Mechanical Engineering Series

The Mechanical Engineering Series presents advanced level treatment of topics on the cutting edge of mechanical engineering. Designed for use by students, researchers and practicing engineers, the series presents modern developments in mechanical engineering and its innovative applications in applied mechanics, bio-engineering, dynamic systems and control, energy, energy conversion and energy systems, fluid mechanics and fluid machinery, heat and mass transfer, manufacturing science and technology, mechanical design, mechanics of materials, micro- and nano-science technology, thermal physics, tribology, and vibration and acoustics. The series features graduate-level texts, professional books, and research monographs in key engineering science concentrations.

More information about this series at http://www.springer.com/series/1161

Afshin J. Ghajar

Single- and Two-Phase Flow Pressure Drop and Heat Transfer in Tubes

Springer

Afshin J. Ghajar
Mechanical and Aerospace Engineering
Oklahoma State University
Stillwater, OK, USA

ISSN 0941-5122 ISSN 2192-063X (electronic)
Mechanical Engineering Series
ISBN 978-3-030-87280-9 ISBN 978-3-030-87281-6 (eBook)
https://doi.org/10.1007/978-3-030-87281-6

This Springer imprint is published by the registered company Springer Nature Switzerland AG
The registered company address is: Gewerbestrasse 11, 6330 Cham, Switzerland

To: Homa, Aarash, Aria, Margo Louise,
Julian Amir, and Noah Jahan

Preface

This book provides in two parts a summary of three decades of pioneering research conducted by Professor Afshin J. Ghajar and his students at Oklahoma State University on single- and two-phase flow pressure drop and heat transfer in tubes. The first part summarizes the research conducted on different aspects of single-phase pressure drop and heat transfer in the transition region in plain horizontal macro, mini, micro tubes, and horizontal micro-fin tubes. The second part describes the different issues related to two-phase gas–liquid flow in pipes with different orientations.

For proper design of thermal systems, such as compact heat exchangers involving horizontal tubes or heating, ventilating, and air conditioning systems involving horizontal micro-fin tubes heated by a uniform wall heat flux boundary condition, knowledge of the pressure drop (friction factor) and the heat transfer coefficient (Nusselt number) in tubes of different sizes for the entrance and fully developed regions with different inlet geometries under different flow regimes, different modes of convection heat transfer (forced and mixed), and different fin geometry is essential.

In the first part of this book, a detailed investigation on the effects of entrance flow, buoyancy (secondary flow), inlet configurations (e.g., square-edged, re-entrant inlets, and bell-mouth), and fin geometries (spiral angle, fin height-to-diameter ratio, and number of starts) on friction factor and heat transfer in the entrance and fully developed regions of different size horizontal plain (macro, mini, and micro) and micro-fin tubes under laminar, transitional, and turbulent regimes is presented. The Reynolds numbers for the ethylene glycol-water mixtures used throughout the experiments ranged from 800 to 25,000.

The reliability of the experimental system was verified by plain tube friction factor and heat transfer experiments. Owing to the lack of information on the entrance flow friction factors in the open literature, the effects of the inlet and heating on the entrance and fully developed flow friction factors are presented. The results showed that the transition range is inlet-dependent, and heating affected friction factors in the laminar and transitional regions. Furthermore, on the basis of their experimental data, they developed practical and easy-to-use correlations for the isothermal and non-isothermal (heating) friction factor (pressure drop) and heat transfer coefficient (Nusselt number) in the entrance and fully developed regions in all flow regimes (laminar, transitional, and turbulent) for different inlet configurations.

From micro-fin tube heat transfer and friction factor test results, the transition from laminar to turbulent flow was clearly established and the transition Reynolds number was shown to be inlet- and fin-geometry dependent. The influence of buoyancy and entrance flow on heat transfer was observed in the laminar region, and the effect of heating on the friction factor in the lower transitional region was clearly established. Correlations to predict forced and mixed convection heat transfer in laminar, transitional, and turbulent regions were developed. Also, correlations for predicting isothermal and non-isothermal entrance and fully developed friction factors in laminar and transitional regions were developed.

As an extension to the aforementioned studies, Ghajar and co-workers also investigated pressure drop (friction factor) and heat transfer in different size mini and micro tubes (2083 μm to 337 μm) with different inlet configurations (straight tube length, 90° bend, and 180° bend) in all flow regimes (laminar, transitional, and turbulent).

The results presented in the first part of this book provide the thermal systems design engineer using liquid single-phase flow in different industrial applications the necessary design tools and fundamental understanding of the variables that influence pressure drop and heat transfer in different size plain (macro, mini, and micro) and micro-fin tubes.

The phenomenon of gas–liquid two-phase flow in inclined systems, although not as common as horizontal or vertical flow, is of great practical significance in several applications such as undulating oil-gas flow lines, chemical process engineering, and inclined flow paths in steam condensers and generators. In these practical applications, accurate determination of two-phase flow variables such as void fraction, pressure drop, and heat transfer is of great importance for system sizing and optimization. It is a well-established fact that these parameters are very sensitive to the spatial and morphological variations of the two-phase flow structure. The two-phase flow structure, commonly termed as flow pattern, depends on the interaction and balance between the buoyancy-inertia-gravity forces, which, in turn, are a function of pipe inclination. Thus, a correct understanding of the effect of change in pipe inclination on the two-phase flow structure is needed, and its effect on the thermo-fluidics of two-phase flow needs to be understood. Thus, the objective of the second part of the book is to present an overview of the pipe inclination effects of the gas–liquid two-phase flow phenomenon.

The pipe inclination effects on non-boiling two-phase flow are studied through the data available in literature and extensive experiments carried out at the Two-Phase Flow Laboratory, Oklahoma State University, Stillwater, Oklahoma. These experiments were carried out using air-water as fluid combination. The experimental setup used for the experiments is unique as it can be rotated to any inclination between +90° to −90° including the horizontal orientation and is capable of flow visualization and simultaneous measurements of void fraction, pressure drop, and heat transfer. The experimental data show the flow pattern and pipe inclination dependency of all two-phase flow variables. At low mass fluxes, this effect is found to be most significant where buoyant forces acting on the gas phase dominate the two-phase flow phenomenon. Experiments also reveal the prevalence and insights about the flow reversal and transient nature of the two-phase flow in upward and downward inclined systems.

Two-phase flow literature reports a plethora of correlations/models for determination of void fraction, pressure drop, and non-boiling heat transfer. Since the two-phase flow is a function of several variables such as flow patterns (phase flow rates), fluid properties, pipe diameter, and inclination angle (pipe orientation), it is quite a challenging task for the end user to select an appropriate flow condition-specific correlation/model. Selection of a correct model also requires some fundamental understanding of the two-phase flow physics and the underlying principles/assumptions/limitations associated with these correlations. To address these issues, some of the top performing two-phase flow models validated against a comprehensive data set are also introduced.

Stillwater, OK, USA Afshin J. Ghajar

Acknowledgments

In Part I of the book, majority of the results presented is based on the work of my former graduate students. In particular, the contributions of Professor Lap Mou Tam and Dr. Hou Kuan Tam of the University of Macau, China, are greatly appreciated.

For Part II of the book, I would like to thank several of my colleagues around the world for sharing their two-phase flow experimental data with me. Majority of the results presented in this part is based on the work of my former graduate students Dr. Swanand M. Bhagwat of Southwest Research Institute, San Antonio, Texas, and Professor Clement C. Tang of the University of North Dakota, Grand Forks. Their contributions are greatly appreciated.

Contents

About the Author

Afshin J. Ghajar is Regents Professor and John Brammer Professor in the School of Mechanical Engineering and Aerospace Engineering at Oklahoma State University. He is a fellow of the American Society of Mechanical Engineers (ASME) and the American Society of Thermal and Fluids Engineers (ASTFE). He is a Registered Professional Engineer in the State of Oklahoma.

Professor Ghajar has received countless teaching/service awards, such as the 75th Anniversary Medal of the ASME Heat Transfer Division, the ASME ICNMM Outstanding Leadership Award, and the Donald Q. Kern Award, among others.

Professor Ghajar's research work has resulted in over 250 publications including professional journals, reports, books, peer-reviewed conference papers, and symposium proceedings. His research achievements have also been documented by a large number of presentations as well as keynote and invited lectures all over the world. As of 2021, his research has been cited more than 7300 time. A 2020 study conducted by Ionnidis et al. of Stanford University [Updated science-wide author databases of standardized citation indicators (plos.org)] ranked nearly 160,000 scientists of all disciplines based on citations to their work over their career, and for the year 2019, Professor Ghajar ranked in the top 1.3% of researchers in mechanical engineering and transports category.

He currently serves as the editor-in-chief of *Heat Transfer Engineering Journal* and is the Heat Transfer Series editor for CRC Press (he has edited 13 books to date). He is the co-author of two popular textbooks, *Heat and Mass Transfer: Fundamentals and Applications, 6th Edition (2020)* and *Fundamentals of Thermal-Fluid Sciences, 6th Edition (2022)*, both published by McGraw-Hill, and the author of *Two-Phase Gas-Liquid Flow in Pipes with Different Orientations*, Springer Briefs in Applied Sciences and Technology, published by Springer 2020.

Nomenclature

Part I

a	Constant, dimensionless
b	Constant, dimensionless
A_c	Cross-sectional area of the tube [= $\pi D^2/4$], m^2
A_{xs}	Actual flow area of the micro-fin tube [= A_c - N_s e t], m^2
c	Constant, dimensionless
c_p	Specific heat of the test fluid evaluated at T_b, J/(kg·K)
C_f	Fully developed friction factor (Fanning friction factor), [= $\Delta PD/2L\rho V^2$], dimensionless
D	Inside diameter of the test section (tube), m
D_o	Outside diameter of the test section (tube), m
e	Internal fin height, m
Gr	Local bulk Grashof number [=$g\beta\rho^2D^3(T_w-T_b)/\mu_b{}^2$], dimensionless
f_{app}	Apparent (developing) Fanning friction factor [=$\Delta P_{0\text{-}x}D/2x\rho V^2$], dimensionless
f	Fully developed friction factor (Darcy friction factor), [= $2\Delta PD/L\rho V^2$], dimensionless
g	Acceleration of gravity, m/s^2
h	Fully developed peripheral heat transfer coefficient, W/(m^2·K)
h_b	Local peripheral heat transfer coefficient at bottom of tube, W/(m^2·K)
h_t	Local peripheral heat transfer coefficient at top of tube, W/(m^2·K)
j	Colburn j-factor [= St $Pr^{0.67}$], dimensionless
k	Thermal conductivity evaluated at T_b, W/(m·K)
l_{csw}	Modified characteristic length for swirling flows, m
L	Length of the test section (tube), m
L_{ent}	Entrance length before the heating length of the test section, mm
L_t	Heating length of the test section, mm
m	Exponent of the bulk-to-wall-viscosity ratio; see Eqs. (4.2) and (4.5), dimensionless
$\dot{m}$	Mass flow rate, kg/s

N_s	Number of starts (fins) per cross-sectional area, dimensionless
Nu	Local average or fully developed peripheral Nusselt number [= hD/k], dimensionless
p	Axial fin pitch [$= \pi D/(N_s \tan \alpha)$], m
Pr	Local bulk Prandtl number [$= c_p \mu_b /k$], dimensionless
Q	Volume flow rate, m^3/s
r_i	Inner radius, mm
R	Radius of bend, mm
R_a	Average roughness from SPM, m
R_{max}	Maximum roughness profile peak height from SPM, m
R_q	Root mean square roughness from SPM, m
Re	Local bulk Reynolds number [$= \rho VD/\mu_b$], dimensionless
SEM	Scanning electron microscope
SPM	Scanning probe microscope
St	Local average or fully developed peripheral Stanton number [= Nu/(PrRe)], dimensionless
SW	Modified non-dimensional axial pitch [$= N_s \sin(\alpha/\pi)$], dimensionless
t	Mean fin width (or fin thickness), m
T	Temperature, °C
T_b	Local bulk temperature of the test fluid, °C
T_w	Local inside wall temperature, °C
V	Average velocity in the test section, m/s
W	Inter-fin flow area [$= (\pi/N_s - t/D)(\cos \alpha)$], dimensionless
X	Local axial distance along the test section from the inlet, m

Part II

a_1	Variable in Xiong and Chung (2006) correlation
A	Cross-sectional area (m^2)
B_1	Variable defined by Eq. (16.28)
B_2	Variable defined by Eq. (16.29)
B_3	Variable defined by Eq. (16.30)
Bo	Bond number defined by Eq. (16.35)
c_1	Variable in Bhagwat and Ghajar (2014) correlation
c_2	Variable in Bhagwat and Ghajar (2014) correlation
c_3	Variable in Bhagwat and Ghajar (2014) correlation
C	Constant in Lockhart and Martinelli (1949) parameter
C_o	Distribution parameter
$C_{o,1}$	Variable used by Bhagwat and Ghajar (2014)
C_1	Variable in Bhagwat and Ghajar (2015a) correlation
C_2	Variable in Bhagwat and Ghajar (2015a) correlation
C_3	Variable in Bhagwat and Ghajar (2015a) correlation
C_4	Variable in Bhagwat and Ghajar (2015a) correlation
d_{def}	Bubble diameter above which bubble is deformed (m)

d_{max}	Maximum bubble diameter (m)
d_{migr}	Bubble diameter below which migration of bubbles is prevented (m)
dp/dz	Pressure gradient (Pa/m)
$(dp/dz)^+$	Non-dimensional pressure gradient
D	Inside pipe diameter (m)
D_h	Hydraulic diameter (m)
D^+	Non-dimensional pipe diameter used by Bhagwat and Ghajar (2015a)
D_h^+	Non-dimensional hydraulic diameter by Kataoka and Ishii (1987)
E	Liquid entrainment fraction
Eo	Eötvös number
f	Fanning friction factor
F	Froude number defined by Eq. (14.10)
F_p	Flow pattern factor
F_s	Shape factor
Fr	Modified Froude number defined by Eq. (14.16)
g	Acceleration due to gravity (m/s^2)
G	Mass flux (kg/m^2-s)
h_L	Liquid level in stratified flow (m), single-phase heat transfer coefficient (W/m^2-K)
h_{TP}	Two-phase heat transfer coefficient (W/m^2-K)
$\overline{h}$	Average heat transfer coefficient (W/m^2-K)
I	Inclination factor
k	Thermal conductivity (W/m-K)
L	Test section length (m)
La	Laplace number
$\dot{m}$	Mass flow rate (kg/s)
$MARD$	Mean absolute relative deviation
MRD	Mean relative deviation
N	Number of data points
N_{30}	Percentage of data points predicted within ±30%
N_{50}	Percentage of data points predicted within ±50%
N_μ	Viscosity number
N_{ST}	Number of thermocouple stations
p	Pressure (Pa)
P	Perimeter (m)
Pr	Prandtl number
$\dot{q}$	Heat transfer rate (W)
Re	Reynolds number
S	Slip ratio
T_b	Bulk temperature (°C)
$\overline{T}_{w,i}$	Average inside wall temperature (°C)
$\overline{T}_{w,o}$	Average outside wall temperature (°C)
U	Phase velocity (m/s)

U_b	Bubble velocity (m/s)
U_{GM}	Drift velocity (m/s)
U_{GM}^{+}	Non-dimensional drift velocity by Kataoka and Ishii (1987)
$U_{SG,\ t}$	Superficial gas velocity at smooth to wavy stratified flow transition (m/s)
V	Volume (m^3)
We	Weber number
x	Two-phase flow quality
X	Lockhart and Martinelli (1949) parameter
Y	Taitel and Dukler (1976) parameter, Chisholm (1973) parameter
Δz	Differential pipe length (between two thermocouple stations)

Greek Symbols: Part I

α	Spiral/helix angle, degree
β	Coefficient of thermal expansion of the test fluid evaluated at T_b, K^{-1}
β'	Fin included angle of Fig. 2.2, degree
ΔP	Pressure difference, Pa
$\Delta P_{0\text{-}x}$	Pressure drop from the inlet to a specific location down the tube, Pa
μ_b	Absolute viscosity of the test fluid evaluated at T_b, Pa·s
μ_w	Absolute viscosity of the test fluid evaluated at T_w, Pa·s
ν	Kinematic viscosity of the test fluid evaluated at T_b, m^2/s
ρ	Density of the test fluid evaluated at T_b, kg/m^3
ζ	Dimensionless axial distance [= x/D/Re], dimensionless

Greek Symbols: Part II

α_G	Void fraction
α_L	Liquid holdup
β	Volume fraction of liquid film
δ	Liquid film thickness in annular flow (m)
$\Delta\rho$	Density difference (kg/m^3)
ε	Roughness (m)
μ	Dynamic viscosity (Pa-s)
ν	Kinematic viscosity (m^2/s)
σ	Surface tension (N/m)
ρ	Density (kg/m^3)
λ	Gas volumetric flow fraction
τ	Shear stress (N/m^2)
θ	Pipe orientation (degree)
ϕ	Fraction of pipe circumference occupied by liquid phase
Φ^2	Two-phase frictional multiplier
Ψ	Angles subtended by liquid phase with pipe centerline (radian)
ξ	Correction factor in Eq. (18.6)
π_1	Non-dimensional variable defined by Eq. (16.31)

π_2	Constant defined by Eq. (16.32)
π_3	Non-dimensional variable defined by Eq. (16.33)

Subscripts: Part I

app	Apparent
cal	Refers to the calculated value
exp	Refers to the experimental value
heat	Refers to non-isothermal (or heating) condition
i	Inlet
iso	Refers to isothermal condition
l	Refers to laminar
lam	Laminar
p	Plain tube
t	Refers to turbulent
tr	Refers to transition
trans	Transition

Subscripts: Part II

a	Accelerational
atm	Atmospheric
c	Gas core
C	Critical
CAL	Calculated
EXP	Experimental
f	Frictional
G	Gas
GO	Gas only
h	Hydrostatic
i	Interface or inside
j	Phase
L	Liquid
LF	Liquid film
LO	Liquid only
max	Maximum
M	Mixture
o	Outside
ref	Reference
S	Superficial
sys	System
t	Total
tt	Turbulent-turbulent

TP	Two-phase
w	Wall

Superscripts: Part II

~	Non-dimensional quantities in Taitel and Dukler (1976) model
+	Non-dimensional quantities in Kataoka and Ishii (1987) model

Part I
Single-Phase Flow Pressure Drop and Heat Transfer in Tubes

Chapter 1
Introduction

Background

For proper design of thermal systems such as compact heat exchangers involving horizontal tubes that are heated by a uniform wall heat flux boundary condition, proper knowledge of the friction factor (pressure drop) and heat transfer coefficient (Nusselt number) in the tube for the entrance and fully developed regions with different inlet geometries under different flow regimes (laminar, transitional, and turbulent) is essential. In practical engineering design, the usual recommendation is to avoid design and operation in this region, but this is not always feasible under design constraints. The usually cited transitional Reynolds number range of about 2300 (onset of turbulence) to 10,000 (fully turbulent condition) applies, strictly speaking, to a very steady and uniform entry flow with a rounded entrance. If the flow has a disturbed entrance typical of heat exchangers in which there is a sudden contraction and possibly even a re-entrant entrance, the transitional Reynolds number range will be very different.

The transition from laminar to turbulent flow has been considered to be one of the most challenging problems in the thermal-fluid sciences since it was described by Osborne Reynolds in 1883. However, not a lot of progress has been made since then in the state of the art. Ghajar and co-workers were the first to break the ground to determine the very important relationship between inlet geometry and transition. In a series of studies, they experimentally investigated the inlet configuration (e.g., square-edged, re-entrant inlets, and bell-mouth) and buoyancy (secondary flow) effects on the developing and fully developed transitional pressure drop (friction factor) under isothermal and heating (uniform wall heat flux boundary) conditions and developing and fully developed transitional forced and mixed convection heat transfer in plain circular horizontal tubes. On the basis of their experimental data, they developed practical and easy to use correlations for the isothermal and

A. J. Ghajar, *Single- and Two-Phase Flow Pressure Drop and Heat Transfer in Tubes*, Mechanical Engineering Series,
https://doi.org/10.1007/978-3-030-87281-6_1

non-isothermal friction factor (pressure drop) and heat transfer coefficient (Nusselt number) in the transition region as well as the laminar and turbulent flow regions for different inlet configurations.

Later on, they extended their macro-tube studies to micro-fin tubes and mini- and micro- tubes. For the micro-fin tube studies, they also looked into the effect of fin geometries (spiral angle, fin height-to-diameter ratio, and number of starts) on the transitional pressure drop and heat transfer. They also developed correlations for the prediction of transitional pressure drop and heat transfer for micro-fin tubes covering a wide range of fin geometries and different inlet configurations. For the mini- and micro-tube studies, they performed a systematic and careful experimental study of the friction factor in the transition region to identify the influence of tube diameter on the start and end of the transition region. They also investigated the influence of inlet configuration on pressure drop and heat transfer transition region in different size mini- and micro-tubes with different inlet configurations (straight tube length, 90° bend, and 180° bend).

The work presented in the Part I of this book provides a detailed overview of the pioneering research done by Professor Ghajar and his students over a period of three decades. The groundbreaking work done by Ghajar and coworkers made it possible for other researchers to follow their work. A brief summary of such efforts is summarized in Ghajar (2019) and Meyer and Everts (2019).

Need for Single-Phase Flow Study in the Transition Region

From a practical standpoint, the transition region in the majority of thermo-hydraulic systems is unavoidable. The transition from laminar to turbulent flow has been considered to be one of the most challenging problems in the thermal-fluid sciences and has not been studied in great detail. Ghajar and his coworkers have systematically and in great detail experimentally studied the numerous factors that influence the start and end of the transition region for pressure drop (friction factor) and heat transfer coefficient (Nusselt number) in different size tubes and micro-fin tubes, with different inlet configurations and different boundary conditions. Based on their carefully obtained experimental data, they have developed practical and easy to use correlations for prediction of friction factor (pressure drop) and heat transfer coefficient (Nusselt number) in all flow regimes (laminar, transitional, and turbulent).

The results of their study should help in addressing the following issues:

1. What is the effect of inlet configuration and wall boundary conditions (isothermal vs heating) on the plain tube developing and fully developed friction factors?
2. What type of correlation to use for prediction of developing and fully developed friction factors in the laminar and transition flow regions of a plain tube with different inlet configurations and wall boundary conditions (isothermal vs heating)?

3. What is the effect of inlet configuration on the plain tube heat transfer?
4. What is the effect of buoyancy (secondary flow) on the plain tube heat transfer with different inlet configurations?
5. How does local Nusselt number vary along the plain tube with different inlet configurations?
6. How does heating influence the transition range for friction factor in comparison to the transition range for heat transfer in a plain tube with different inlet configurations?
7. What type of correlation to use for prediction of developing and fully developed heat transfer in different flow regions of a plain tube with different inlet configurations?
8. How to select an appropriate forced or mixed convection heat transfer correlation for a plain tube in different flow regions?
9. What is the effect of inlet configuration and wall boundary conditions (isothermal vs heating) on micro-fin tubes' developing and fully developed friction factors?
10. What is the effect of fin geometry on micro-fin tubes' friction factor?
11. What is the effect of change in spiral angle, fin height-to-diameter ratio, and number of starts on micro-fin tubes' friction factors?
12. What type of correlation to use for prediction of developing and fully developed friction factors in the laminar and transition flow regions of micro-fin tubes with different inlet configurations and wall boundary conditions (isothermal vs heating)?
13. What is the effect of inlet configuration and buoyancy (secondary flow) on micro-fin tubes' heat transfer?
14. What is the effect of fin geometry on micro-fin tube heat transfer?
15. What is the effect of change in spiral angle, fin height-to diameter ratio, and number of starts on micro-fin tubes' heat transfer?
16. How does heating influence the transition range for friction factor in comparison to the transition range for heat transfer in micro-fin tubes with different inlet configurations?
17. What type of correlation to use for prediction of developing and fully developed heat transfer in all flow regions of micro-fin tubes with different inlet configurations?
18. What is the effect of tube size on the friction factor of mini- and micro-tubes?
19. What is the effect of tube size on the friction factor transition region of mini- and micro-tubes?
20. What is the effect of inner surface roughness and heating on the friction factor of mini- and micro-tubes?
21. What is the effect of bending inlets with different angles on the heat transfer of mini tubes?

To answer the above and many other questions related to the friction factor (pressure drop) and heat transfer in plain, micro-fin, mini-, and micro-tubes with different inlet configurations in all flow regimes (laminar, transitional, and turbulent), dedicated unique experimental setups were designed to collect appropriate experimental data. The experimental data were also used to develop practical correlations for the prediction of friction factor and heat transfer coefficient under different operating conditions. The application of some of the developed correlations will be illustrated through a few worked out example problems.

Chapter 2
Single-Phase Flow Experimental Setup for Plain and Micro-fin Tubes

Background

As mentioned in Chap. 1, a dedicated unique experimental setup is needed to provide detailed information on the effect of entrance flow, buoyancy (secondary flow), inlet configurations (square-edged, re-entrant, and bell-mouth), and fin geometries (spiral angle, fin height-to-diameter ratio, and number of starts) on friction factor and heat transfer in the entrance and fully developed regions of horizontal plain and micro-fin tubes in all flow regions (laminar, transitional, and turbulent). No such experimental setup exists in the literature. The experimental setup detailed in this chapter will allow the necessary detailed and systematic measurements of friction factor (pressure drop) and heat transfer in response to the issues raised in Chap. 1.

Details of the Single-Phase Flow Experimental Setup

A schematic diagram of the overall experimental setup used for heat transfer and pressure drop experiments in plain and micro-fin tubes is shown in Fig. 2.1. The test section was composed of horizontal copper circular tubes with an inside diameter of 1.49 cm and an outside diameter of 1.59 cm. One plain tube and six micro-fin tubes were used in this study. Figure 2.2 depicts the sectional view of the micro-fin tube. In this study, the magnitude of included angle β' shown in the figure was 42°. Table 2.1 provides the specifications of the fin geometric parameters of the six micro-fin tubes used for the experiments. The enhanced area of the micro-fin tube was calculated by taking the surface area of the maximum inside diameter, subtracting the surface area of the base thickness of the fins, and adding the surface area from the sides of the fins. The total length of the test section was 6 m, providing a maximum length-to-

A. J. Ghajar, *Single- and Two-Phase Flow Pressure Drop and Heat Transfer in Tubes*, Mechanical Engineering Series,
https://doi.org/10.1007/978-3-030-87281-6_2

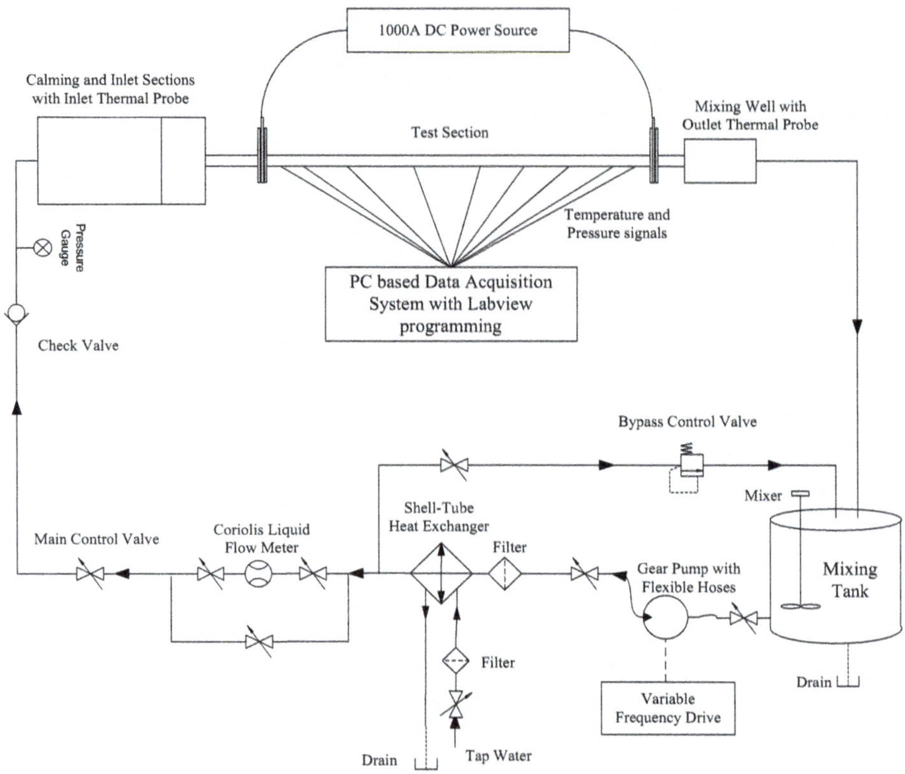

Fig. 2.1 Schematic of experimental setup used for friction factor and heat transfer measurements in plain and micro-fin tubes

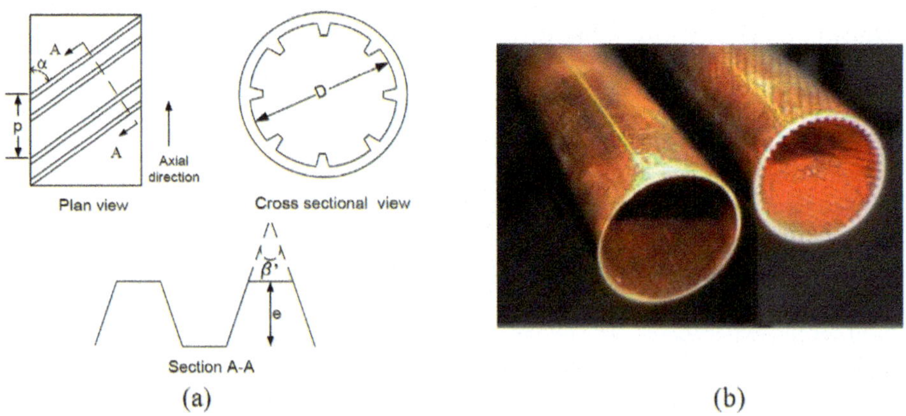

Fig. 2.2 (**a**) Sectional view of the micro-fin tube and (**b**) plain and micro-fin tubes

Table 2.1 Specifications of the test sections

Tube number	Outer diameter, D_o (mm)	Inner diameter, D (mm)	Fin height, e (mm)	Axial pitch, p (mm)	Number of starts, N_s (degree)	Spiral angle, α (degree)	e/D	p/e	p/D	Area enhanced ratio, A_c/A_p
0	15.88	14.88	Plain	–	–	–	–	–	–	–
1	15.88	14.88	0.5	5.75	25	18°	0.0336	11.5	0.39	1.34
2	15.88	14.88	0.5	4.01	25	25°	0.0336	8.02	0.27	1.34
3	15.88	14.88	0.5	2.67	25	35°	0.0336	5.34	0.18	1.34
4	15.88	14.88	0.3	2.67	25	35°	0.0202	8.9	0.18	1.20
5	15.88	14.88	0.5	3.20	45	18°	0.0336	6.4	0.22	1.61
6	15.88	14.88	0.5	2.21	65	18°	0.0336	4.42	0.15	1.89

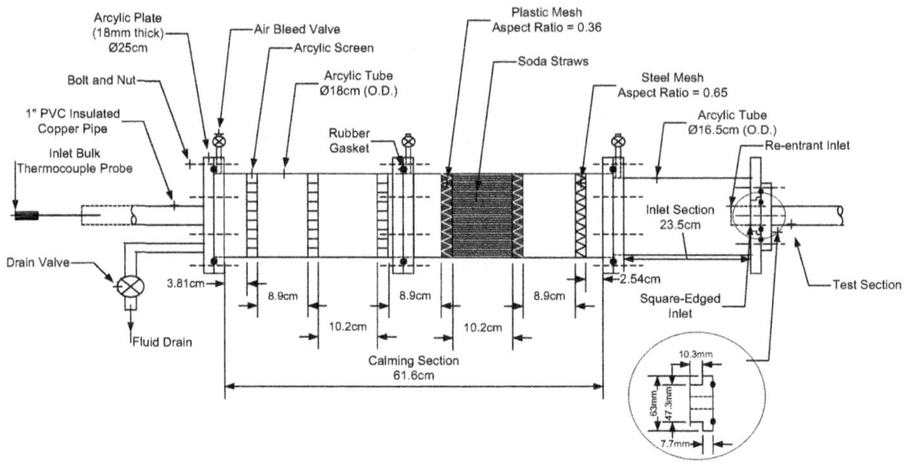

Fig. 2.3 Calming and inlet sections

inside diameter ratio (L/D) of 405. The end connections of the test section consisted of copper plates that were arc soldered with silver to the ends of the test section to secure a well-defined electric circuit through the end plates (Fig. 2.1). The heated length of the test section was 5.8 m, which was the distance between the copper plates. The test section was insulated from the environment by using foamed nitrile rubber pipe insulation and vaporproof pipe tape. The total thickness of the Armaflex® insulation materials was about 2.5 cm. The test section was placed on a 10-m long I-beam platform. To ensure a uniform fluid bulk temperature at the exit of the test section, a static mixer that consisted of several mixing elements was used. A one-shell, two-tube pass heat exchanger was used to cool the fluid from the test section to a desirable inlet bulk temperature.

To ensure a uniform velocity distribution in the test fluid before it entered the test section, the flow passed through a calming section (Fig. 2.3) which consisted of an 18-cm diameter acrylic plastic cylinder with three perforated acrylic plastic plates with an open area ratio of 0.31 (73 holes per plate, diameter 1.1 cm) followed by tightly packed soda straws (inside diameter 0.6 cm, length 10.2 cm, open area ratio 0.92) sandwiched between galvanized steel mesh screens (wire diameter 0.07 cm, mesh width 0.28 cm, open area ratio 0.65). Before leaving the calming section, test fluid passed through a fine mesh screen (wire diameter 0.03 cm, mesh width 0.14 cm, open area ratio 0.65). The total length of the calming section was 61.6 cm. Test fluid leaving the calming section entered the inlet section and flowed undisturbed through 23.5 cm of a 16.5-cm diameter acrylic plastic tube before it entered the test section. This was done to ensure a uniform velocity distribution upon entering the test section. The calming and inlet sections were also equipped with air bleed valves that were used to evacuate air during startup. As seen in Fig. 2.3, the inlet section had the versatility of being modified to incorporate different type of inlet configurations.

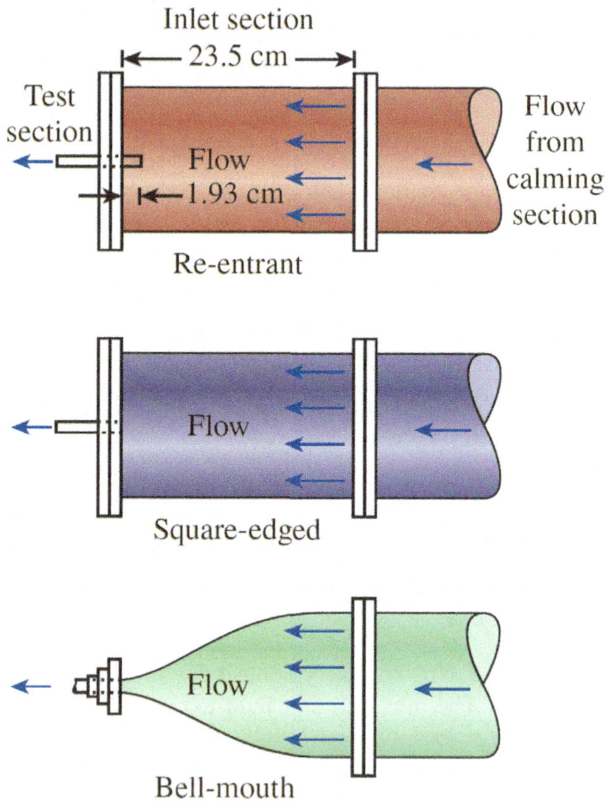

Fig. 2.4 Schematic of the three different inlet configurations

Figure 2.4 shows the schematic of the three different inlet configurations used. The re-entrant inlet was simulated by sliding 1.5 cm of the tube entrance length into the inlet section which was otherwise the square-edged (sudden contraction) inlet. In a heat exchanger, the re-entrant inlet represents the tube extending beyond the tube sheet face and into the head of the distributor, the square-edged inlet represents the tube end being flush with the tube sheet face, and the bell-mouth inlet represents a tapered entrance of tube from the tube sheet face. It should be noted that the bell-mouth inlet was only used for selected plain tube experiments and none of the micro-fin tube experiments.

For the uniform wall heat flux boundary condition, the heat flux was maintained by a Miller Dimension 1250 DC CC/CV power source along the 5.8 m length of the test section. Thermocouples were placed on the outer surface of the tube wall at close intervals near the entrance and at greater intervals further downstream (see Fig. 2.5 for details). Sixteen thermocouple stations were designated with four thermocouples at each station. The thermocouples were placed 90° apart around the periphery. Omega SA1XL-T-72 fiberglass-insulated T-type surface thermocouples were used with Omega EXPP-T-TWSH-SLE-1000 extension wire for relay to the data acquisition system. The thermocouples were attached to the outside of the tube wall with

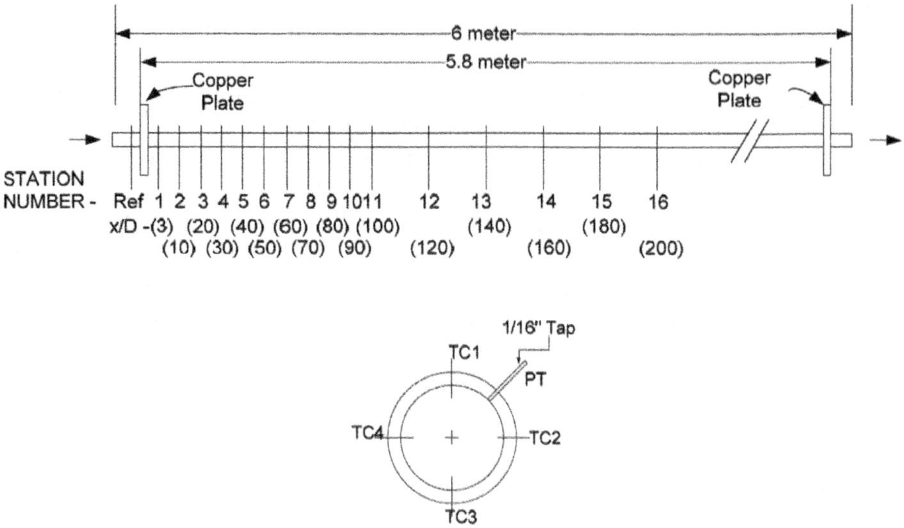

Fig. 2.5 Arrangement of the thermocouples and pressure tap on the test section. ("Ref" indicates the reference pressure tap station)

Omegabond 101, an epoxy adhesive with high thermal conductivity and electrical resistivity. The inlet and exit bulk temperatures were measured by means of thermocouple probes (Omega TMQSS-125U-6) inserted in the calming section and the mixing well, respectively. All the thermocouples and thermocouple probes were calibrated by a NIST-calibrated thermocouple probe (±0.22 °C) and the Omega HCTB-3030 constant temperature circulating bath. They were accurate within ±0.22 °C. The data acquisition system used for temperature measurements consisted of two sets of National Instrument SCXI systems with the maximum of 256 thermocouple input channels. The flow rate was measured by a Micro Motion Coriolis flow meter CMF100 equipped with a model 2200S transmitter. The flow meter had a maximum mass flow accuracy of ±0.77% in the flow range from 0.3 to 30 USGPM and had a density accuracy of 0.2 kg/m^3. The calibration was done by the manufacturer. The voltage drop across the test section was directly input to the SCXI system with the high-voltage attenuator terminal block. The current carried by the test section was measured by a Powertek CTH 1000A current transducer with an accuracy of ±0.5%. The voltage signal generated from the current transducer was also input to the SCXI system.

To prevent the power source and pumps from transmitting noise and vibration through the experimental setup, they were mounted inside a plywood box lined with insulation. The power source was placed far away from the test section. In addition, rubber hoses were connected to the pump box to prevent the transmission of vibration to the fluid return tubing. All the equipment was placed on damping pads to prevent vibration from being transmitted through the floor.

The heat transfer measurements at uniform wall heat flux boundary condition were carried out by measuring the local outside wall temperatures at 16 stations along the axis of the tube and the inlet and outlet bulk temperatures in addition to other measurements such as the flow rate, room temperature, voltage drop across the test section, and current carried by the test section. The local peripheral heat transfer coefficient and Nusselt number were then calculated for each of these stations based on the knowledge of pipe inside wall temperature and inside wall heat flux obtained from a data reduction computer program developed exclusively for this type of experiment by Ghajar and Zurigat (1991) and Ghajar and Kim (2006). The local average peripheral values for inside wall temperature, inside wall heat flux, heat transfer coefficient, and Nusselt number were then obtained by averaging all the appropriate individual local peripheral values at each axial location. The computer program used a finite-difference formulation to determine the inside wall temperature and the inside wall heat flux from measurements of the outside wall temperature, the heat generation within the pipe wall, and the thermophysical properties of the pipe material (electrical resistivity and thermal conductivity). In these calculations, axial conduction was assumed negligible (RePr > 41,000 in all cases), but peripheral and radial conduction of heat in the tube wall were included. In addition, for the entrance and fully developed regions, the bulk fluid temperature was assumed to increase *linearly* from the inlet to the outlet according to Çengel and Ghajar (2020).

Two test fluids of different concentrations were used in the experiments. Full-strength concentrations of distilled water and ethylene glycol were followed by mixtures of 35–75% ethylene glycol and water (by mass fraction). Expressions for the physical properties of water, ethylene glycol, and ethylene glycol-water mixtures along with their accuracy and range of application are reported in Ghajar and Zurigat (1991) and Ghajar and Kim (2006). In the experiments, the local bulk dimensionless variables (Reynolds number, Prandtl number, friction factor, etc.) were calculated with the physical properties of the test fluid at each station along the test section. Those physical properties were obtained based on the fluid temperature and wall temperature measured at each station. For the experiments (plain and micro-fin tubes), the local bulk Reynolds number ranged from 800 to 25,000, the local bulk Prandtl number varied from 4.3 to 56.1, the local bulk Grashof number range was from 1800 to 112,000, the local bulk Nusselt number varied from 10.2 to 383.1, and the friction factor varied from 6.5×10^{-3} to 6.9×10^{-2}. The uniform wall average heat flux for the experiments ranged from 3.4 to 8.9 kW/m^2. Heat balance errors were calculated for all experimental runs by taking a percent difference between two methods of calculating the heat addition. The product of the voltage drop across the test section and the current carried by the tube was the primary method, while the fluid enthalpy rise from inlet to exit was the second method. In all cases, the heat balance error was less than $\pm 10\%$. The primary method was the one used in the computer program developed by Ghajar and Zurigat (1991) and Ghajar and Kim (2006) for all heat flux and heat transfer coefficient calculations.

For simultaneous measurement of the heat transfer and the pressure drop at the same station, the pressure taps should be arranged at the same axial location where the four surface thermocouples were placed. To determine the appropriate number of

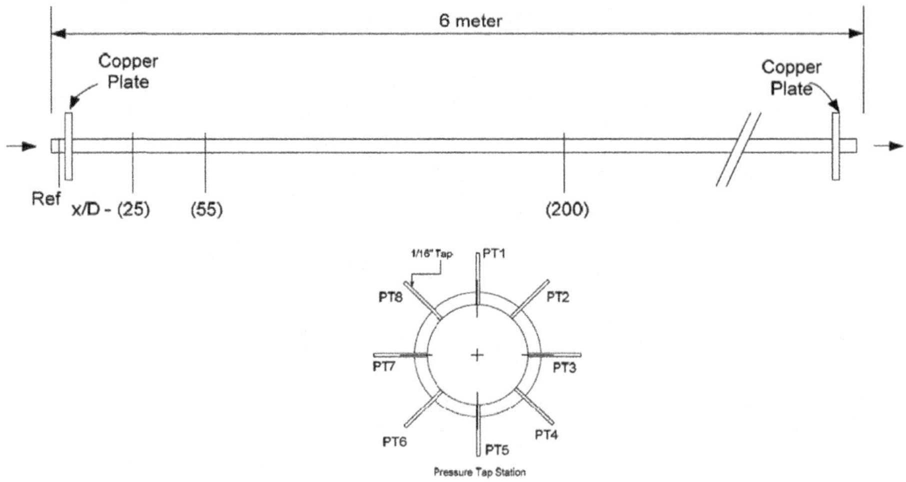

Fig. 2.6 Arrangement of the pressure taps on the test section for verification. ("Ref" indicates the reference pressure tap station)

peripheral pressure taps to be used for determination of friction factor, the variations of the pressure readings were first verified from the different pressure taps around the periphery of the tube wall.

For verifying the variation of peripheral pressure readings in the entrance and fully developed flow regions, eight pressure taps with the arrangement shown in Fig. 2.6 were used. As shown in the figure, the reference pressure tap was located between the tube inlet and copper plate, and the other pressure tap stations were arranged at three axial locations (x/D = 25, 55, 210). In each axial location, eight pressure taps (PT1 to PT8) at equal intervals were placed around the periphery of the tube wall. The verification was conducted for isothermal and non-isothermal (heating) boundary conditions for a range of Reynolds number that varied from 1000 to 4000.

Under the isothermal condition, as seen from Fig. 2.7, for the entire range of Reynolds numbers, the friction factors deduced from all eight pressure taps were similar. For x/D of 25, 55, and 210, the deviation between the maximum and minimum value was 6.3%, 3.6%, and 5.6%, respectively.

Under the heating (non-isothermal) condition, as seen from Fig. 2.7, for the entire range of Reynolds numbers, the friction factors deduced from all eight pressure taps were similar. For x/D of 25, 55, and 210, the deviation between the maximum and minimum value was 6.6%, 4.7%, and 4.8%, respectively.

Based on the above results, for both isothermal and non-isothermal (heating) boundary conditions in the entrance and fully developed regions, the pressure readings from peripheral pressure taps were similar and did not show substantial deviation. Therefore, for the simultaneous heat transfer and pressure drop experiments, one pressure tap was arbitrarily placed near the TC1 and TC2 thermocouples on the outer surface of the wall at close intervals near the entrance and at greater

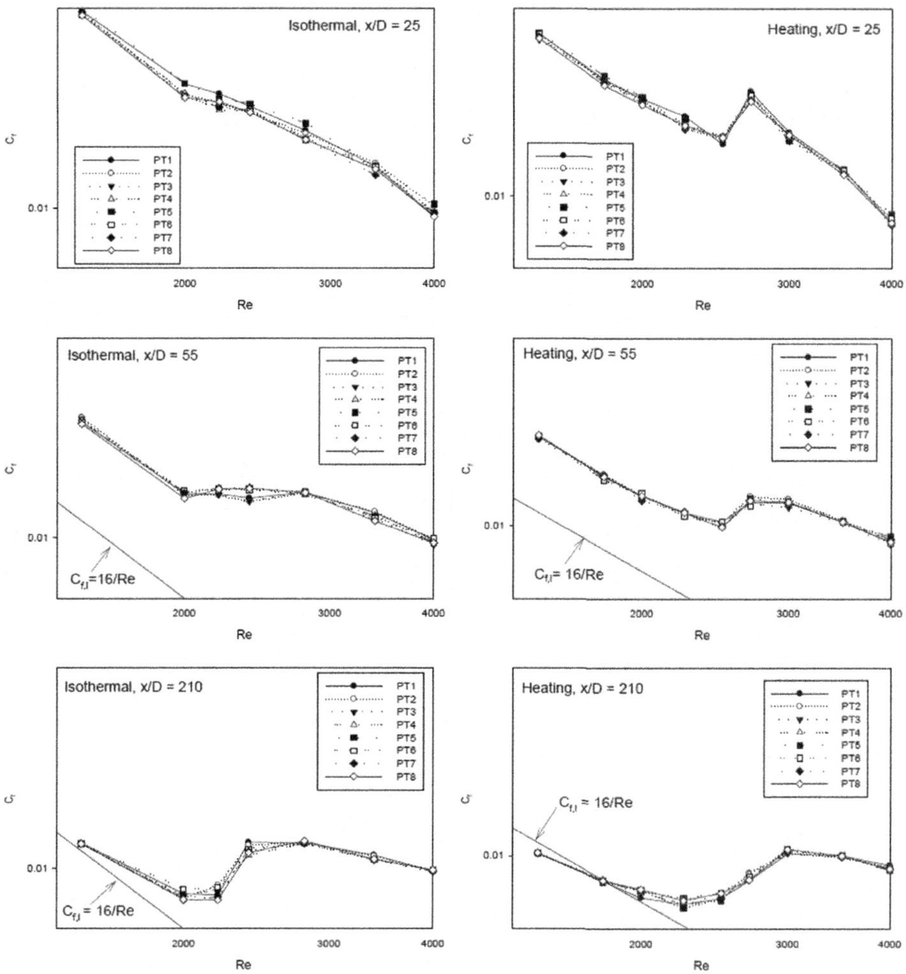

Fig. 2.7 Isothermal and non-isothermal (heating) friction factors of the peripheral eight pressure taps (PT1 to PT8) at the three stations (*x/D* = 25, 55, 210)

intervals further downstream (see Fig. 2.5 for details). A reference pressure station and 16 pressure stations were designated with one 0.16 cm diameter hole drilled at each station. Each of the pressure tap holes was deburred with a special tool to ensure the characteristics of a smooth tube. Pressure measurements were made with three differential pressure transducers with ranges of 0.86 kPa, 1.4 kPa, and 14 kPa. The differential pressure transducers had an accuracy of ±0.25% reading, and their calibration was checked regularly by the Furness PPC500 pressure calibrator (an accuracy of ±0.1% reading) during the experiments. In order to take sequential pressure drop measurements along the test section, a Scanivalve electronic rotary system was used. This system made it possible to measure pressure drop from the reference pressure tap to any of the other 16 pressure taps.

Table 2.2 Uncertainties in experimental data

Measured variables		Calculated variables	
Variable	Uncertainty	Variable	Uncertainty
Temperature	0.22 °C	C_f (friction factor)	2.4%
Mass flow rate	0.77% of full range (0.3-30USGPM)	h (heat transfer coefficient)	22.6%
Density	0.2 kg/m^3	Nu (Nusselt number)	22.6%
Diameter	0.02 mm		
Length	0.1 mm		
Voltage	1%		
Current	1%		

To establish the uncertainty of the experimental data, all the thermocouples and the pressure transducers were calibrated before the experimental runs. The high-precision Coriolis flow meter was calibrated by the manufacturer. The uncertainty analyses of the overall experimental procedures using the method of Kline and McClintock (1953) showed that there is a maximum of 22.6% uncertainty for the heat transfer coefficient calculations, a maximum of 2.4% uncertainty for the friction factor calculations, and a maximum 1.3% uncertainty for Reynolds number calculations. Table 2.2 lists the uncertainties of the measured variables and calculated variables over the entire range of Reynolds numbers studied.

The reliability of the flow circulation system and of the experimental procedures was checked by making several calibration runs with distilled water, and the heat transfer results were compared with the well-established heat transfer correlations by Gnielinski (1976), Sieder and Tate (1936), and Dittus and Boelter (1930) for fully developed turbulent pipe flow. The experimental data were within ±18.3% of the predicted values. Therefore, the experimental setup for providing the heating condition was confirmed to be reliable.

Moreover, the isothermal friction factor results were also compared with the well-established friction factor correlation of Blasius (1913) ($C_{f,t} = 0.0791\ \mathrm{Re}^{-0.25}$) for the fully developed turbulent pipe flow and the classic fully developed laminar flow equation ($C_{f,l} = 16/\mathrm{Re}$). The experimental data were within ±10% of the predicted values, again confirming the reliability of the experimental setup for friction factor (pressure drop) measurements.

For further detailed information regarding experimental setup and uncertainties in the measurements of pressure drop and heat transfer coefficient for the plain tube, refer to the work of Tam (1995) and for the micro-fin tubes to the work of Tam (2013).

Chapter 3
Effect of Inlet Configuration and Heating on Plain Tube Friction Factor

Background

For proper design of thermal systems, for example, compact heat exchangers involving horizontal tubes being heated by uniform wall heat flux boundary condition, proper knowledge of the friction factors in the tube for the entrance and fully developed regions with different inlet configurations (square-edged and re-entrant) under different flow regimes (laminar, transitional, and turbulent) is essential. For this purpose, using the experimental setup detailed in Chap. 2, experiments were conducted to obtain accurate and broad pressure drop (friction factor) database across all flow regimes. This chapter presents isothermal and non-isothermal (heating) friction factor results in the entrance and fully developed regions of the plain tube, the effect of inlet configuration on heat transfer modes (forced and mixed convection) and on the start and end of transition of the fully developed friction factor, the effect of inlet configuration and heating on the entrance and fully developed friction factors in all flow regimes, a discussion of the role of inlet configuration and heating, and the effect of entrance flow on pressure drop.

Fully Developed Friction Factor

To demonstrate the influence of entrance disturbances on the fully developed friction factor (pressure drop), the isothermal and non-isothermal (heating) friction factor data of Tam et al. (2013) for re-entrant and square-edged inlets were used. Figure 2.4 shows the schematic of the two inlet configurations (square-edged and re-entrant) used. In a heat exchanger, the square-edged inlet represents the tube end being flush with the tube sheet face, and the re-entrant inlet represents the tube extending beyond the tube sheet face and into the head of the distributor.

A. J. Ghajar, *Single- and Two-Phase Flow Pressure Drop and Heat Transfer in Tubes*, Mechanical Engineering Series,
https://doi.org/10.1007/978-3-030-87281-6_3

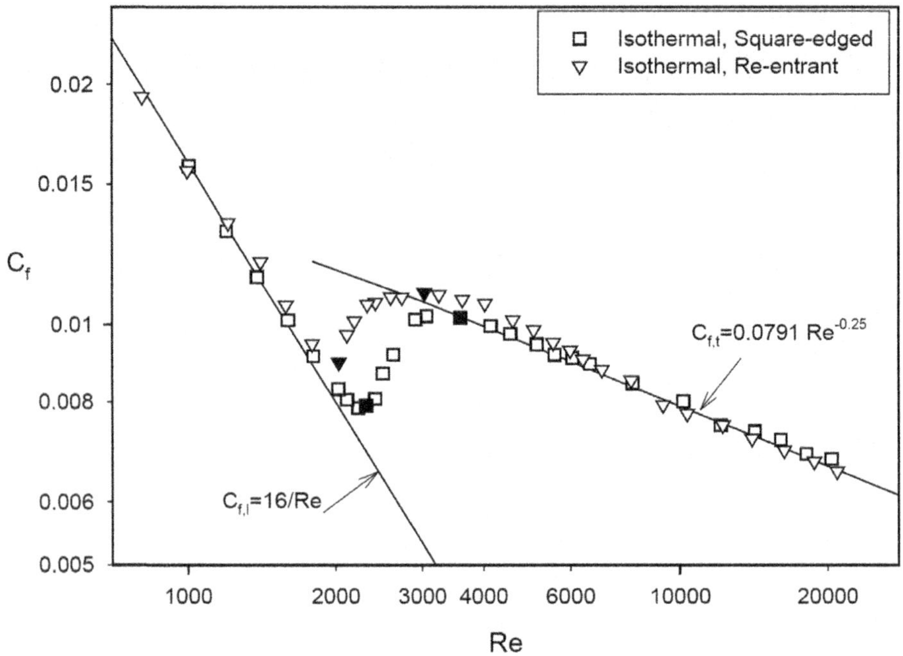

Fig. 3.1 Influence of different inlet configurations on the friction factor in the transition region for a plain tube at 200 diameters from the tube entrance under isothermal boundary condition (solid symbols designate the start and end of the transition region for each inlet)

Figure 3.1 clearly shows the influence of inlet configuration on the start and end of the friction factor transition region under isothermal condition. This figure plots the fully developed friction factors versus the Reynolds numbers for the two inlets in all flow regimes. For the purpose of comparisons, the figure also shows the classic fully developed laminar pipe flow friction factor equation ($C_{f,l} = 16/Re$) and the established Blasius (1913) friction factor correlation for fully developed turbulent pipe flow ($C_{f,t} = 0.0791Re^{-0.25}$). The solid symbols in the figure represent the start and end of the fully developed transition region for the two different inlets. As shown by the solid symbols, the lower and upper limits of the friction factor (pressure drop) transition Reynolds number range are dependent on the inlet configuration. The Reynolds number for the start of the transition region is defined as the Reynolds number corresponding to the first abrupt change in the friction factor, and the Reynolds number for the end of the transition region corresponds to the Reynolds number of the friction factor that first reaches the fully developed turbulent friction factor line. From these data, the limits for the transition Reynolds number range for these two inlet configurations can be summarized as follows: square-edged ($2222 < Re < 3588$) and re-entrant ($2032 < Re < 3031$).

Because the calming section is a common factor for the two inlets (see Fig. 2.3) and the flow is isothermal, the difference between the transitional Reynolds number ranges is due only to the effect of different inlet configuration. The preceding limits for the friction factor transition Reynolds numbers indicate that the inlet that caused the most disturbance (re-entrant) produced an early transition (Re = 2032) and that the inlet with less disturbance (square-edged) did not experience transition until a Reynolds number of about 2222. From these observations, it can be concluded that the transition Reynolds number range can be manipulated by using different inlet configurations.

Tam et al. (2013) used a DC welder to approximate the uniform wall heat flux boundary condition. The results shown in Fig. 3.2 clearly establish the influence of the heating condition on the friction factor. In the laminar and transition regions, heating seems to have a significant influence on the value of the friction factor. However, in the turbulent region, heating did not affect the magnitude of the friction factor. In the turbulent pipe flow region, the Blasius (1913) type fully developed correlation ($C_{f,t} = 0.0791Re^{-0.25}$) for both isothermal and non-isothermal (heating) conditions could be used. According to Ghajar and Tam (1994), the significant influence of heating on the values of the friction factor in the laminar and transition regions is directly due to the effect of secondary flow (the occurrence of the mixed convection). In the turbulent region, the secondary flow effect is suppressed by the turbulent motion (the occurrence of the pure forced convection). Ghajar and Tam (1994) stated that the ratio of the heat transfer coefficient of the top and bottom (h_t/h_b) of the heated tube should be close to unity (0.8–1.0) for forced convection and

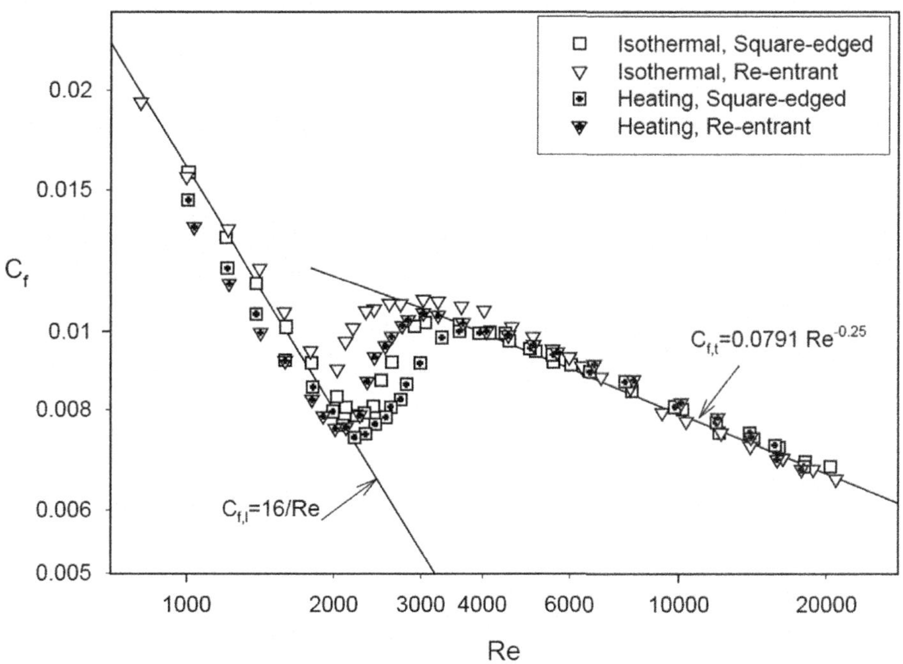

Fig. 3.2 Friction factor characteristics at 200 diameters from the tube entrance under isothermal and heating boundary conditions for different inlet configurations and all flow regimes (laminar, transition, and turbulent)

is much less than unity (<0.8) for a case in which mixed convection exists. They established that when h_t/h_b is less than 0.8, a strong buoyancy effect is present. Figure 3.3, which is based on the more recent experimental data of Tam et al. (2013), supports their findings and shows that the buoyancy effect is significant ($h_t/h_b < 0.8$) when the Reynolds numbers are lower than 2600 (square-edged inlet) and 2400 (re-entrant inlet), respectively.

In their study of the effect of heating on the transition Reynolds numbers, Ghajar and Tam (1994) observed that the increase of heat flux delayed the flow transition or stabilized the flow and caused the flow to go into transition at a higher Reynolds

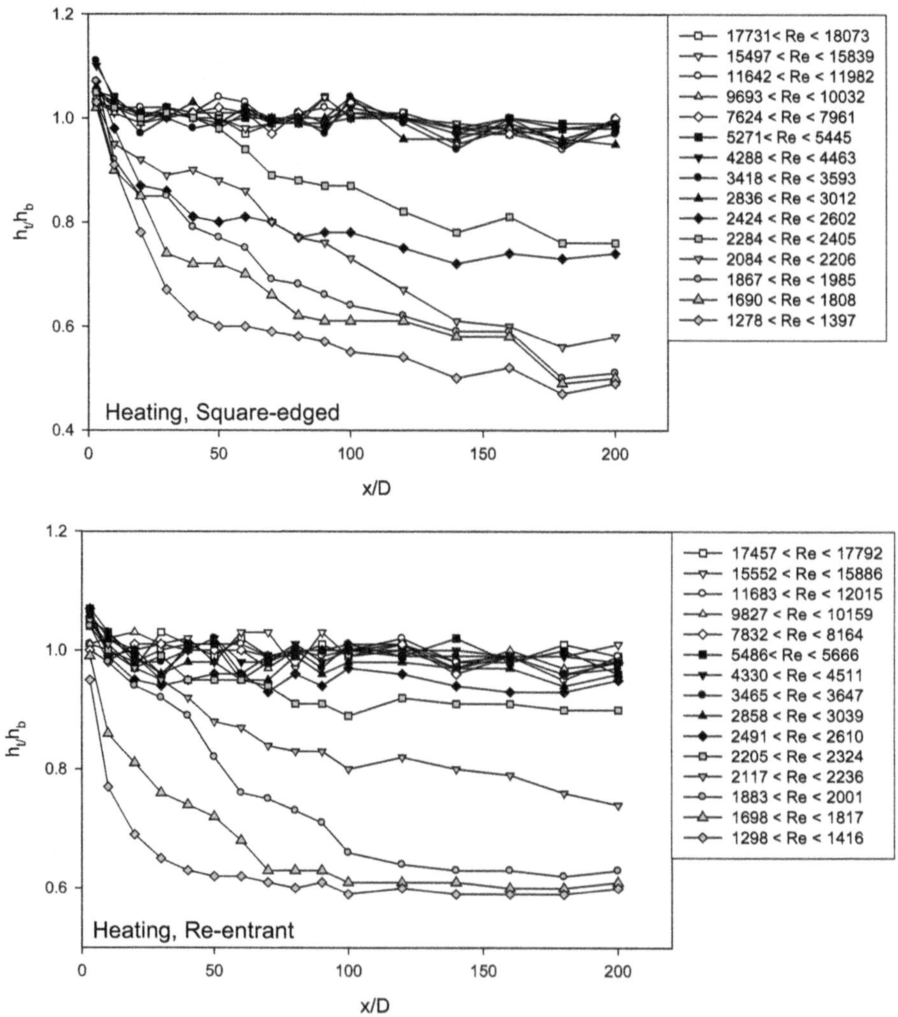

Fig. 3.3 The ratio of heat transfer coefficients at the top and bottom on the tested tube with square-edged and re-entrant inlets and all flow regimes (laminar, transition, and turbulent)

Table 3.1 Start and end of transition of the fully developed friction factor at 200 diameters from the tube entrance

	Friction Factor			
Tube, Condition, Inlet	Re_{start}	C_f	Re_{end}	C_f
Plain, isothermal (square-edged)	2222	0.0079	3588	0.0102
Plain, heating (square-edged)	2316	0.0074	3941	0.0100
Plain, isothermal (re-entrant)	2032	0.0090	3031	0.0110
Plain, heating (re-entrant)	2257	0.0079	3250	0.0105

number. Table 3.1 summarizes the start and end of transition of the fully developed friction factor under isothermal and non-isothermal (heating) conditions. For both inlets, the effect of heating is to delay the start and end of the transition. For the square-edged inlet, the start and end of transition are delayed from Re = 2222 to 2316 and from Re = 3588 to 3941, respectively. For the re-entrant inlet, the start and end of transition are delayed from Re = 2032 to 2257 and from Re = 3031 to 3250, respectively. Thus, both entrance disturbance and heating influence the start and end of the transition region of the fully developed friction factor.

Entrance and Fully Developed Friction Factors

To analyze the effects of inlet geometries and heating on the entrance and fully developed friction factors, the friction factor data of Tam et al. (2013) are arranged into the *apparent friction factor* plots and are shown in Fig. 3.4 for the isothermal and non-isothermal (heating) conditions. The apparent (developing) friction factor accounts for the combined effects of flow acceleration (variation in momentum flux) and surface shear stress. For comparison purposes, the isothermal entrance and fully developed correlation of Shah (1978) given by Eq. (3.1) are also shown in Fig. 3.4.

$$f_{app} = \frac{16}{Re}\left(\frac{3.44}{\zeta^{1/2}} + \frac{0.31/\zeta + 16 - 3.44/\zeta^{1/2}}{1 + 0.00021/\zeta^2}\right) \tag{3.1}$$

where $\zeta = \frac{x/D}{Re}$.

For the isothermal condition, Fig. 3.4 displays the laminar apparent friction factors for the two inlets (re-entrant and square-edged). The figure presents the apparent friction factor (f_{app}) multiplied by the Reynolds number (Re) versus ζ, which is the dimensionless axial location of the pressure taps (x/D) divided by the Reynolds number. In Fig. 3.4, it can be observed that for the square-edged inlet when Re is less 2300, the apparent friction factor data collapse into a single curve ($f_{app}Re$ vs ζ) and that after a certain length, it approaches the fully developed isothermal constant value of the apparent fraction factor ($f_{app}Re = 16$). For Re values greater than 2300, the apparent friction factor becomes a strong function of the Reynolds number. Similar behavior is shown for the re-entrant inlet. When Re is less than

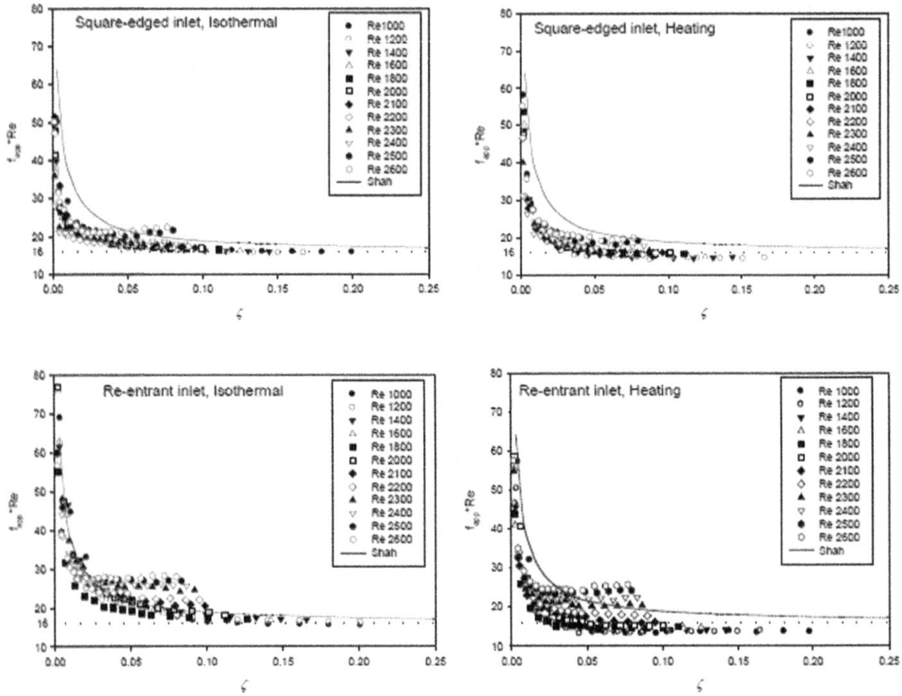

Fig. 3.4 Comparison between experimental laminar flow apparent friction factors with square-edged and re-entrant inlets and correlation of Shah (1978) under isothermal and heating conditions

2100, the apparent friction factor data collapse into a single curve ($f_{app}Re$ vs ζ), and after a certain length, it approaches the fully developed isothermal constant value of the apparent fraction factor ($f_{app}Re = 16$). For Re values greater than 2100, the apparent friction factor becomes a strong function of the Reynolds number. Comparison of the results given in Fig. 3.4 for the two inlets at comparable Reynolds numbers shows that the length required for the square-edged inlet to reach the fully developed isothermal apparent friction factor ($f_{app}Re = 16$) is shorter than what is required for the re-entrant inlet. A longer entrance flow region could be attributed to the additional disturbance caused by the re-entrant inlet. In Fig. 3.4, it also can be observed that the square-edged and re-entrant data do not follow the correlation of Shah (1978), Eq. (3.1). This observation also was made by Ghajar and Madon (1992). Those authors also observed that the Shah correlation was applicable to the bell-mouth inlet (smooth inlet with no appreciable disturbance, see Fig. 2.4) but was not applicable to the square-edged and re-entrant inlets.

For the non-isothermal condition, Fig. 3.4 also provides the laminar apparent friction factors for the two inlets (re-entrant and square-edged), similar to the presentation for the isothermal condition. For comparison purposes, the isothermal entrance and fully developed correlation of Shah (1978) given by Eq. (3.1) are also shown in Fig. 3.4.

As is shown in Fig. 3.4, the trend of the apparent friction factor under the non-isothermal condition is similar to that of the isothermal condition. For Re values less than 2400 (square-edged inlet) and 2200 (re-entrant inlet), the apparent friction factor data collapse into a single curve, and after a certain length, it approaches the fully developed non-isothermal constant value of the apparent fraction factor. However, compared with the isothermal data, the entire non-isothermal apparent friction factor trend is shifted downward as a result of the effect of heating. The apparent friction factor then approaches a constant line ($f_{app}Re = 14$) that is less than the isothermal value of 16. Therefore, it can also be observed that in the fully developed friction factor region (Fig. 3.2), the non-isothermal laminar data do not follow the classic fully developed laminar friction factor ($C_{f,l} = 16/Re$). As can be seen in Fig. 3.4, for Re values greater than 2400 (square-edged inlet) and 2200 (re-entrant inlet), the non-isothermal apparent friction factor becomes a strong function of the Reynolds number. In Fig. 3.4, it is clear that the Shah (1978) isothermal correlation (Eq. 3.1) predicts friction factor values much higher than do Tam et al. (2013) non-isothermal laminar flow apparent friction factor data.

Chapter 4
Proposed Correlations for Laminar and Transition Friction Factors in Plain Tubes with Different Inlet Configurations

Background

An important design problem in industrial heat exchangers arises when flow inside the horizontal tubes fall into the transition region. For proper design of these systems, the experimental data presented in Chap. 3 was used to develop accurate friction factor correlations for the laminar and transitional friction factors in the entrance and fully developed regions of plain tubes under isothermal and non-isothermal (heating) conditions and different inlet configurations (square-edged and re-entrant). For the turbulent region, as discussed in Chap. 3, the Blasius (1913) type fully developed correlation ($C_{f,t} = 0.0791\mathrm{Re}^{-0.25}$) could be used for prediction of isothermal and non-isothermal (heating) friction factors. At the conclusion of this chapter, an illustrative example is presented to demonstrate the use of the proposed correlations.

Laminar Region

Tam et al. (2013) proposed a friction factor correlation for the entrance region that is similar to the correlation of Shah (1978) given by Eq. (3.1). They introduced a correction factor in terms of ζ in the classic relation, $C_{f,l} = 16/\mathrm{Re}$, to account for the variation of the friction factor in the entrance region. In the development of the correlation, they used a total of 223 isothermal experimental data points (130 data points for the square-edged inlet and 93 data points for the re-entrant inlet). For the *isothermal* laminar flow, their proposed correlation is of the following form:

A. J. Ghajar, *Single- and Two-Phase Flow Pressure Drop and Heat Transfer in Tubes*, Mechanical Engineering Series,
https://doi.org/10.1007/978-3-030-87281-6_4

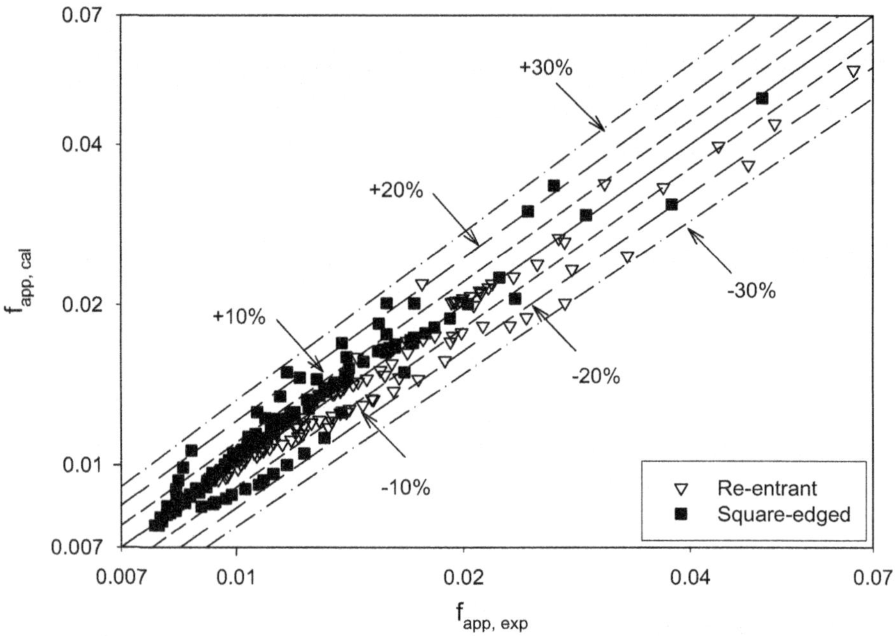

Fig. 4.1 Comparison between experimental entrance and fully developed friction factors and the proposed laminar region correlation (Eq. 4.1) under the isothermal condition

$$f_{app,lam,iso} = \frac{1}{Re}\left(16 + \frac{0.00314}{0.00004836 + 0.0609\,\zeta^{1.28}}\right) \tag{4.1}$$

where $\zeta = \frac{x/D}{Re}$ and the range of application of Eq. (4.1) is 799 < Re < 2240 and 3 < x/D < 200.

Equation (4.1) gives a representation of the experimental data to within +28.1% to −26.1%. The average deviation between the results predicted by the correlation and the experimental data is 6.4%; 6% (13 data points) were predicted with ±20–30% deviation, 17% (38 data points) were predicted with ±10–20% deviation, and 77% (172 data points) were predicted with less than ±10% deviation. Figure 4.1 compares the predicted entrance and fully developed laminar friction factors obtained from the proposed correlation with measurements. Compared with the correlation of Shah (1978), Eq. (4.1) can predict the isothermal apparent friction factor data with much better accuracy.

For the non-isothermal condition, referring to Fig. 3.4, the effect of heating influences the entrance and fully developed friction factors in the laminar region. Therefore, to account for the effect of heating, Tam et al. (2013) introduced a correction factor in terms of the bulk-to-wall-viscosity ratio raised to power m in Eq. (4.1). Similar to the work of Tam and Ghajar (1997), the power m is a function of Prandtl and Grashof numbers. The proposed correlation for *non-isothermal (heated)* laminar flow is of the following form:

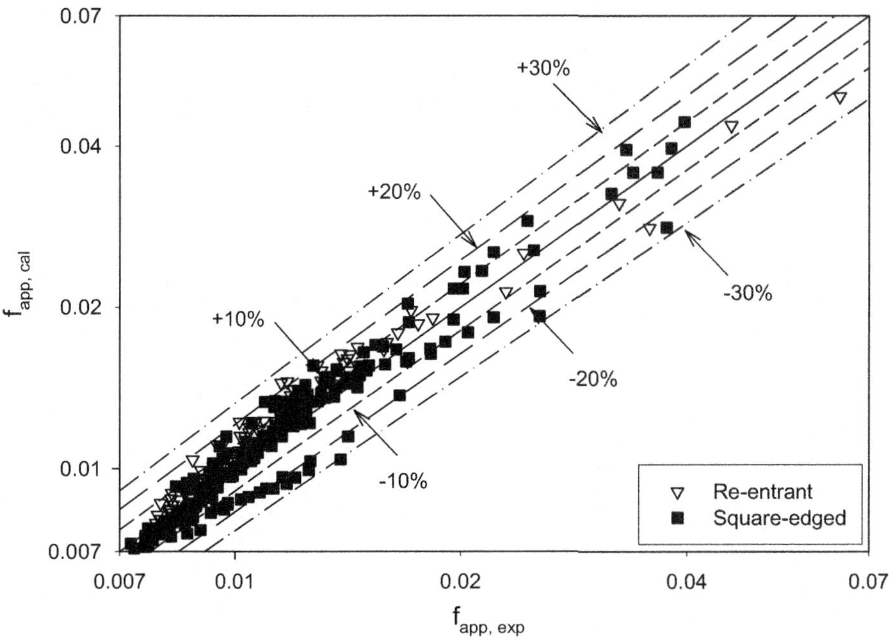

Fig. 4.2 Comparison between experimental entrance and fully developed friction factors and the proposed laminar region correlation (Eq. 4.2) under the non-isothermal condition

$$f_{app,lam,heat} = f_{app,lam,iso}\left(\frac{\mu_b}{\mu_w}\right)^m \tag{4.2}$$

where $m = -5.06 + 0.84\ Pr^{0.23}\ Gr^{0.09}$ and the range of application of Eq. (4.2) is $897 < Re < 2189$, $7141 < Gr < 18{,}224$, $1.27 < \mu_b/\mu_w < 1.56$, and $39 < Pr < 47$.

Equation (4.2) provides a representation of the experimental data to within +25.2% to −29.0%. In the development of the correlation, Tam et al. (2013) used a total of 301 experimental data points (211 data points for the square-edged inlet and 90 data points for the re-entrant inlet). The average deviation between the results predicted by the correlation and the experimental data is 7.6%; 5% (16 data points) were predicted with ±20–30% deviation, 24% (73 data points) were predicted with ±10–20% deviation, and 70% (212 data points) were predicted with less than ±10% deviation. Figure 4.2 compares the predicted apparent friction factors obtained from the proposed correlation with measurements for the non-isothermal condition.

Transition Region

As is shown in Fig. 3.1, the type of inlet configuration influences the start and end of the transition region. Thus, a single correlation cannot predict the data, and a specific correlation for each inlet should be developed. The correlation for the fully developed flow is presented first, and then the correlation for the entrance flow is proposed by adding a correction term. Since the fully developed friction factor in the transition region is an asymptotic problem from the laminar region to the turbulent region, under the isothermal condition, the fully developed correlation proposed by Tam et al. (2013) was based on the asymptotic correlating technique proposed by Churchill and Usagi (1972). The proposed correlation for the isothermal fully developed flow in the transition region is of the following form:

$$C_{f,tr,iso} = \left(\frac{16}{Re}\right)\left\{\left[1 + \left(0.0049\ Re^{0.75}\right)^{a}\right]^{1/a} + b\right\} \tag{4.3}$$

where the coefficients a and b are inlet dependent and are obtained separately for each inlet. The coefficients, the exponents, and their range of application for each inlet are as follows:

- Re-entrant: $a = 0.52$, $b = -3.47$; for $2026 < Re < 3257$; $3 < x/D < 200$.
- Square-edged: $a = 0.50$, $b = -4.0$; for $2111 < Re < 4141$; $3 < x/D < 200$.

Equation (4.3) is applicable to an isothermal and fully developed transition region and should be used with the a and b values for the correct geometry (re-entrant vs square-edged). It gives a representation of the experimental data to within +7.9% to −8.6%. In the development of the correlation, Tam et al. (2013) used a total of 30 data points for each inlet. The average deviation between the results predicted by the correlation and the experimental data is 3.6%. All the data points were predicted with less than 10% deviation. Figure 4.3 compares the predicted fully developed transition region friction factors obtained from the proposed correlation with isothermal measurements.

To account for the entrance effect, Tam et al. (2013) introduced a correction factor in terms of length-to-diameter ratio (x/D) into Eq. (4.3). Therefore, the proposed correlation for the prediction of entrance and fully developed flow in the transition region under isothermal condition is of the following form:

$$f_{app,tr,iso} = C_{f,tr,iso}\left[1 + \left(\frac{c}{x/D}\right)\right] \tag{4.4}$$

where the coefficient c is inlet dependent and is obtained separately for each inlet. The coefficient for each inlet and its range of application are as follows:

- Re-entrant: $c = 4.8$; for $2019 < Re < 3257$; $3 < x/D < 200$.
- Square-edged: $c = 3.0$; for $2109 < Re < 4184$; $3 < x/D < 200$.

Equation (4.4) is applicable to an isothermal entrance and fully developed transition region and should be used with the c value for the correct geometry (re-entrant vs square-edged). For the development of the transition region correlation for the re-entrant inlet, Tam et al. (2013) used 150 experimental data points. The correlation gave a representation of the experimental data to within +21.9% to

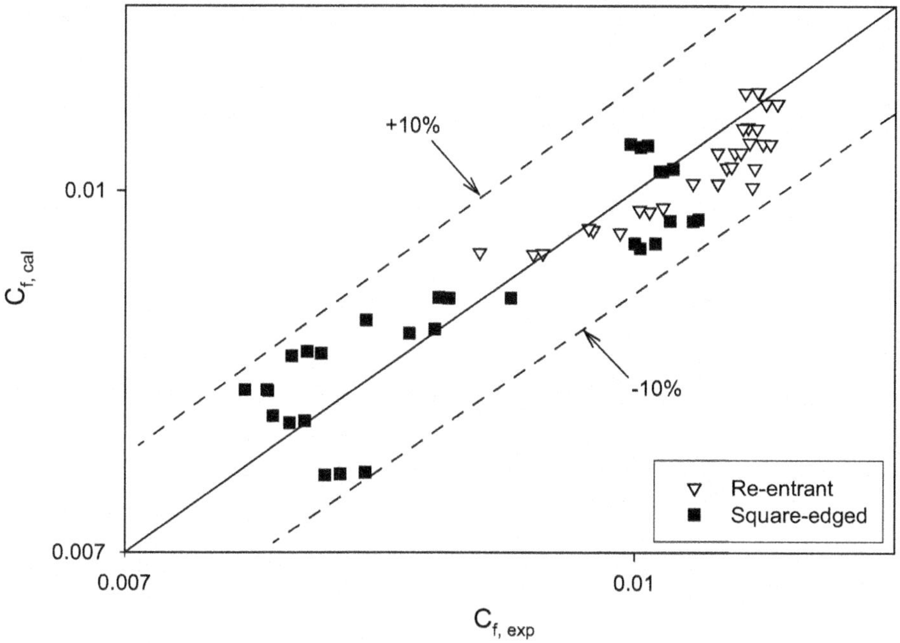

Fig. 4.3 Comparison between experimental fully developed friction factors and the proposed transition region correlation (Eq. 4.3) under the isothermal condition

−25.9% and had an average deviation of 5.9%; 3% (5 data points) were predicted with ±20–30% deviation, 14% (21 data points) were predicted with ±10–20% deviation, and 83% (124 data points) were predicted with less than ±10% deviation. For the square-edged inlet, Tam et al. (2013) used 153 experimental data points for the development of the correlation. The equation correlated the experimental data to within +21.4% to −28.7% and had an average deviation of 7.1%; 4% (6 data points) were predicted with ±20–30% deviation, 15% (23 data points) were predicted with ±10–20% deviation, and 81% (124 data points) were predicted with less than ±10% deviation. Figure 4.4 compares the predicted entrance and fully developed friction factors obtained from the proposed Eq. (4.4) for each inlet with isothermal measurements.

To account for the effect of heating, Tam et al. (2013) applied the viscosity ratio correction factor proposed by Tam and Ghajar (1997) expressed as a function of the Grashof and Prandtl numbers to the isothermal correlation (Eq. 4.4). The proposed correlation is of the following form:

$$f_{app,tr,heat} = f_{app,tr,iso}\left(\frac{\mu_b}{\mu_w}\right)^m \tag{4.5}$$

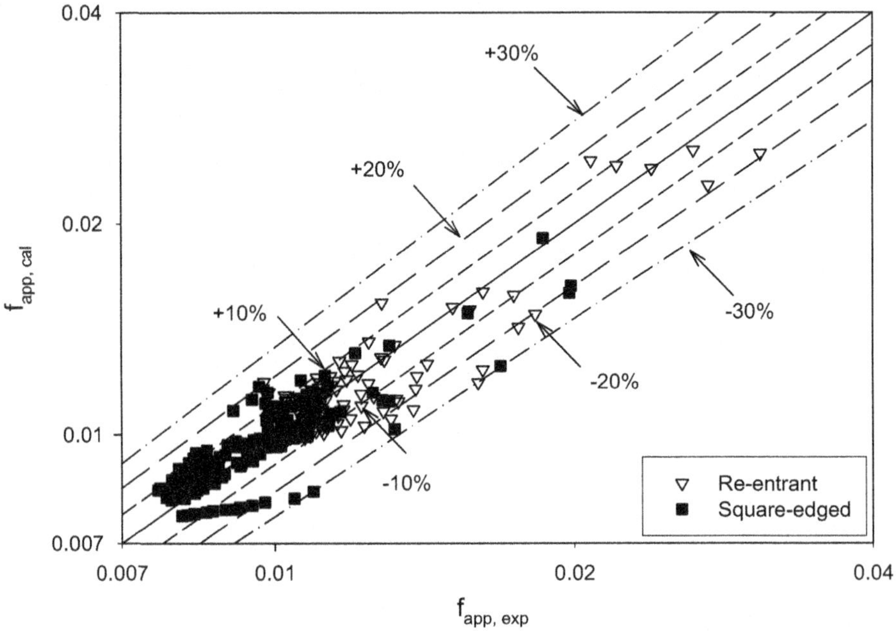

Fig. 4.4 Comparison between experimental entrance and fully developed friction factors and the proposed transition region correlation (Eq. 4.4) under the isothermal condition

where the exponent m is inlet dependent and is obtained separately for each inlet. The exponent of the viscosity ratio term and its range of application for each inlet are as follows:

- Re-entrant: $m = -1.8 + 0.46\ Gr^{-0.13}\ Pr^{0.41}$; for $1883 < Re < 3262$, $19.1 < Pr < 46.5$, $4560 < Gr < 24{,}339$, $1.12 < \mu_b/\mu_w < 1.54$.
- Square-edged: $m = -1.13 + 0.48\ Gr^{-0.15}\ Pr^{0.55}$; for $2084 < Re < 3980$, $19.6 < Pr < 47.3$, $6169 < Gr < 35{,}892$, $1.10 < \mu_b/\mu_w < 1.54$.

Equation (4.5) is applicable to the non-isothermal entrance and fully developed transition regions and should be used with the m value for the correct geometry (re-entrant vs square-edged). For the development of the transition region correlation for the re-entrant inlet, Tam et al. (2013) used 212 experimental data points. The correlation gave a representation of the experimental data to within +27.6% to −28.0% and had an average deviation of 10.2%; 15% (32 data points) were predicted with ±20–30% deviation, 29% (61 data points) were predicted with ±10–20% deviation, and 56% (119 data points) were predicted with less than ±10% deviation. For the square-edged inlet, Tam et al. (2013) used 226 experimental data points for the development of the correlation. The equation correlated the experimental data to within +18.6% to −29.0% and had an average deviation of 9.2%; 8% (19 data points) were predicted with ±20–30% deviation, 31% (69 data points) were predicted with ±10–20% deviation, and 61% (138 data points) were

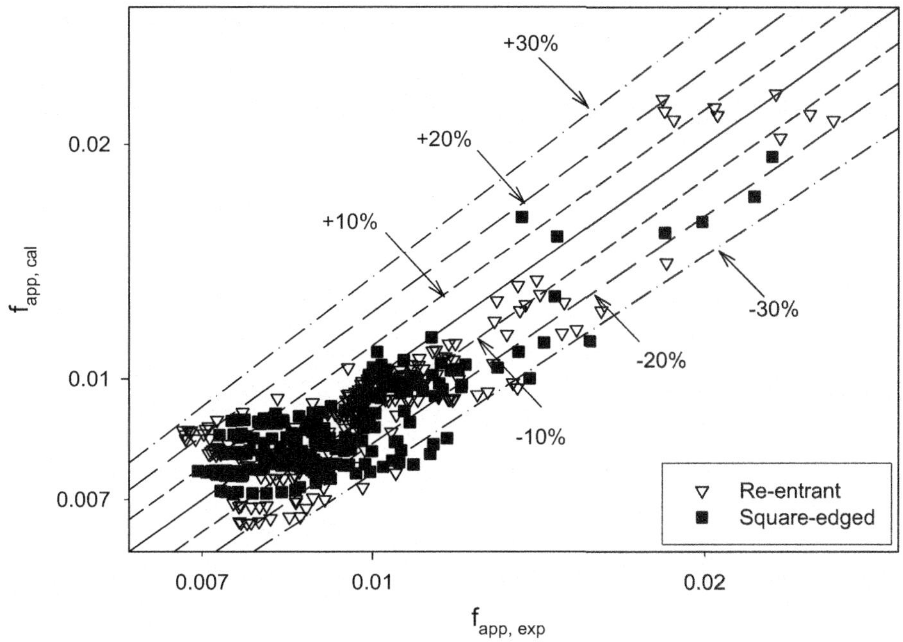

Fig. 4.5 Comparison between experimental entrance and fully developed friction factors and the proposed transition region correlation (Eq. 4.5) under the non-isothermal condition

predicted with less than ±10% deviation. Figure 4.5 compares the predicted entrance and fully developed friction factors obtained from the proposed Eq. (4.5) for each inlet with non-isothermal measurements.

An Illustrative Example

Use of the Non-isothermal Apparent Friction Factor in the Transition Region A 4-m long tube is subjected to uniform wall heat flux. The tube has an inside diameter of $D = 0.0149$ m and a volume flow rate of $Q = 7.8 \times 10^{-5}$ m^3/s. The liquid flowing inside the tube is ethylene glycol-distilled water mixture with a mass fraction of 0.5. Determine the apparent (developing) friction factor at the location $x/D = 20$ if the inlet configuration of the tube is (a) re-entrant and (b) square-edged. At this location, the local Grashof number is $Gr = 28{,}090$, and the properties of the mixture of distilled water and ethylene glycol (with a mass fraction of 0.5) are $Pr = 20.9$, $\nu = 2.33 \times 10^{-6}$ m^2/s, and $\mu_b/\mu_w = 1.25$.

Solution

For the calculation of the non-isothermal apparent (developing) friction factor, it is necessary to determine the flow regime before deciding about which friction factor relation to use. The Reynolds number at the specified location is:

$$Re = \frac{(Q/A_c)D}{\nu} = \frac{\left[\left(7.8 \times 10^{-5} m^3/s\right)/\left(1.744 \times 10^{-4}\ m^2\right)\right](0.0149\ m)}{2.33 \times 10^{-6} m^2/s} = 2860$$

since $A_c = \pi D^2/4 = \pi(0.0149\ m)^2/4 = 1.744 \times 10^{-4}\ m^2$.

From Table 3.1, we see that for both inlets, with heating the flow is in the transition region. Therefore, Eq. (4.5) applies.

$$f_{app,tr,heat} = f_{app,tr,iso}\left(\frac{\mu_b}{\mu_w}\right)^m$$

where $f_{app,\ tr,\ iso}$ is calculated from Eq. (4.4) and $C_{f,\ tr,\ iso}$ is based on Eq. (4.3):

$$f_{app,tr,iso} = C_{f,tr,iso}\left[1 + \left(\frac{c}{x/D}\right)\right]$$

$$C_{f,tr,iso} = \left(\frac{16}{Re}\right)\left\{\left[1 + \left(0.0049\ Re^{0.75}\right)^a\right]^{1/a} + b\right\}$$

Introducing Eqs. (4.3) and (4.4) into Eq. (4.5), the final equation for the calculation of friction factor in this case becomes:

$$f_{\text{app,tr,heat}} = \left\{\left(\frac{16}{Re}\right)\left[\left(1 + \left(0.0049\ Re^{0.75}\right)^a\right)^{1/a} + b\right]\left[1 + \left(\frac{c}{x/D}\right)\right]\right\}\left(\frac{\mu_b}{\mu_w}\right)^m$$

Note that the values of m and the constants a, b, and c are inlet dependent and are provided with the appropriate equations.

(a) Re-entrant inlet:

$$\begin{aligned} \text{m} &= -1.8 + 0.46\ \text{Gr}^{-0.13}\ \text{Pr}^{0.41} = -1.8 + 0.46\ (28,090)^{-0.13}\ (20.9)^{0.41} \\ &= -1.3776 \end{aligned}$$

$$\text{a} = 0.52, \text{b} = -3.47, \text{c} = 4.8$$

$$f_{\text{app,tr,heat}} =$$

$$\left\{\left(\frac{16}{2860}\right)\left[\left(1 + \left(0.0049 \times 2860^{0.75}\right)^{0.52}\right)^{1/0.52} - 3.47\right]\left[1 + \left(\frac{4.8}{20}\right)\right]\right\}(1.25)^{-1.3776}$$

$$f_{app,tr,heat} = 0.009820$$

(b) Square-edged inlet:

$$\begin{aligned} m &= -1.13 + 0.48\ Gr^{-0.15}\ Pr^{0.55} = -1.13 + 0.48\ (28,090)^{-0.15}\ (20.9)^{0.55} \\ &= -0.58041 \end{aligned}$$

$$a = 0.50, b = -4.0, c = 3.0$$

$$f_{app,tr,heat} = \left\{\left(\frac{16}{2860}\right)\left[\left(1+\left(0.0049\times 2860^{0.75}\right)^{0.5}\right)^{1/0.5} - 4\right]\left[1+\left(\frac{3}{20}\right)\right]\right\}(1.25)^{-0.58041}$$

$$f_{app,tr,heat} = 0.009523$$

Discussion If the flow is considered to be isothermal, in the above calculations, the viscosity ratio should be set to unity ($m = 0$), or one should use in Eq. (4.5). The apparent friction factor for the re-entrant inlet would be $f_{app,tr,iso} = 0.01335$ (about a 36% increase), and for the square-edged inlet, it would be $f_{app,tr,iso} = 0.01084$ (about a 14% increase). Heating causes a decrease in the friction factor.

Chapter 5
Heat Transfer Results in Plain Tubes with Different Inlet Configurations

Background

Heating a fluid in a horizontal tube produces a secondary flow (buoyancy). Buoyancy influences forced convection heat transfer in a horizontal tube in ways that depend on the Reynolds, Prandtl, and Grashof numbers. The type of inlet configuration significantly influences the establishment of secondary flow, the development of the heat transfer coefficient along the pipe, and the start and end of the heat transfer transition region. These types of information are critical for the proper design of heat exchangers. For this purpose, using the experimental setup detailed in Chap. 2, detailed experiments were conducted to obtain accurate and broad heat transfer database across all flow regimes (laminar, transitional, and turbulent). The effect of inlet configuration and buoyancy on heat transfer in different flow regimes is presented in this chapter. In addition, the variation of local Nusselt number along the tube in the laminar, transitional, and turbulent regions is discussed. In particular the unusual variation of local heat transfer coefficient in a circular tube with a bell-mouth inlet is highlighted.

Effect of Inlet Configuration on Plain Tube Heat Transfer

Ghajar and coworkers also experimentally investigated the inlet configuration effects on heat transfer in the transition region between laminar and turbulent flows in plain circular horizontal tubes. For this investigation, they used the same two inlet configurations (re-entrant and square-edged) used for the friction factor (pressure drop) studies and added a third inlet (bell-mouth), as shown in Fig. 2.4. In a heat exchanger, the bell-mouth inlet represents a tapered entrance of tube from the tube sheet face.

A. J. Ghajar, *Single- and Two-Phase Flow Pressure Drop and Heat Transfer in Tubes*, Mechanical Engineering Series,
https://doi.org/10.1007/978-3-030-87281-6_5

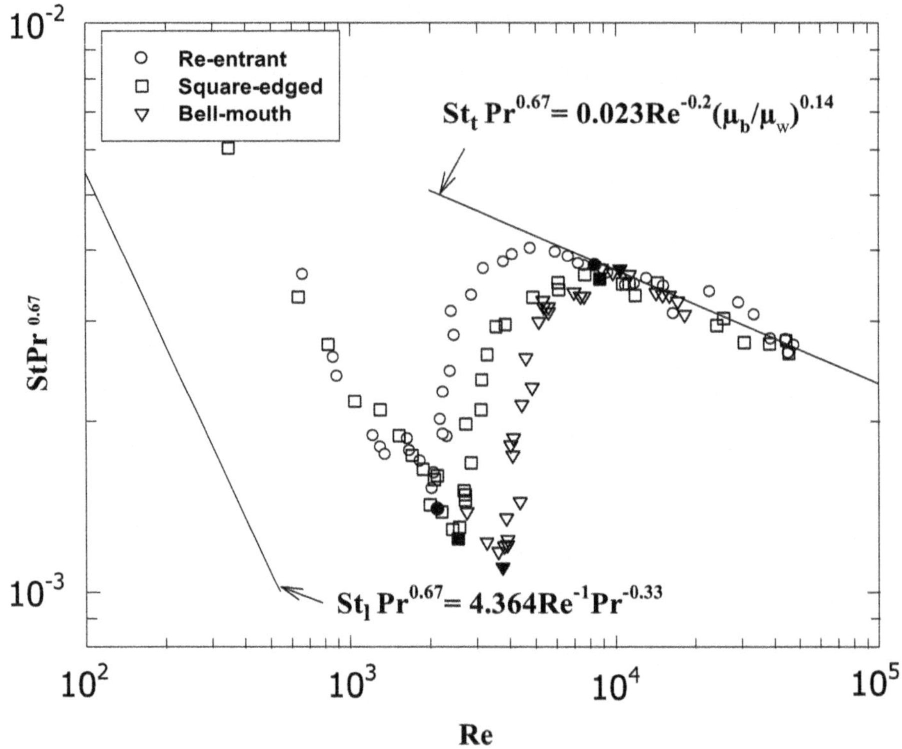

Fig. 5.1 Influence of different inlet configurations on heat transfer in the transition region for a plain tube at 200 tube diameters from the tube entrance (filled symbols designate the start and end of the transition region for each inlet)

Figure 5.1 clearly shows the influence of inlet configuration on the start and end of the heat transfer transition region. This figure plots the local average peripheral heat transfer coefficients in terms of the Colburn j factor ($j = St\ Pr^{0.67}$) versus the local Reynolds number for all flow regimes at the length-to-diameter ratio of 200. The filled symbols in Fig. 5.1 represent the start and end of the heat transfer transition region for each inlet configuration. The presented results show how heat transfer in the transition region varies between the limits of the Dittus and Boelter (1930) correlation with Sieder and Tate (1936) viscosity property correction factor [$Nu_t = 0.023Re^{0.8}Pr^{0.33}(\mu_b/\mu_w)^{0.14}$] for fully developed turbulent flow and $Nu_l = 4.364$ for fully developed laminar flow with a uniform wall heat flux boundary condition. Note the large influence of natural convection (buoyancy effect) superimposed on the forced convective laminar flow heat transfer process ($Nu_l = 4.364$ for a fully developed laminar flow with a uniform heat flux boundary condition without a buoyancy effect), yielding a much larger mixed convection value of about $Nu = 14.5$.

Table 5.1 Lower and upper limits of heat transfer transition Reynolds numbers

Inlet Geometry	Lower Limit	Upper Limit
Re-entrant	$Re_{lower} = 2157 - 0.65[192 - (x/D)]$	$Re_{upper} = 8475 - 9.28[192 - (x/D)]$
Square-edged	$Re_{lower} = 2524 - 0.82[192 - (x/D)]$	$Re_{upper} = 8791 - 7.69[192 - (x/D)]$
Bell-mouth	$Re_{lower} = 3787 - 1.80[192 - (x/D)]$	$Re_{upper} = 10481 - 5.47[192 - (x/D)]$

As shown by the filled symbols in Fig. 5.1, the lower and upper limits of the heat transfer transition Reynolds number range depend on the inlet configuration. In addition, these transition Reynolds number limits are *x*/D dependent, and they linearly increase with an increase in *x*/D, as reported by Ghajar and Tam (1995). To determine the range of heat transfer transition Reynolds numbers along the tube, Ghajar and Tam (1995) used their experimental data and developed figures similar to Fig. 5.1 for 20 other *x*/D locations. From those figures, the heat transfer transition Reynolds number range for each inlet was determined to be about 2000–8500 for the re-entrant inlet, 2400–8800 for the square-edged inlet, and 3400–10,500 for the bell-mouth inlet. The lower and upper limits of the heat transfer transition Reynolds number ranges for the three different inlets along the tube ($3 \leq x/D \leq 200$) are summarized in Table 5.1. The results in this table indicate that the re-entrant inlet configuration causes the earliest transition from laminar flow into the transition regime (at about 2000), whereas the bell-mouth entrance retards this regime change (at about 3400). The square-edged entrance falls in between (at about 2400), which is close to the frequently quoted value of 2300 in most textbooks.

Effect of Buoyancy (Secondary Flow) on Plain Tube Heat Transfer Coefficient

Application of heat to the tube wall produces a temperature difference in the fluid. The fluid near the tube wall has a higher temperature and a lower density than the fluid close to the centerline of the tube. This temperature difference may produce a secondary flow caused by free convection. In the laminar flow region, the effect of free convection (or buoyancy) on forced convection can be seen clearly in Fig. 5.1, and it resulted in an upward parallel shift of the Colburn j factors from their fully developed forced convection laminar values. In the transition region, the effect of mixed convection cannot be seen easily unless the local heat transfer information, which is the ratio of the local peripheral heat transfer coefficient at the top of the tube to the local peripheral heat transfer coefficient at the bottom of the tube (h_t/h_b), is examined carefully. To account for the effect of mixed convection properly, in their experiments, Ghajar and Tam (1994) used four thermocouples around the periphery of the tube (90° apart) at each of the 26 designated axial locations along the tube. The thermocouples were at close intervals near the entrance of the tube and at greater intervals farther downstream as shown in Fig. 5.2.

According to Ghajar and Tam (1994) (as mentioned in Chap. 3 in the context of the influence of heating on the friction factor; see Fig. 3.3), h_t/h_b should be close to

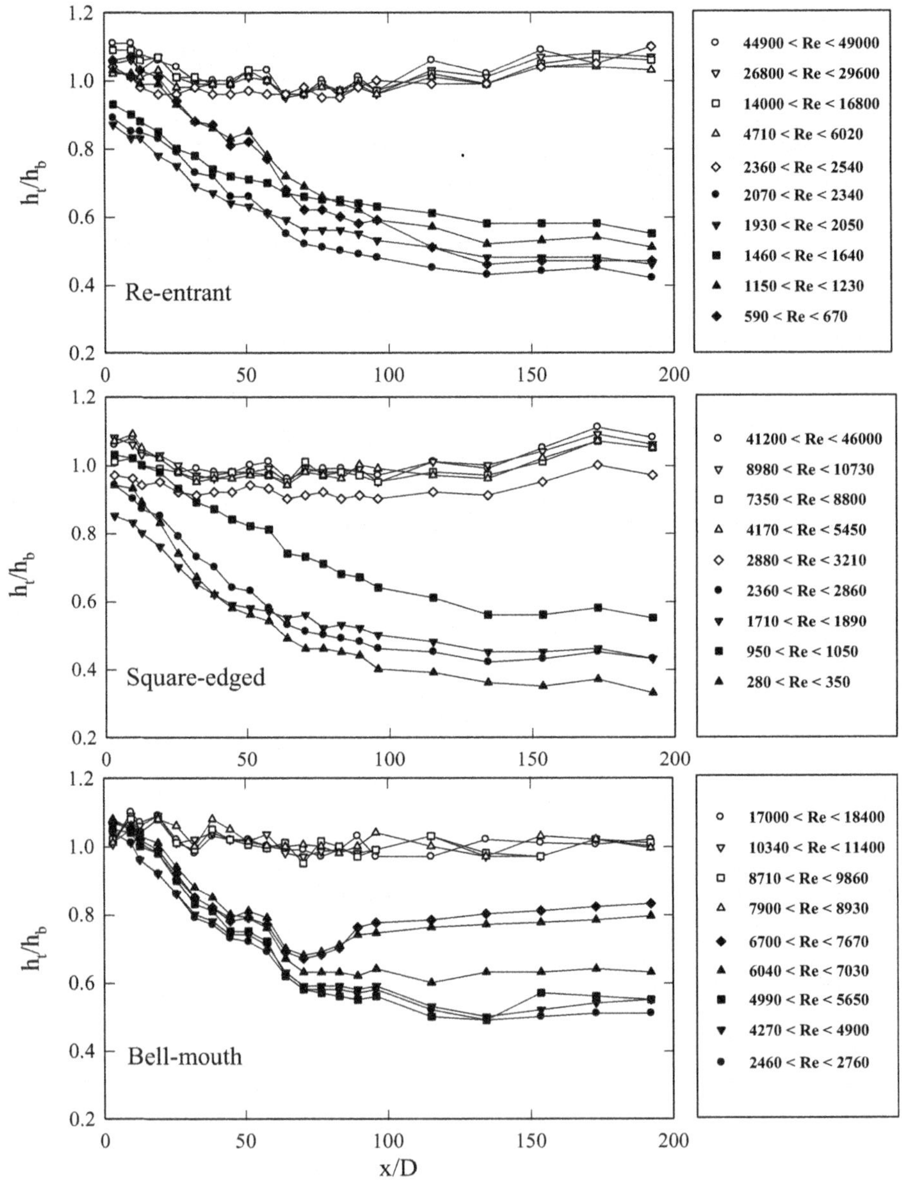

Fig. 5.2 Effect of secondary flow on heat transfer coefficient for different inlets and flow regimes

unity (0.8–1.0) for forced convection and is much less than unity (< 0.8) for a case in which mixed convection exists. Figure 5.2, which is similar to Fig. 3.3, shows the effect of secondary flow on the heat transfer coefficient for three different inlets and all flow regimes. For the re-entrant, square-edged, and bell-mouth inlets when the Reynolds numbers were greater than 2500, 3000, and 8000, respectively, the flows were dominated by forced convection heat transfer, and the heat transfer coefficient ratios (h_t/h_b) did not fall below 0.8–0.9 and at times exceeded unity as a result of rounding-off errors in the property evaluation calculation of Ghajar and Tam (1994) data reduction program. The flows dominated by mixed convection heat transfer had h_t/h_b ratios start near 1 but dropping off rapidly as the length-to-diameter ratio (*x*/D) increased. Beyond about 125 diameters from the tube entrance, the ratio was almost invariant with *x*/D, indicating a much less dominant role for forced convection heat transfer and increased free convection activity. In Fig. 5.2, it is interesting to observe that the starting length necessary for the establishment of the free convection effect for low Reynolds number flows is also inlet dependent. When the secondary flow is established, a sharp decrease in h_t/h_b occurs.

Depending on the type of inlet configuration, for low Reynolds number flows (Re < 2500 for re-entrant, Re < 3000 for square-edged, and Re < 8000 for bell-mouth), the flow is dominated by forced convection over the first 20 to 70 diameters from the entrance to the tube. It should be noted that the reported lower and upper limits of heat transfer transition Reynolds numbers in Table 5.1 are not influenced by the presence of mixed convection. However, as the flow travels the tube length required for the establishment of secondary flow, the lower transition region for all three inlets will be influenced by the presence of mixed convection. It should be pointed out that for the bell-mouth inlet, the mixed convection effect will influence not only the lower transition region but also the upper part of the transition region, as shown by Tam and Ghajar (1998). These results will be discussed next.

Variation of Local Nusselt Number Along the Plain Tube with Different Inlet Configuration in Different Flow Regimes

In a focused experimental study, Tam and Ghajar (1998) further explored the unusual behavior of the local heat transfer coefficients in the transition region for a tube with a bell-mouth inlet. This type of inlet is used in some heat exchangers mainly to avoid the presence of eddies, which are believed to be one of the causes of erosion in the tube inlet region. Figure 5.3 shows the variation of the local Nusselt number along the tube length (*x/D*) in the transition region for the three inlet configurations at comparable Reynolds numbers. As is shown in Fig. 5.3, the re-entrant and square-edged inlets show no influence on the local heat transfer coefficients. For these inlets, the local Nusselt number has a minimum value at an *x/D* that is approximately equal to 25 and increases monotonically along the tube rather than staying at a relatively constant value, as is the expected behavior in the

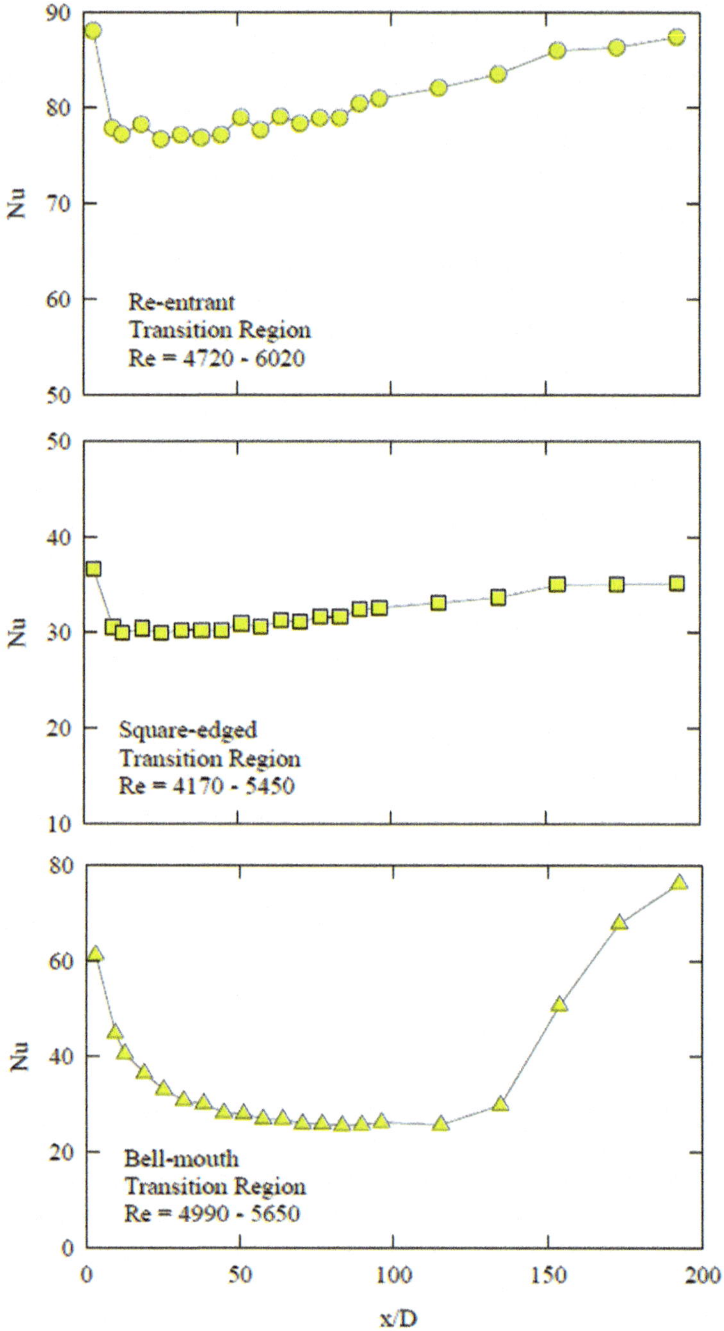

Fig. 5.3 Variation of local Nusselt number with length for re-entrant, square-edged, and bell-mouth inlets in the transition region

fully developed laminar and turbulent flows (Fig. 5.4). Considering Newton's law of cooling for a uniform wall heat flux boundary condition, the increase in the Nu values with the tube length is associated with the decrease in the difference between the inside wall and fluid bulk temperatures ($T_w - T_b$) along the tube. The inside wall temperature and the fluid bulk temperature increase monotonically after they have reached their fully developed values, but the fluid bulk temperature has a larger rate of increase. This causes the heat transfer coefficient to increase along the tube.

However, for the bell-mouth inlet, the variation of the local heat transfer coefficient with tube length in the transition (Fig. 5.3) and turbulent (Fig. 5.4) flow regions is very unusual. Tam and Ghajar (1998) in their work further investigated this unusual behavior of the bell-mouth inlet by looking into the influence of Reynolds number and inlet disturbance on the length of the dip in the transition region. To investigate the effect of inlet disturbance on the bell-mouth entrance, three different screen sizes were used in the last section of the calming section before the fluid entered the bell-mouth entrance (see Fig. 2.3). The three different plastic mesh screens were referred to as coarse, medium, and fine mesh screens with open area ratios of 0.825, 0.759, and 0.650, respectively. For their experiments, the length of the dip varied from $x/D = 100$ to 175 depending on the level of disturbance at the inlet and the transition flow Reynolds number.

The results of the study showed that the boundary layer along the tube wall is first laminar and then changes through a transition to the turbulent condition, causing a dip in the Nu versus x/D curve. The length of the dip decreases with the increase of Reynolds number. The length of the dip in the transition region is much longer ($100 < x/D < 175$) than it is in the turbulent region ($x/D < 25$), as is shown in Figs. 5.3 and 5.4, respectively. Hence, a mixed convection effect is strongly present even in the high Reynolds numbers' upper transition region. The presence of the dip in the transition region causes a significant influence in both the local and the average heat transfer coefficients. This is particularly important for heat transfer calculations in short tube heat exchangers with a bell-mouth inlet.

Simultaneous Heat Transfer and Friction Factor for Plain Tube with Different Inlet Configurations

In most studies, heat transfer and friction factor characteristics for plain tubes usually are represented and discussed independently. The simultaneous representation for the heat transfer and friction factor characteristics may not only help in observing the difference of the transition range between the heat transfer and the friction factor but also inspire the researchers to study the relationship between the heat transfer and the friction factor. Therefore, the plain tube heat transfer and pressure drop data (friction factor), which were measured simultaneously in the experiments of Tam et al. (2013), were plotted together in a single graph, Fig. 5.5. The figure includes the fully developed heat transfer data (with uniform wall heat flux boundary condition)

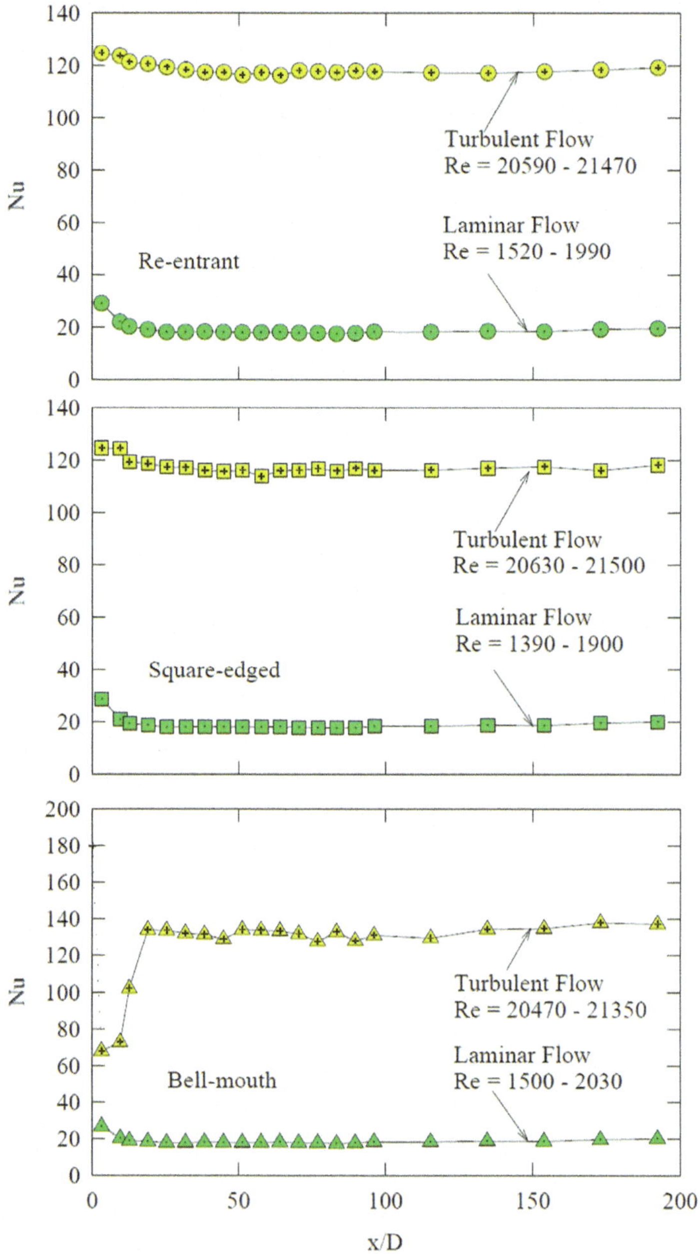

Fig. 5.4 Variation of local Nusselt number with length for re-entrant, square-edged, and bell-mouth inlets in the laminar and turbulent regions

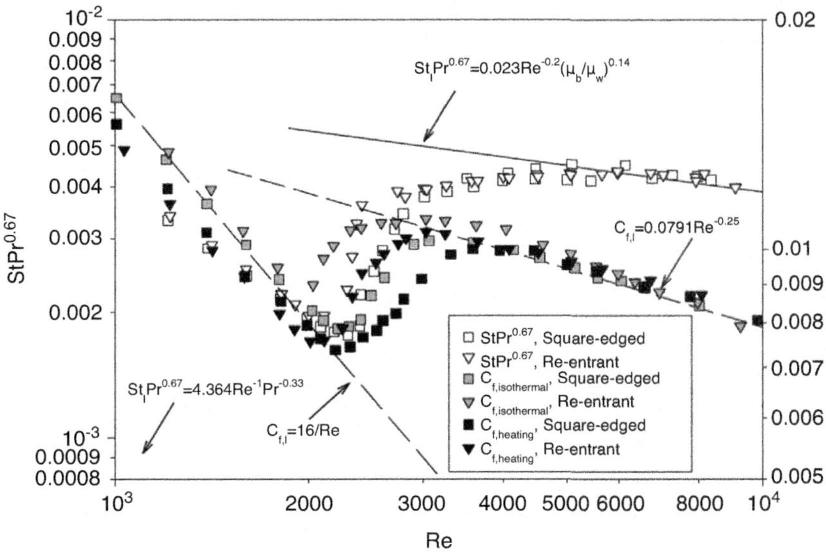

Fig. 5.5 Simultaneous heat transfer and friction factor characteristics of plain tube at 200 diameters from the tube entrance for square-edged and re-entrant inlets under different wall boundary conditions

Table 5.2 Start and end of the transition of the fully developed heat transfer and friction factor at 200 diameters from the plain tube entrance for square-edged and re-entrant inlets

	Heat Transfer				Friction Factor			
Condition, Inlet	Re_{start}	$StPr^{0.7}$	Re_{end}	$StPr^{0.7}$	Re_{start}	C_f	Re_{end}	C_f
Isothermal (square-edged)	–	–	–	–	2222	0.0079	3588	0.0102
Heating (square-edged)	2298	0.0017	8357	0.0041	2316	0.0074	3941	0.0100
Isothermal (re-entrant)	–	–	–	–	2032	0.0090	3031	0.0110
Heating (re-entrant)	2001	0.0020	7919	0.00412	2257	0.0079	3250	0.0105

and the fully developed friction factor data (with isothermal and uniform wall heat flux boundary conditions) for the square-edged and re-entrant inlets. On the basis of Fig. 5.5, the transition Reynolds numbers for heat transfer and the friction factor are summarized in Table 5.2.

In the laminar region, it can be observed from Fig. 5.5 that the heating friction factor is slightly lower than the isothermal friction factor as a result of the effect of heating. In the same plot, the laminar heat transfer data have a downward trend similar to that of the laminar friction factor data. When the Reynolds number reaches about 2300, the heat transfer and the friction factor start to move away from the laminar region (start of the transition region). As was discussed in Chap. 3, the start of transition for the friction factor is inlet dependent. Moreover, the heating condition also influences the start of transition of the friction factor. As shown in Table 5.2, the start of transition for heat transfer is also inlet dependent. The square-edged inlet (Re = 2298) will delay the start of transition compared to the re-entrant inlet (Re = 2001). For the heating condition (Table 5.2), it can be observed that the transition Reynolds number for heat transfer (Re = 2298 for square-edged and Re = 2001 for re-entrant inlets) is slightly earlier than that of the heating (non-isothermal) friction factor (Re = 2316 for square-edged and Re = 2257 for re-entrant inlets).

As shown in Fig. 5.5, when the Reynolds number reaches approximately 3000, all the friction factor data start to decrease and follow the fully developed turbulent friction factor line ($Cf = 0.0791Re^{-0.25}$). However, the heat transfer data continue to increase until a Reynolds number of about 8000. Around the Reynolds number of 8000, the heat transfer data start to follow the fully developed turbulent pipe flow heat transfer correlation [$St_t Pr^{0.67} = 0.023Re^{-0.2}(\mu_b/\mu_w)^{0.14}$]. It can be observed clearly in Table 5.2 that the end of transition for the heat transfer (Re = 8357 for square-edged and Re = 7919 for re-entrant inlets) is much later than that of the heating friction factor (Re = 3941 for square-edged and Re = 3250 for re-entrant inlets). On the basis of the start and end of transition, it can be observed that heat transfer has a much wider transition range than the friction factor. For the turbulent region, the heat transfer and the friction factor data will continue to decrease along the fully developed turbulent heat transfer and friction factor lines.

Chapter 6
Proposed Correlations and Flow Regime Map for Laminar, Transition, and Turbulent Heat Transfer in Plain Tubes with Different Inlet Configurations

Background

An important design problem in industrial heat exchangers arises when flow inside the horizontal tubes fall into the transition region. For proper design of these systems, the experimental data presented in Chap. 3 was used to develop accurate heat transfer correlations for the laminar, transitional, and turbulent dimensionless heat transfer coefficients (Nusselt number) in the entrance and fully developed regions of plain tubes under uniform wall heat flux boundary condition and different inlet configurations (square-edged and re-entrant). The proposed correlations were used to develop a flow regime map. The flow regime map provided in this study can be used to assist the heat exchanger designer in determining the influence of buoyancy in a horizontal circular tube with uniform wall heat flux boundary condition for a specified inlet configuration for all flow regimes. For the identified pure or mixed convection heat transfer regimes, heat transfer calculations can be made based on the recommended correlations. In addition, an illustrative example is presented to demonstrate the use of the proposed correlations.

Proposed Correlations for All Flow Regimes

In developing their transition region heat transfer correlations, Ghajar and Tam (1994) took an approach completely different from that of the previous investigators. They first performed very careful experiments in the transition region, paying particular attention to the role of secondary flow (free convection superimposed on the forced convection or mixed convection) and inlet configuration effects on the start and end of the transition region and the magnitude of heat transfer in this region. After careful analysis of their transitional heat transfer data, following the general

A. J. Ghajar, *Single- and Two-Phase Flow Pressure Drop and Heat Transfer in Tubes*, Mechanical Engineering Series,
https://doi.org/10.1007/978-3-030-87281-6_6

form of Churchill's (1977a) correlation, they proposed some prediction methods for this region to bridge between laminar correlations and turbulent correlations, applicable to forced and mixed convection in the entrance and fully developed regions for three types of inlet configurations (re-entrant, square-edged, and bell-mouth), which are presented below. The local heat transfer coefficient in transitional flow is obtained from the transitional Nusselt number, $\mathrm{Nu}_{\mathrm{trans}}$, which is calculated as follows at a distance x from the tube entrance:

$$\mathrm{Nu}_{\mathrm{tr}} = \mathrm{Nu}_l + \left\{ \exp\left[\left(a - \mathrm{Re}\right)/b\right] + \mathrm{Nu}_t^c \right\}^c \tag{6.1}$$

where Nu_l is the laminar flow Nusselt number and Nu_t is the turbulent flow Nusselt number given by Eqs. (6.2) and (6.3), respectively.

Ghajar and Tam (1994) used a total of 546 experimental data points in the entrance and fully developed laminar regions with natural convection effects and proposed the following correlation for the laminar flow Nusselt number (Nu_l), which represented 86% of experimental data with less than ±10% deviation and 100% of measured data with less than ±17% deviation:

$$\mathrm{Nu}_l = 1.24\left[\left(\frac{\mathrm{RePr}D}{x}\right) + 0.025(\mathrm{GrPr})^{0.75}\right]^{1/3}\left(\frac{\mu_b}{\mu_w}\right)^{0.14} \tag{6.2}$$

where $3 \leq x/D \leq 200$, $280 \leq \mathrm{Re} \leq 3800$, $40 \leq \mathrm{Pr} \leq 160$, $1000 \leq \mathrm{Gr} \leq 28{,}000$, and $1.2 \leq \mu_b/\mu_w \leq 3.8$.

For the turbulent flow Nusselt number (Nu_t), Ghajar and Tam (1994) used a total of 604 experimental data points in the entrance and fully developed turbulent regions and developed the following correlation, which correlated 100% of experimental data with less than ±11% deviation and 73% of measured data with less than ±5% deviation:

$$\mathrm{Nu}_t = 0.023\,\mathrm{Re}^{0.8}\mathrm{Pr}^{0.385}\left(\frac{x}{D}\right)^{-0.0054}\left(\frac{\mu_b}{\mu_w}\right)^{0.14} \tag{6.3}$$

where $3 \leq x/D \leq 200$, $7000 \leq \mathrm{Re} \leq 49{,}000$, $4 \leq \mathrm{Pr} \leq 34$, and $1.1 \leq \mu_b/\mu_w \leq 1.7$.

The physical properties (k, μ, ρ, c_p, β) appearing in the dimensionless numbers (Nu, Re, Pr, Gr) are all evaluated at the bulk fluid temperature (T_b) unless otherwise specified. The values of the empirical constants a, b, and c in Eq. (6.1) depend on the inlet configuration and are given in Table 6.1. The viscosity ratio accounts for the temperature effect on the process. The range of application of the heat transfer correlation based on Ghajar and Tam's (1994) database of 1290 data points (441 points for re-entrant inlet, 416 points for square-edged inlet, and 433 points for bell-mouth inlet) is as follows:

Re-entrant: $3 \leq x/D \leq 200$, $1700 \leq \mathrm{Re} \leq 9100$, $5 \leq \mathrm{Pr} \leq 51$, $4000 \leq \mathrm{Gr} \leq 210{,}000$, $1.2 \leq \mu_b/\mu_w \leq 2.2$

Table 6.1 Constants for transition heat transfer correlation Eq. (6.1) for plain tube and each inlet type

Inlet geometry	a	b	c
Re-entrant	1766	276	−0.955
Square-edged	2617	207	−0.950
Bell-mouth	6628	237	−0.980

Square-edged: $3 \le x/D \le 200$, $1600 \le \text{Re} \le 10{,}700$, $5 \le \text{Pr} \le 55$, $4000 \le \text{Gr} \le 250{,}000$, $1.2 \le \mu_b/\mu_w \le 2.6$

Bell-mouth: $3 \le x/D \le 200$, $3300 \le \text{Re} \le 11{,}100$, $13 \le \text{Pr} \le 77$, $6000 \le \text{Gr} \le 110{,}000$, $1.2 \le \mu_b/\mu_w \le 3.1$

These correlations capture about 70% of measured data within a deviation band of ±10% and 97% of measured data within ±20%, which is remarkable for transitional flows. The individual expressions above for Nu_l and Nu_t can be used alone for developing and fully developed flows in those respective regimes. The lower and upper limits of the heat transfer transition Reynolds number ranges for the three different inlets are summarized in Table 5.1.

A detailed comparison of the performance of Eq. (6.1) with the experimental data of Ghajar and Tam (1994) is shown in Figs. 6.1, 6.2, and 6.3 for the re-entrant, square-edged, and bell-mouth inlets, respectively. Equation (6.1) predicted 96.9%, 99.7%, and 98.9% of the data in the mixed convection region with less than 20% deviation for the re-entrant, square-edged, and bell-mouth inlets, respectively. In the forced convection region, Eq. (6.1) predicted 97.2% (re-entrant inlet), 90.8% (square-edged inlet), and 100% (bell-mouth inlet) of the experimental data with less than 20% deviation.

In a subsequent study, Tam and Ghajar (2006) conducted a detailed study comparing the performance of their proposed correlation Eq. (6.1) with all the available correlations in the transition region using the experimental data of Ghajar and Tam (1994). The study showed that all the existing correlations failed to account properly for the existence of the mixed convection in the laminar and transition regions. In addition, none of the correlations could account properly for the effect of inlet configuration on the start and end of the transition region.

Flow Regime Map for Forced and Mixed Convection

Heating a fluid flowing in a horizontal tube produces a secondary flow (buoyancy). The fluid near the tube wall due to its higher temperature and lower density circulates upward, while the fluid near the central region of the tube, having a lower temperature and a higher density, circulates downward. These counterrotating transverse vortices that are superimposed on the streamwise main flow due to free convection effects (buoyancy influences) can significantly increase the forced convection heat transfer. It is important to realize that heat transfer in combined forced and free (mixed) convection can be significantly different from its values in both pure free

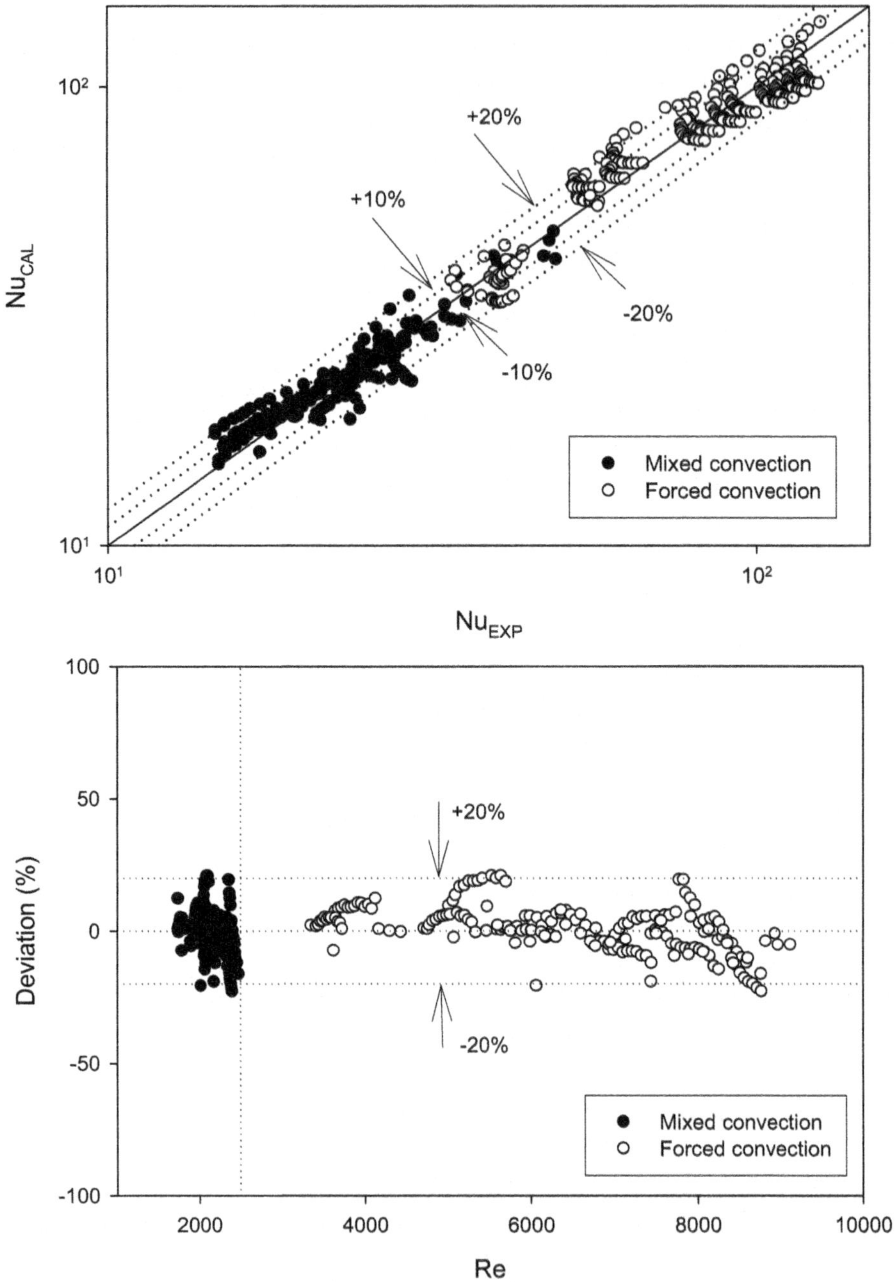

Fig. 6.1 Comparison between predictions of Eq. (6.1) for the transition region and heat transfer experimental data of Ghajar and Tam (1994) for the re-entrant inlet

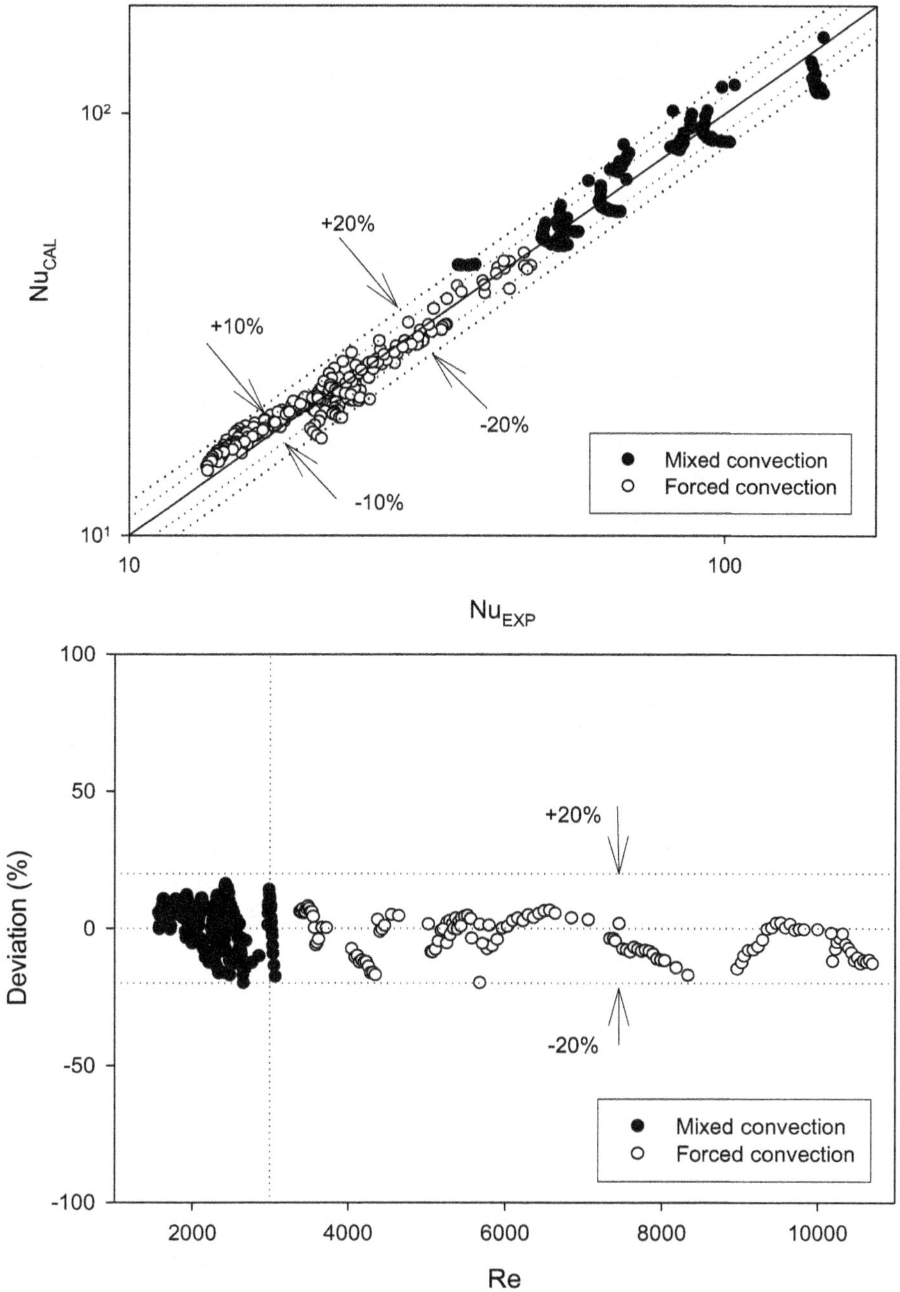

Fig. 6.2 Comparison between predictions of Eq. (6.1) for the transition region and heat transfer experimental data of Ghajar and Tam (1994) for the square-edged inlet

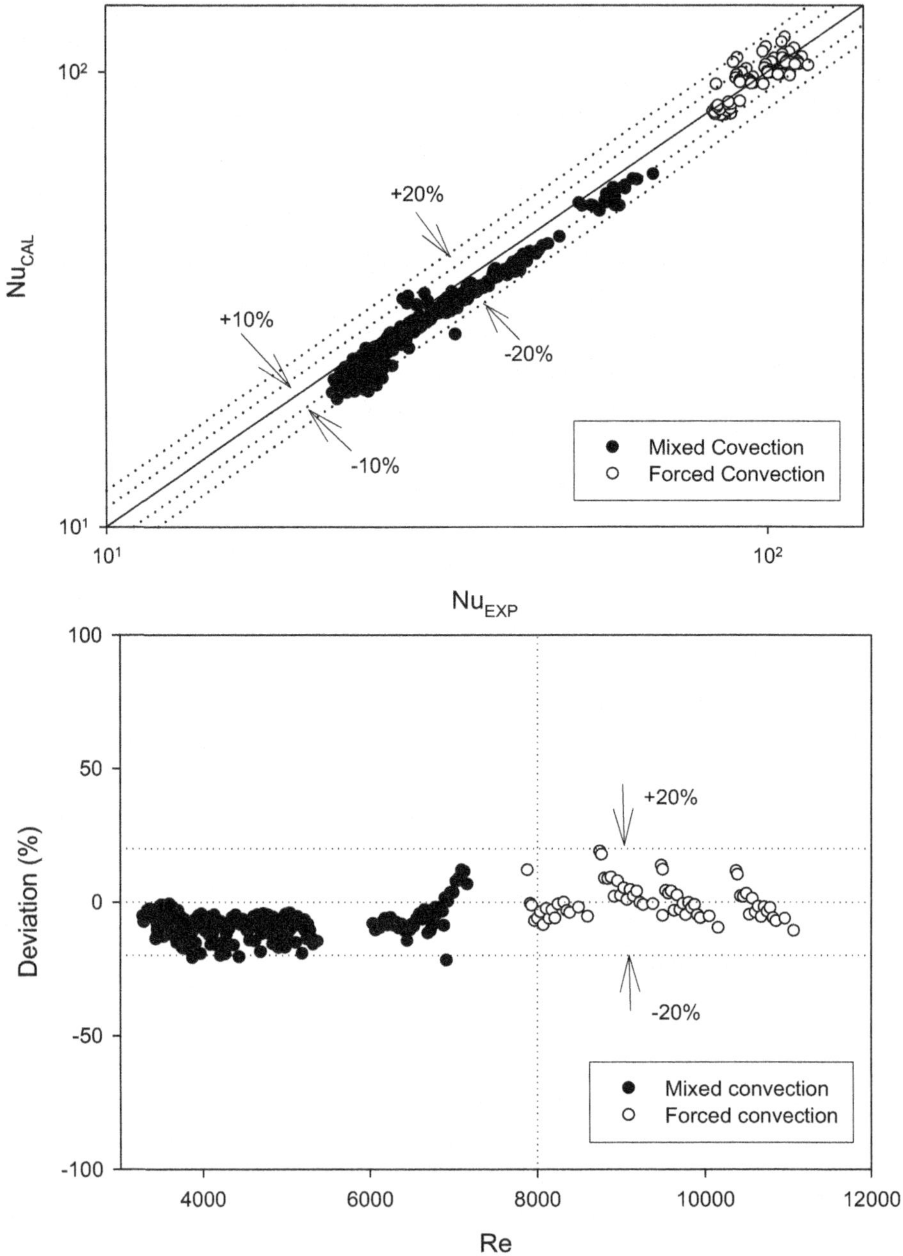

Fig. 6.3 Comparison between predictions of Eq. (6.1) for the transition region and heat transfer experimental data of Ghajar and Tam (1994) for the bell-mouth inlet

and pure forced convection. Actually, buoyancy forces (free convection effects) are present in any forced convection flow, and for design purposes, it is of interest to know when they can be neglected and when they have to be accounted for. For this purpose, Ghajar and Tam (1995) used their uniform wall heat flux heat transfer data and their proposed correlations for laminar Eq. (6.2), transitional Eq. (6.1), and turbulent Eq. (6.3) flows for the three different inlets (re-entrant, square-edged, and bell-mouth) and constructed a flow regime map (see Fig. 6.4), similar to the one developed for uniform wall temperature boundary condition by Metais and Eckert (1964). Their flow regime map is unique in the sense that it was the first attempt to develop such a map for the case of horizontal tube with uniform wall heat flux boundary condition.

In the development of their flow regime map, they paid particular attention to the influence of inlet configuration on the start and end of the heat transfer transition region and the development of secondary flow along the tube. Their flow regime map determines the boundary between forced and mixed convection in a horizontal circular straight tube with three different inlets under uniform wall heat flux boundary condition. Their map is applicable to all flow regimes and both developing and fully developed flows. From the flow regime map, for any forced flow represented by a given Reynolds number, the value of the parameter GrPr at a particular *x/D* location indicates where it is necessary to consider buoyancy effects (secondary flow). Once the convection heat transfer flow regime (pure forced or mixed) for any

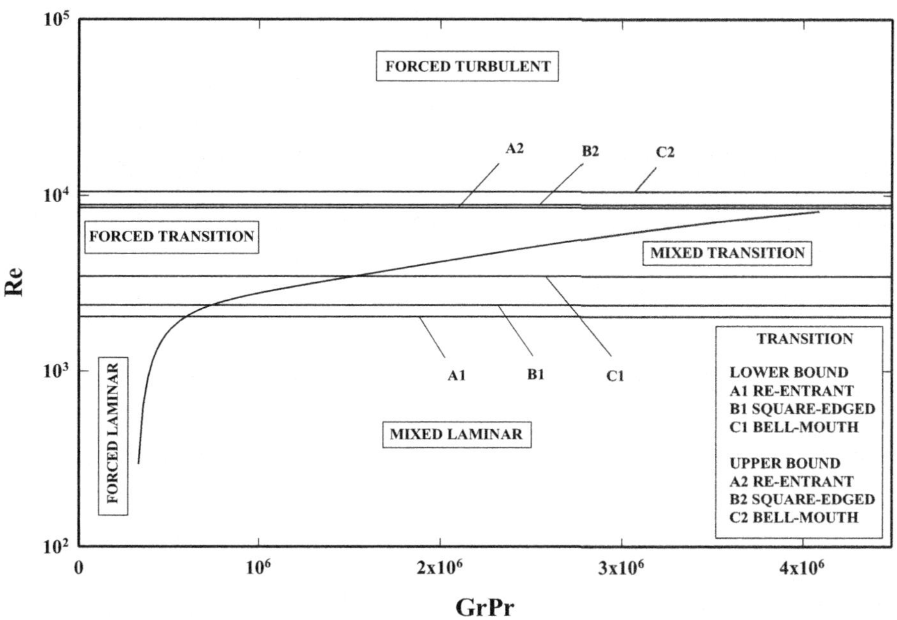

Fig. 6.4 Flow regime map for flow in horizontal tubes with three different inlet configurations and uniform wall heat flux

of the three inlets has been established, correlations for calculation of heat transfer coefficient in the laminar, transition, and turbulent regimes are offered. Equation (6.2) is recommended for laminar forced and mixed convection in the entrance and fully developed regions and can be used for all three inlets. For turbulent forced convection in the entrance and fully developed regions for all three inlets, Eq. (6.3) is recommended. Equation (6.1) is applicable to transition forced and mixed convection in the entrance and fully developed regions and should be used with the *a*, *b*, and *c* constant values for the correct inlet geometry (see Table 6.1).

Ghajar and Tam (1995) developed and used the following equation as shown in Fig. 6.4, to separate the forced and mixed convection regions in their flow regime map:

$$\mathrm{Re} = 2674 + 5.35 \times 10^{-13}\,(\mathrm{GrPr})^{2.5} - 1.85 \times 10^{-16}\,(\mathrm{GrPr})^{3} - 2.64 \times 10^{14}\,(\mathrm{GrPr})^{-2} \tag{6.4}$$

Equation (6.4) correlates the 51 data points with a correlation coefficient of 0.96. These data points represented the boundary between laminar and transition forced and mixed convection for the three inlets in their experimental data. In identifying these points, they used the criterion that the ratio of the local peripheral heat transfer coefficient at the bottom of the tube (h_t/h_b) should be greater than or equal to 0.8 for the forced convection and less than 0.8 for the mixed convection (see Fig. 5.2).

An Illustrative Example

Use of the Mixed Convection Heat Transfer Correlation in the Transition Region An ethylene glycol-distilled water mixture with a mass fraction of 0.6 and a volume flow rate of $Q = 2.6 \times 10^{-4}$ m^3/s flows inside a tube with an inside diameter of $D = 0.0158$ m with a uniform wall heat flux boundary condition. For this flow, determine the Nusselt number at the location $x/D = 90$ if the inlet configuration of the tube is (a) re-entrant, (b) square-edged, and (c) bell-mouth. At this location, the local Grashof number is Gr = 51,770. The properties of the ethylene glycol-distilled water mixture (with a mass fraction of 0.6) at the location of interest are Pr = 29.2, $\nu = 3.12 \times 10^{-6}$ m^2/s, and $\mu_b/\mu_w = 1.77$.

Solution

For a tube with a known diameter and volume flow rate, the type of flow regime is determined before any decision about which Nusselt number correlation to use is made.

The Reynolds number at the specified location is:

$$\mathrm{Re} = \frac{(Q/A_c)D}{\nu} = \frac{\left[(2.6\times10^{-4}\ \mathrm{m^3/s})/(1.961\times10^{-4}\ \mathrm{m^2})\right](0.0158\ \mathrm{m})}{3.12\times10^{-6}\ \mathrm{m^2/s}} = 6713$$

since $A_c = \pi D^2/4 = \pi(0.0158\ \mathrm{m})^2/4 = 1.961\times10^{-4}\ \mathrm{m^2}$.

From Table 5.1 with $x/D = 90$ and Re = 6713, the flow for all three inlet configurations is in the transition region. Therefore, Eq. (6.1) should be used with the constants a, b, and c given in Table 6.1. However, Nu_l and Nu_t are the inputs to Eq. (6.1), and they have to be evaluated first from Eqs. (6.2) and (6.3), respectively. It should be mentioned that the correlations for Nu_l and Nu_t have no inlet configuration dependency.

From Eq. (6.2), the laminar Nusselt number is:

$$\mathrm{Nu}_l = 1.24\left[\left(\frac{\mathrm{RePr}D}{x}\right) + 0.025(\mathrm{GrPr})^{0.75}\right]^{1/3}\left(\frac{\mu_b}{\mu_w}\right)^{0.14}$$

$$= 1.24\left[\left(\frac{(6713)(29.2)}{90}\right) + 0.025\left[(51770)(29.2)\right]^{0.75}\right]^{1/3}(1.77)^{0.14} = 19.9$$

From Eq. (6.3), the turbulent Nusselt number is:

$$\mathrm{Nu}_t = 0.023\,\mathrm{Re}^{0.8}\mathrm{Pr}^{0.385}\left(\frac{x}{D}\right)^{-0.0054}\left(\frac{\mu_b}{\mu_w}\right)^{0.14}$$

$$= 0.023(6713)^{0.8}(29.2)^{0.385}(90)^{-0.0054}(1.77)^{0.14} = 102.7$$

Then the transition Nusselt number can be determined from Eq. (6.1):

$$\mathrm{Nu_{tr}} = \mathrm{Nu}_l + \left\{\exp\left[(a - \mathrm{Re})/b\right] + \mathrm{Nu}_t^c\right\}^c$$

(a) Re-entrant inlet:

$$\mathrm{Nu_{tr}} = 19.9 + \left\{\exp\left[(1766 - 6713)/276\right] + 102.7^{-0.955}\right\}^{-0.955} = 88.2$$

(b) Square-edged inlet:

$$\mathrm{Nu_{tr}} = 19.9 + \left\{\exp\left[(2617 - 6713)/207\right] + 102.7^{-0.950}\right\}^{-0.950} = 85.3$$

(c) Bell-mouth inlet:

$$\mathrm{Nu_{tr}} = 19.9 + \left\{\exp\left[(6628 - 6713)/237\right] + 102.7^{-0.980}\right\}^{-0.980} = 21.2$$

Discussion It is worth mentioning that for the re-entrant and square-edged inlets, the flow behaves normally. For the bell-mouth inlet, the Nusselt number is low in comparison to the other two inlets. This is the case because of the unusual behavior of the bell-mouth inlet noted earlier (Fig. 5.3), that is, the boundary layer along the tube wall is at first laminar and then changes through a transition region to the turbulent condition.

Chapter 7
Friction Factor Results for Micro-fin Tubes

Background

For heat transfer enhancement, one of the general methods is to use enhanced tubes instead of plain tubes. Among the enhanced tubes, micro-fin tubes are often applied in commercial heating, ventilating, and air-conditioning (HVAC) applications because the cost of production of these tubes is relatively low and the manufacturing process is simple. For micro-fin tubes, tube side roughening increases the heat transfer surface area, resulting in an increase in heat transfer for the micro-fin tube over the plain tube. However, there is also an increase in the pressure drop. The increase in the surface area causes low flow rates in the heat exchanger tubes, resulting in the flow being at Reynolds numbers that are between laminar and fully turbulent regions, that is, in the transition region. Owing to the high heat transfer rates, it is likely that more HVAC units will operate in the transition region where the understanding of friction factor (pressure drop) and heat transfer is limited. In particular, there is limited information on the effect of different inlet configurations and different fin geometries on the friction factor in the entrance and fully developed regions of the micro-fin tubes. Refer to Tam et al. (2012) and Tam (2013) for a detailed review of related studies on micro-fin tubes.

For this purpose, using the experimental setup detailed in Chap. 2, experiments were conducted to obtain accurate and broad pressure drop (friction factor) database across all flow regimes. This chapter presents isothermal and non-isothermal (heating) friction factor results in the entrance and fully developed regions of the micro-fin tubes, the effect of inlet configuration and heating on the entrance and fully developed friction factors in all flow regimes (laminar, transitional, and turbulent), the effect of inlet configuration on the start and end of transition of the fully developed friction factor, and the effect of fin geometries (spiral angle, fin height-to-diameter ratio, and number of starts) on friction factors. Detailed information on these effects may be obtained from Tam et al. (2012) and Tam (2013).

A. J. Ghajar, *Single- and Two-Phase Flow Pressure Drop and Heat Transfer in Tubes*, Mechanical Engineering Series,
https://doi.org/10.1007/978-3-030-87281-6_7

Effect of Inlet Configuration and Heating on Micro-fin Tubes' Fully Developed Friction Factors

Figure 7.1 shows the influence of inlet configuration on the start and end of the friction factor (pressure drop) transition region for the plain and micro-fin tube #1 (see Table 2.1) under isothermal conditions. The isothermal plain tube results are also shown in Fig. 7.1 for comparison with the isothermal micro-fin tube results, and they are the same as those presented in Fig. 3.1. Figure 7.1 plots the fully developed friction factors versus Reynolds numbers for the square-edged and re-entrant inlets in all flow regimes. Similar to Fig. 3.1, Fig. 7.1 also shows the established plain tube fully developed laminar and turbulent pipe flow friction factor correlations. The solid symbols in the figure represent the start and end of the fully developed transitional region for the plain and micro-fin tubes with different inlets.

A parallel shift from the classic laminar flow equation ($C_{f,l} = 16/\text{Re}$) is observed for the micro-fin tube in the laminar region. The Reynolds number for the start of the transitional region for the micro-fin tube (refer to the solid symbols in Fig. 7.1) is defined as the point of sudden departure from the isothermal laminar flow friction factor line parallel to the classic $C_{f,l} = 16/\text{Re}$ line, and the end of the transition in the micro-fin tube is defined as the first point where the isothermal friction factor

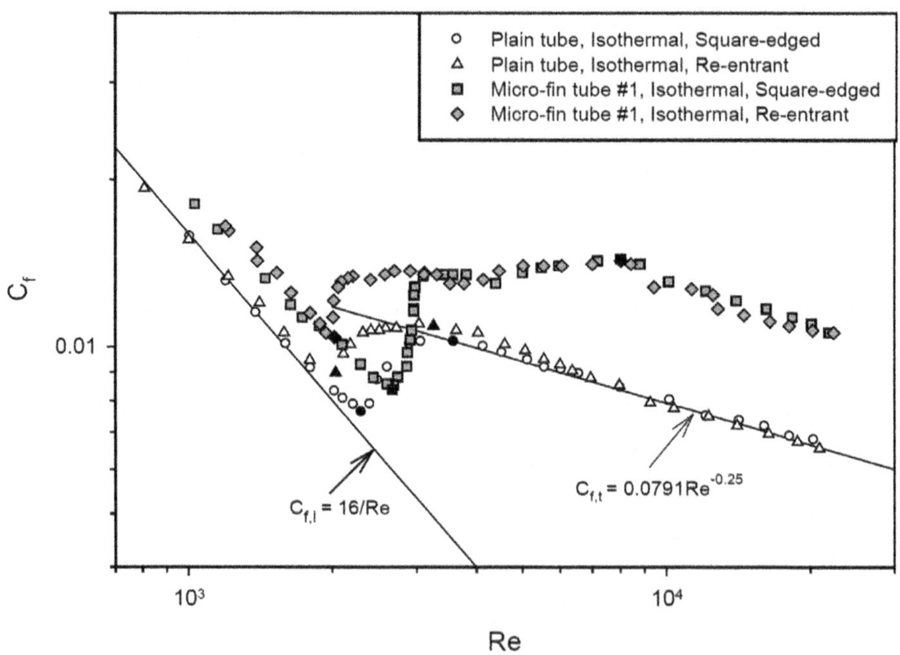

Fig. 7.1 Friction factor characteristics for the plain and micro-fin tube #1 at 200 diameters from the tube entrance under isothermal boundary conditions (solid symbols indicate the start and end of the transition region for each inlet)

data lands on a line parallel to the Blasius turbulent flow correlation ($C_{f,t} = 0.0791\text{Re}^{-0.25}$). Flow is considered turbulent beyond the end of the transition. Micro-fin tube isothermal friction factors in the transitional region go through a steep increase followed by a relatively constant C_f section and then an upward parallel shift (64.8% increase for micro-fin tube #1) from the Blasius turbulent friction factor correlation.

The lower limit of the friction factor (pressure drop) transition Reynolds number range is dependent on the inlet configuration (solid symbols in Fig. 7.1), although the effect of inlet configuration on the upper limit is not significant. From the results presented in Fig. 7.1 for the isothermal micro-fin tube, the limits for the transition Reynolds number range for the two inlet configurations can be summarized as follows: square-edged ($2675 < \text{Re} < 8007$) and re-entrant ($2021 < \text{Re} < 8023$). Comparing these limits to those presented in Table 3.1 for the isothermal plain tube and the same two inlets (square-edged, $2222 < \text{Re} < 3588$, and re-entrant, $2032 < \text{Re} < 3031$), the transition range for the isothermal micro-fin tube is much wider than that of the isothermal plain tube.

Because the calming section is a common factor for the two inlets (Fig. 2.3) and the flow is isothermal, the differences between transitional Reynolds number ranges are due only to the effect of different inlet configurations. For the isothermal friction factor in the micro-fin tube, the preceding limits for friction factor transition Reynolds numbers also indicate that the inlet that caused the most disturbance (re-entrant) produced an early transition ($\text{Re} = 2021$) and the inlet with the least disturbance (square-edged) did not experience transition until a Reynolds number of about 2675. Therefore, the transition Reynolds number range for micro-fin tubes can be manipulated by using different inlet configurations.

For the non-isothermal friction factors, a DC welder was used to approximate the uniform wall heat flux boundary condition (Fig. 2.1). The results presented in Fig. 7.2 clearly establish the influence of the heating (non-isothermal) condition for the plain and micro-fin tube #1.

A parallel shift from the classic laminar flow equation was observed for the micro-fin tube (Fig. 7.2). The laminar friction factors for heated and isothermal micro-fin tubes have the same trend as for the laminar region. In other words, the friction factor in that region is insensitive to different boundary conditions. Unlike the plain tube, the significant influence of heating on friction factor values in the laminar region is due to the fact that the effect of secondary flow could not be observed because buoyant flow (or secondary flow) only occurs above the fin tip, and therefore the friction factor (at the fin base) could not be influenced. For the micro-fin tube, heating seems to have a significant influence on the value of the friction factor in the lower transitional region because the start of transition for the isothermal and heating friction factors is different. However, in the upper transitional and turbulent regions, heating did not affect the magnitude of the friction factor.

Comparing isothermal and heated friction factors for the micro-fin tube, it can be observed in Fig. 7.2 that heating delays the start and end of the transition. Table 7.1 summarizes the start and end of the transition of the fully developed friction factors for the plain tube and micro-fin tube #1 under isothermal and non-isothermal

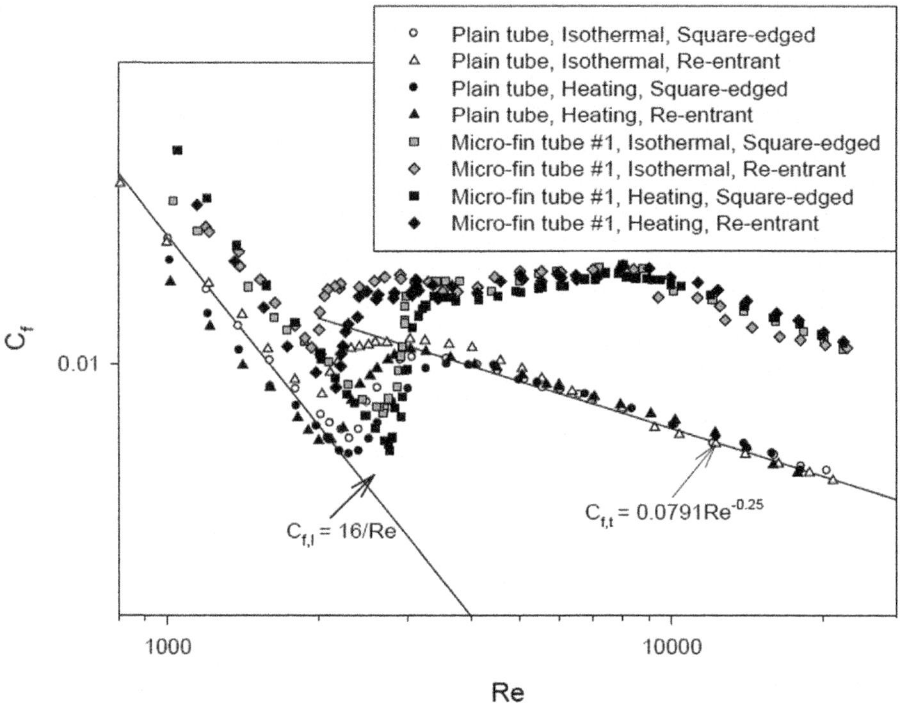

Fig. 7.2 Friction factor characteristics of the plain and micro-fin tube #1 at 200 diameters from the tube entrance under isothermal and heating boundary conditions for different inlet configurations and all flow regimes (laminar, transition, and turbulent)

Table 7.1 Start and end of transition of the fully developed friction factor at 200 diameters from the plain tube and micro-fin tube #1 entrance

	Friction factor			
Tube, condition, inlet	Re_{start}	C_f	Re_{end}	C_f
Plain, Isothermal (square-edged)	2222	0.0079	3588	0.0102
Plain, Heating (square-edged)	2316	0.0074	3941	0.0100
Plain, Isothermal (re-entrant)	2032	0.0090	3031	0.0110
Plain, Heating (re-entrant)	2257	0.0079	3250	0.0105
Micro-fin #1, Isothermal (square-edged)	2675	0.0084	8007	0.0144
Micro-fin #1, Heating (square-edged)	2765	0.0073	8925	0.0136
Micro-fin #1, Isothermal (re-entrant)	2021	0.0104	8023	0.0143
Micro-fin #1, Heating (re-entrant)	2141	0.0092	9027	0.0142

(heating) conditions. The plain tube results are the same as those presented in Table 3.1 and are repeated here for comparison with the micro-fin tube results. For both inlets, the effect of heating is to delay the start and end of the transition in the micro-fin tube. For the square-edged inlet, the start and end of transition are delayed from Re = 2675 to 2765 and from Re = 8007 to 8925, respectively. For the re-entrant inlet, the start and end of the transitions are delayed from Re = 2021 to 2141 and from Re = 8023 to 9027, respectively. Regardless of which inlet was used, heating delayed the transitional range of the micro-fin tube. Thus, both entrance disturbance and heating influence the start and end of the transition region of the fully developed friction factor of the micro-fin tube.

Effect of Inlet Configuration and Heating on Micro-fin Tubes' Entrance and Fully Developed Friction Factors

For analyzing the effect of inlet geometries and heating on the entrance and fully developed friction factors of micro-fin tubes, the micro-fin tube #1 friction factor data is arranged into apparent friction factor plots as shown in Fig. 7.3 for isothermal and non-isothermal conditions. For comparison, the proposed plain tube isothermal entrance and fully developed laminar friction factor correlation, Eq. (4.1), is also shown in Fig. 7.3.

Figure 7.3 displays the laminar apparent friction factors for the two inlet types (re-entrant and square-edged) under isothermal condition, presenting the apparent friction factor multiplied by the Reynolds number versus ζ, which is the

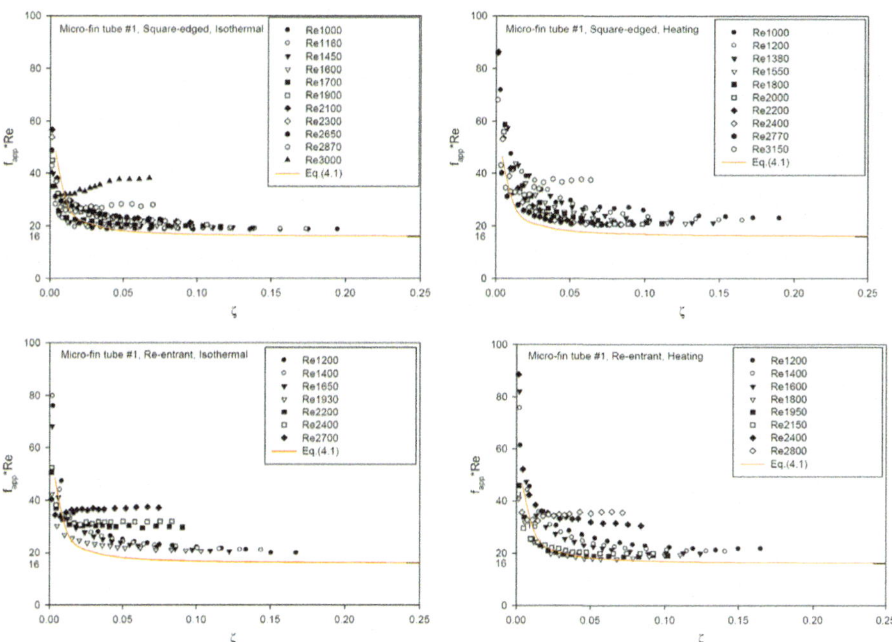

Fig. 7.3 Comparison between experimental laminar flow apparent micro-fin tube #1 friction factors with square-edged and re-entrant inlets and the proposed laminar region correlation (Eq. 4.1) under isothermal and heating conditions

dimensionless axial location of the pressure taps (*x*/*D*, see Fig. 2.5), divided by the Reynolds number. The apparent friction factor data collapse into a single curve ($f_{app}Re$ vs ζ) for the square-edged inlet when Re is less than 2600, and after a certain length, it approaches the fully developed isothermal constant value of the apparent friction factor ($f_{app}Re = 20$), which is higher than the fully developed plain tube classic laminar flow line ($f_{app}Re = 16$) (see Fig. 7.3). For Re values greater than 2600, the apparent friction factor becomes a strong function of the Reynolds number. Similar behavior is shown for the re-entrant inlet. When Re is less than 2000, the apparent friction factor data collapse into a single curve ($f_{app}Re$ vs ζ), and after a certain length, it approaches the fully developed isothermal constant value of the apparent friction factor ($f_{app}Re = 20$). For Re values greater than 2000, the apparent friction factor becomes a strong function of Reynolds number. Comparison of the results given in Fig. 7.3 for the two inlets at comparable Reynolds numbers shows that the length required for the square-edged inlet to reach the fully developed isothermal apparent friction factor ($f_{app}Re = 20$) is similar to that required for the re-entrant inlet. Referring to Fig. 7.3, it can also be observed that the square-edged and re-entrant apparent friction factor data of micro-fin tubes follow the trend of the proposed plain tube correlation, Eq. (4.1). Therefore, in Chap. 8, the form of the proposed plain tube correlation was used to develop the isothermal apparent friction factor correlation for the micro-fin tubes.

Figure 7.3 also displays the laminar flow apparent friction factors for re-entrant and square-edged inlets for non-isothermal (heating) conditions, similar to the presentation for isothermal condition. For comparison purposes, the proposed plain tube isothermal entrance and fully developed friction factor correlation, Eq. (4.1), is also shown in Fig. 7.3.

The trend of the apparent friction factor under the non-isothermal condition is similar to that of the isothermal condition (Fig. 7.3). For Re values less than 2700 (square-edged inlet) and 2200 (re-entrant inlet), the apparent friction factor data collapse into a single curve, and after a certain length, it approaches the fully developed non-isothermal constant value of the apparent fraction factor ($f_{app}Re = 20$). Compared to isothermal data, the entire non-isothermal apparent friction factor trend is identical and then approaches a constant line ($f_{app}Re = 20$), which is less than the isothermal value of 20. Therefore, in the fully developed friction factor region (Fig. 7.3), isothermal and non-isothermal laminar flow data follow the same line parallel to the traditional laminar flow line ($C_{f,l} = 16/\mathrm{Re}$). For Re values greater than 2700 (square-edged inlet) and 2200 (re-entrant inlet), the non-isothermal apparent friction factor becomes a strong function of their Reynolds numbers. The square-edged and re-entrant data of the micro-fin tubes follows the trend of the proposed isothermal plain tube correlation, Eq. (4.1). Therefore, in Chap. 8, the form of the proposed plain tube correlation was used to develop the non-isothermal apparent friction factor correlation for the micro-fin tubes.

Effect of Fin Geometry of Micro-fin Tubes on Friction Factor

Figures 7.4 and 7.5 demonstrate the friction factor characteristics of all micro-fin tubes (#1 to #6, see Table 2.1) with square-edged and re-entrant inlets under isothermal and heating (non-isothermal) boundary conditions. Regardless of boundary condition, the friction factors of all micro-fin tubes throughout their entire flow regimes were basically higher than the plain tube friction factors, and the increase in friction factors for the micro-fin tubes compared to the plain tube ranged from 91% to 212%. In the laminar region, regardless of inlet type and fin geometry, the friction factors of all the micro-fin tubes are parallel shifted from the classic laminar flow line ($C_{f,l} = 16/\text{Re}$). From the transitional to turbulent regions, all the micro-fin tubes have a similar data trend but with different characteristics. It is difficult to observe the influence of fin geometry on friction factor characteristics in Figs. 7.4 and 7.5.

Fin geometry parameters such as fin spiral angle, fin height-to-diameter ratio, and number of starts according to Jensen and Vlakancic (1999) and Zdaniuk et al. (2008) are the major factors influencing micro-fin tube friction factor in the turbulent region. However, the effects of those parameters on friction factor characteristics in the laminar and transitional regions with different inlets are not fully understood. Therefore, a qualitative analysis of the influence of each parameter on friction factor is presented next.

Effect of Change in Spiral Angle on Friction Factor

Micro-fin tubes #1, #2, and #3 have the same fin height-to-diameter ratio ($e/D = 0.0336$) and number of starts ($N_s = 25$) but different spiral angles ($\alpha = 18°$, 25°, and 35°) (Table 2.1). Therefore, those particular micro-fin tubes were selected for analyzing the effect of spiral angle on the friction factor.

Figure 7.6 represents fully developed friction factor characteristics for plain and micro-fin tubes with different spiral angles (#1, #2, and #3) and different inlet configurations (square-edged and re-entrant) under isothermal boundary condition. Isothermal friction factors and Reynolds numbers for the start and end of the transitional region of the plain tube and each micro-fin tube, fitted with two different types of inlet configuration under isothermal boundary condition, are summarized in Table 7.2. The plain tube and micro-fin tube #1 results are the same as those presented in Tables 3.1 and 7.1. They are repeated here for comparison with the micro-fin tubes #2 and #3 results. The parallel shift from the classic laminar flow equation ($C_{f,l} = 16/\text{Re}$) is observed in micro-fin tubes in the laminar region. Laminar friction factor data are insensitive to the inlet type and fin spiral angle. The isothermal lower transition Reynolds numbers for square-edged (Re = 2284) and re-entrant (Re = 1973) inlets in micro-fin tube #2 are different from micro-fin tube #1 (2675 for square-edged and 2021 for re-entrant) and the early transition for micro-fin tube #2. In comparison to micro-fin tubes #1 and #2, micro-fin tube #3 with a larger spiral

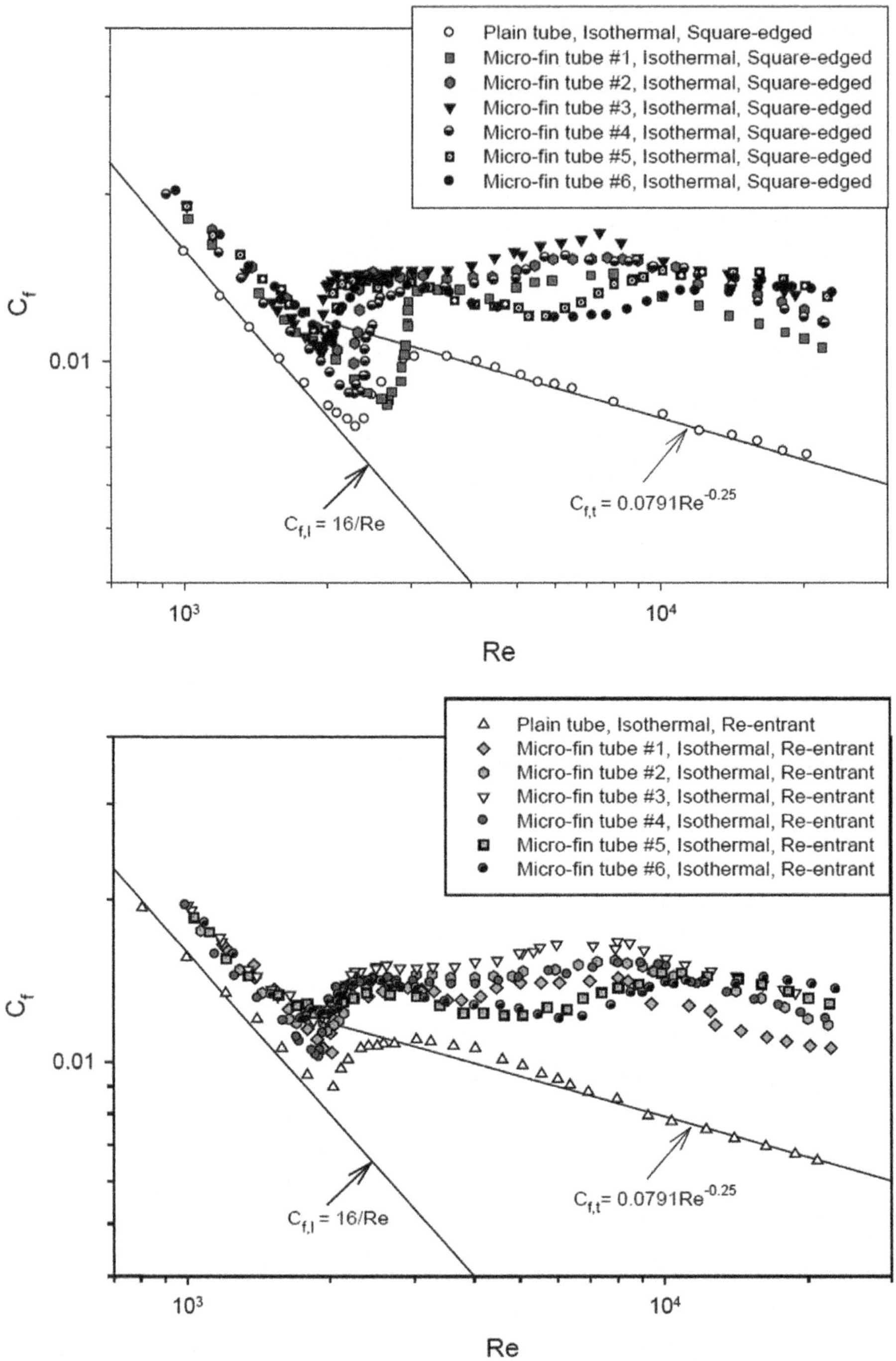

Fig. 7.4 Friction factor characteristics of the plain tube and all micro-fin tubes (#1 to #6) with square-edged and re-entrant inlets at 200 diameters from the tube entrance under isothermal boundary conditions

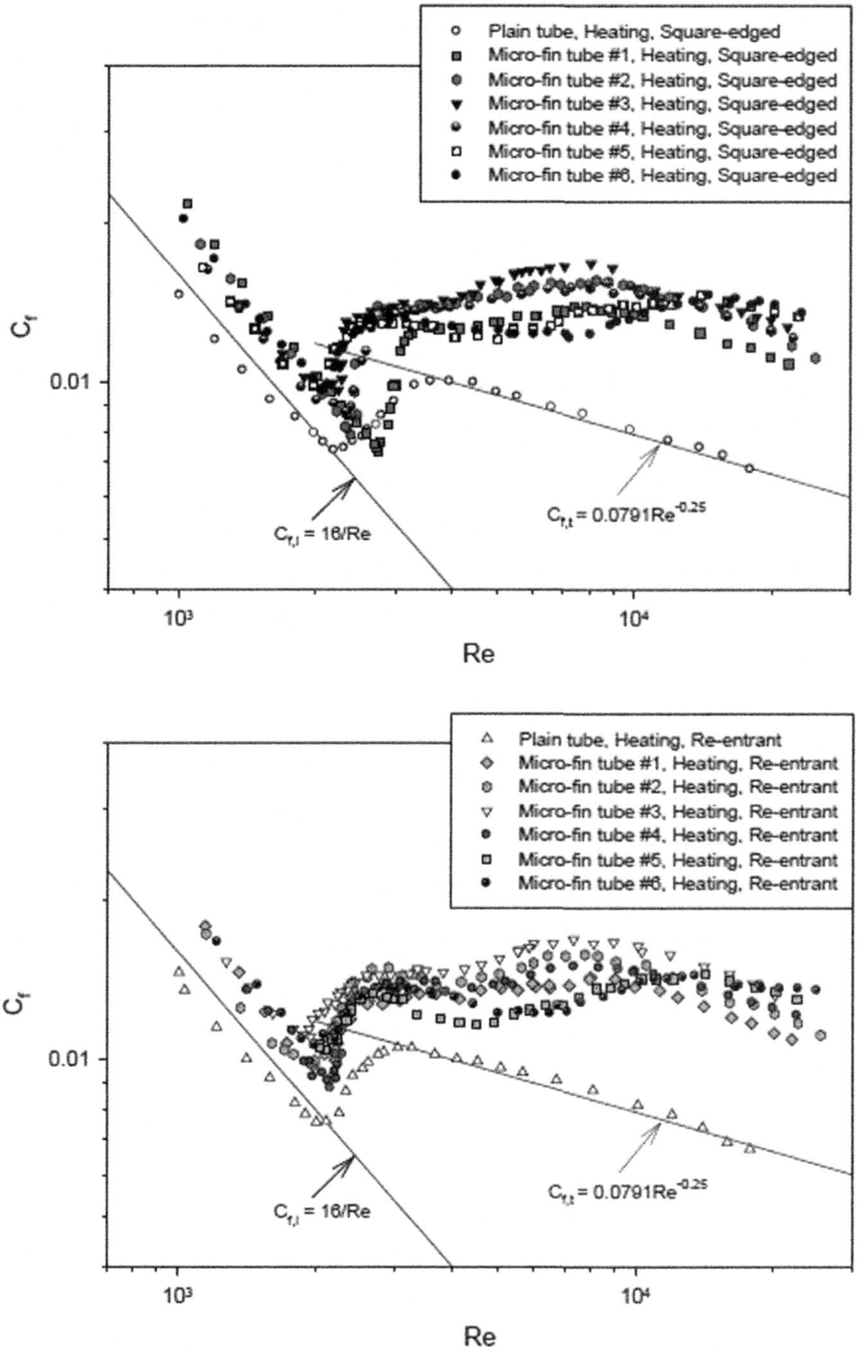

Fig. 7.5 Friction factor characteristics of the plain tube and all micro-fin tubes (#1 to #6) with square-edged and re-entrant inlets at 200 diameters from the tube entrance under heating boundary condition

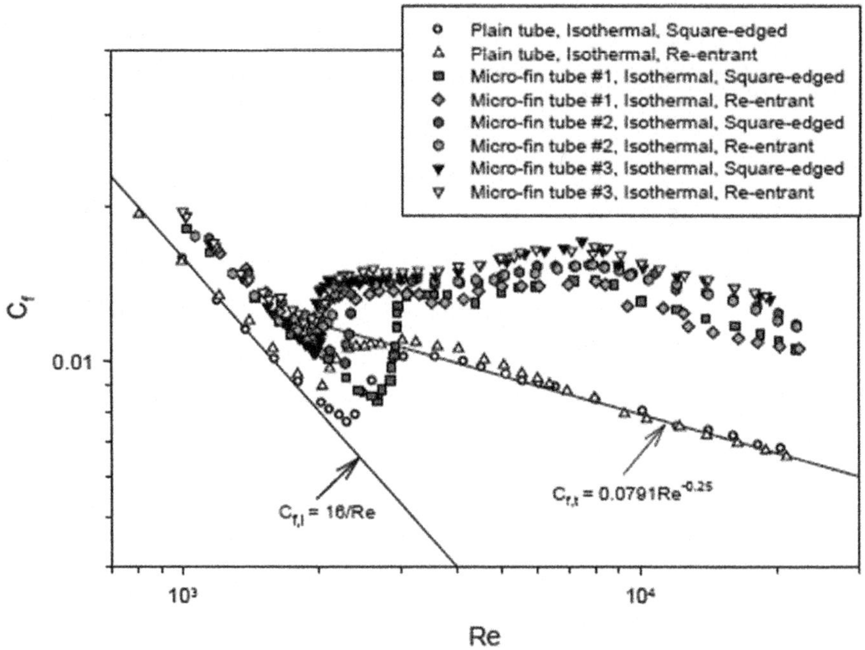

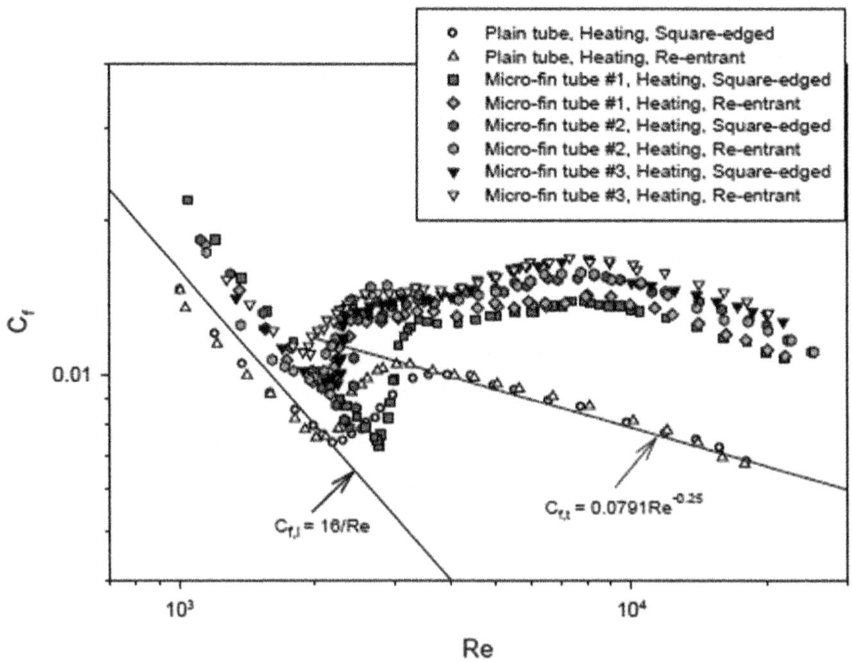

Fig. 7.6 Fully developed friction factor characteristics for plain tube and micro-fin tubes #1, #2, and #3 with different spiral angles and different inlet configurations under isothermal and heating boundary conditions

Table 7.2 Start and end of fully developed friction factor transition in plain and micro-fin tubes #1 ($\alpha = 18°$), #2 ($\alpha = 25°$), and #3 ($\alpha = 35°$) at 200 diameters from the plain and micro-fin tubes' entrance

	Friction factor			
Tube, condition, inlet	Re_{start}	C_f	Re_{end}	C_f
Plain, Isothermal (square-edged)	2222	0.0079	3588	0.0102
Plain, Heating (square-edged)	2316	0.0074	3941	0.0100
Plain, Isothermal (re-entrant)	2032	0.0090	3031	0.0110
Plain, Heating (re-entrant)	2257	0.0079	3250	0.0105
Micro-fin #1, Isothermal (square-edged)	2675	0.0084	8007	0.0144
Micro-fin #1, Heating (square-edged)	2765	0.0073	8925	0.0136
Micro-fin #1, Isothermal (re-entrant)	2021	0.0104	8023	0.0143
Micro-fin #1, Heating (re-entrant)	2141	0.0092	9027	0.0142
Micro-fin #2, Isothermal (square-edged)	2284	0.0099	7762	0.0153
Micro-fin #2, Heating (square-edged)	2402	0.0079	8274	0.0156
Micro-fin #2, Isothermal (re-entrant)	1973	0.0114	7905	0.0152
Micro-fin #2, Heating (re-entrant)	1997	0.0099	8279	0.0157
Micro-fin #3, Isothermal (square-edged)	1962	0.0104	7489	0.0171
Micro-fin #3, Heating (square-edged)	2250	0.0096	8050	0.0168
Micro-fin #3, Isothermal (re-entrant)	1849	0.0119	7594	0.0167
Micro-fin #3, Heating (re-entrant)	1954	0.0110	8106	0.0166

angle has an earlier start of transition (1962 for square-edged and 1849 for re-entrant). Micro-fin tube #3 also has an earlier transition when compared to the plain tube. It can also be seen that the square-edged inlet delays the start of transition relative to the re-entrant inlet. This observation is typically obvious for micro-fin tube #1, which has a smaller spiral angle. In the transitional region, friction factors for those micro-fin tubes go through a steep increase followed by a relatively constant C_f section and then a parallel shift from the classic Blasius turbulent friction factor correlation ($C_{f,t} = 0.0791\mathrm{Re}^{-0.25}$). Friction factors for transitional and turbulent regions increase with larger spiral angles in micro-fin tubes because of their stronger drag. Higher fin spiral angles also lead to higher turbulent friction factors. This was also observed by Jensen and Vlakancic (1999), Brognaux et al. (1997), Webb et al. (2000), and Zdaniuk et al. (2008). A larger spiral angle in the micro-fin tube advances the end of the transition under both inlet configurations (Fig. 7.6, Table 7.2).

Figure 7.6 also represents the fully developed friction factor characteristics for plain and micro-fin tubes #1, #2, and #3 with different inlet configurations (square-edged and re-entrant) under heated (non-isothermal) boundary condition. A parallel shift from the classic laminar equation ($C_{f,l} = 16/\mathrm{Re}$) occurs in micro-fin tubes with both inlets. The laminar friction factors for the heated micro-fin tubes show the same trend in the laminar region. In the other words, the friction factor in that region is insensitive to the boundary condition and fin spiral angle. Lower transition Reynolds numbers for micro-fin tube #2 (2402 and 1997) with square-edged and re-entrant inlets under heated condition are different from micro-fin tube #1 (2765 and 2141), and an early transition is observed in micro-fin tube #2 with a larger spiral angle (25°). Furthermore, an early transition also occurs in micro-fin tube #3, having even a larger spiral angle (35°), when compared with micro-fin tubes #1 and #2 (Table 7.2). The square-edged inlet delays the start of transition relative to the re-entrant inlet. In the transitional region, the heated friction factors for micro-fin tubes #1, #2, and #3 go through a steep increase followed by a relatively constant C_f section and then a parallel shift from the classic Blasius turbulent friction factor correlation ($C_{f,t} = 0.0791\text{Re}^{-0.25}$). With a larger spiral angle in the micro-fin tube, an increased friction factor is observed in the transitional and turbulent regions because of the stronger drag near the tube wall. The larger spiral angle in the micro-fin tube advances the end of the transition under heating condition (Fig. 7.6, Table 7.2).

Thus, regardless of isothermal or heating boundary condition, larger spiral angles cause an earlier transition range and a higher friction factor in the transitional and turbulent regions.

Effect of Change in Fin Height-to-Diameter Ratio on Friction Factor

Micro-fin tubes #3 and #4 have the same spiral angle ($\alpha = 35°$) and number of starts ($N_s = 25$) but have different fin height-to-diameter ratios ($e/D = 0.0336$ and 0.0202) (Table 2.1). Therefore, those two were selected to analyze the effect of fin height-to-diameter ratio on friction factor.

Figure 7.7 represents the fully developed friction factor characteristics for plain and micro-fin tubes (#3 and #4) with different fin height-to-diameter ratio and different inlet configurations (square-edged and re-entrant) under isothermal boundary condition. The figure shows the start and end of transition friction factors and transition Reynolds numbers for isothermal friction factors, which are summarized in Table 7.3. The plain tube and micro-fin tube #3 results are the same as those presented in Tables 3.1 and 7.2. They are repeated here for comparison with the micro-fin tube #4 results. For the micro-fin tubes in the laminar region, a parallel shift from the classic laminar equation ($C_{f,l} = 16/\text{Re}$) is observed. Those isothermal laminar friction factor data are insensitive to inlet type and the fin height-to-diameter ratio. For micro-fin tube #3, the lower transition Reynolds numbers for the square-edged (Re = 1962) and re-entrant (Re = 1849) inlets are different from micro-fin tube #4 (2300 and 1901), and an early transition for micro-fin tube #3 with a larger fin height-to-diameter ratio ($e/D = 0.0336$) is observed. Micro-fin tube #3 also has an earlier transition when compared to the plain tube. Square-edged inlets delay the start of transition relative to re-entrant inlets in micro-tubes #3 and #4. In the transitional region, the friction factors for those micro-fin tubes go through a steep increase followed by a relatively constant C_f section and then a parallel shift from the classic Blasius turbulent friction factor correlation ($C_{f,t} = 0.0791\text{Re}^{-0.25}$). With a

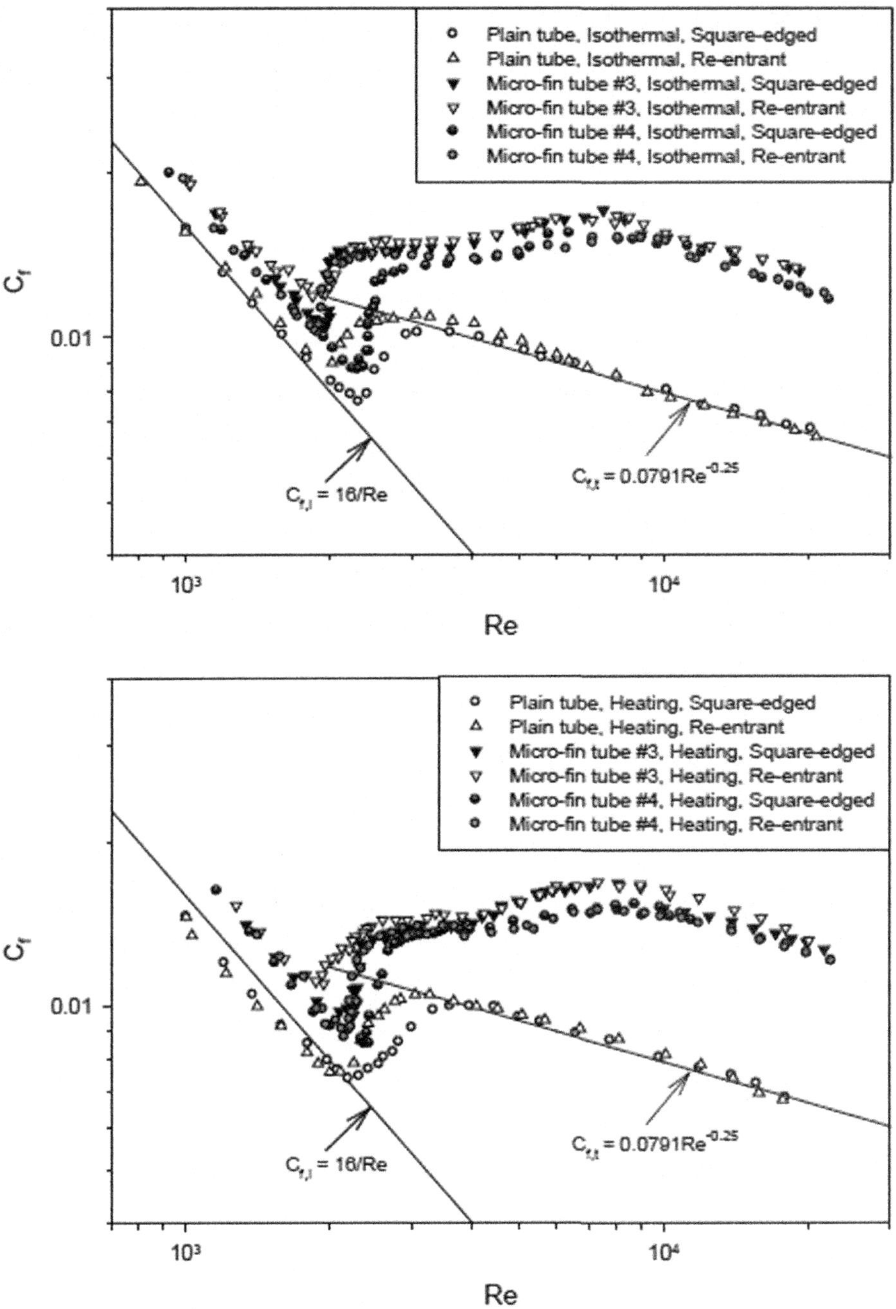

Fig. 7.7 Fully developed friction factor characteristics of plain tube and micro-fin tubes #3 and #4 with different fin height-to-diameter ratios and different inlet configurations under isothermal and heating boundary conditions

Table 7.3 Start and end of fully developed friction factor transition in plain and micro-fin tubes #3 ($e/D = 0.0336$) and #4 ($e/D = 0.0202$) at 200 diameters from the plain and micro-fin tubes' entrance

	Friction factor			
Tube, condition, inlet	Re_{start}	C_f	Re_{end}	C_f
Plain, Isothermal (square-edged)	2222	0.0079	3588	0.0102
Plain, Heating (square-edged)	2316	0.0074	3941	0.0100
Plain, Isothermal (re-entrant)	2032	0.0090	3031	0.0110
Plain, Heating (re-entrant)	2257	0.0079	3250	0.0105
Micro-fin #3, Isothermal (square-edged)	1962	0.0104	7489	0.0171
Micro-fin #3, Heating (square-edged)	2250	0.0096	8050	0.0168
Micro-fin #3, Isothermal (re-entrant)	1849	0.0119	7594	0.0167
Micro-fin #3, Heating (re-entrant)	1954	0.0110	8106	0.0166
Micro-fin #4, Isothermal (square-edged)	2300	0.0087	8076	0.0151
Micro-fin #4, Heating (square-edged)	2497	0.0082	8727	0.0154
Micro-fin #4, Isothermal (re-entrant)	1901	0.0102	8044	0.0152
Micro-fin #4, Heating (re-entrant)	2159	0.0088	8332	0.0149

larger fin height-to-diameter ratio in the micro-fin tube, an increase in friction factors occurs in transitional and turbulent regions. A larger fin height-to-diameter ratio leads to a higher turbulent friction factor because of the greater roughness associated with the higher fin height. This was also observed by Jensen and Vlakancic (1999), Webb et al. (2000), and Zdaniuk et al. (2008). Therefore, a larger fin height-to-diameter ratio in the micro-fin tube advances the end of transition with both inlet configurations (Fig. 7.7, Table 7.3).

Figure 7.7 also represents the fully developed friction factor characteristics for plain and micro-fin tubes #3 and #4 with different inlet configurations under heated (non-isothermal) boundary condition. A parallel shift from the classic laminar equation ($C_{f,l} = 16/\mathrm{Re}$) is also observed in those micro-fin tubes for both inlets. The laminar friction factor for heated micro-fin tubes shows the same trend. In other words, the friction factor in that region is also insensitive to boundary condition and fin height-to-diameter ratio. For the heated data, the lower transition Reynolds numbers of micro-fin tube #3 (2250 and 1954) with square-edged and re-entrant inlets are different from micro-fin #4 tube (2497 and 2159), and an early transition is observed in micro-fin tube #3 with a larger fin height-to-diameter ratio ($e/D = 0.0336$). The start of transition is delayed in the square-edged inlet relative to the re-entrant inlet. In the transitional region, the heated friction factors of micro-fin tubes #3 and #4 also go through steep increases followed by relatively constant C_f sections and then a parallel shift from the classic Blasius turbulent friction factor correlation ($C_{f,t} = 0.0791\mathrm{Re}^{-0.25}$). With a larger fin height-to-diameter ratio in the micro-fin tube, increased heated friction factors are observed in transitional and turbulent regions because the higher fin height increases roughness. A larger fin height-to-diameter ratio in the micro-fin tube advances the end of transition under heated condition for both inlets (Fig. 7.7, Table 7.3).

Thus, regardless of either isothermal or non-isothermal (heating) boundary condition, the larger fin height-to-diameter ratio results in an earlier transition range and a higher friction factor in transitional and turbulent regions.

Effect of Change in Number of Starts on Friction Factor

Micro-fin tubes #1, #5, and #6 have the same fin height-to-diameter ratio ($e/D = 0.0336$) and spiral angles ($\alpha = 18°$) but have different number of starts ($N_s = 25$, 45, and 65) (Table 2.1). Therefore, those three micro-fin tubes were selected to analyze the effect of the number of starts on the friction factor.

Figure 7.8 represents fully developed friction factor characteristics for plain and micro-fin tubes #1, #5, and #6 and different inlet configurations under isothermal boundary condition. The figure shows the start and end of transition friction factors and transition Reynolds numbers for isothermal friction factors, which are summarized in Table 7.4. The plain tube and micro-fin tube #1 results are the same as those presented in Tables 3.1 and 7.1. They are repeated here for comparison with the micro-fin tubes #5 and #6 results. For micro-fin tubes in the laminar region, a parallel shift from the classic laminar equation ($C_{f,l} = 16/\text{Re}$) occurs. Those laminar friction factor data are insensitive to inlet type and the number of starts. For isothermal micro-fin tube #1 ($N_s = 25$), the lower transition Reynolds numbers for square-edged ($\text{Re} = 2675$) and re-entrant ($\text{Re} = 2021$) inlets are different from micro-fin tube #5 (2068 and 1949), and an early transition in micro-fin tube #5 with a larger number of starts ($N_s = 45$) is observed. Micro-fin tube #6, with even a larger number of starts ($N_s = 65$), has an earlier start of transition (2016 for square-edged and 1739 for re-entrant) than micro-fin tubes #1 and #5. Micro-fin tube #6 also has an earlier transition when compared to the plain tube. The square-edged inlet delays the start of transition relative to the re-entrant inlet in these micro-fin tubes and is obvious with micro-fin tube #1, which has a lower number of starts ($N_s = 25$). In the transitional region, the friction factors for those micro-fin tubes go through a steep increase followed by a relatively constant C_f section and then a parallel shift from the classic Blasius turbulent friction factor correlation ($C_{f,t} = 0.0791\text{Re}^{-0.25}$). Interestingly, for micro-fin tubes #5 and #6, which have larger number of starts, Fig. 7.8 shows an unusual behavior (a dip of friction factor) in the relatively constant C_f section. In the turbulent region, micro-fin tubes #5 and #6 have larger number of starts, resulting in higher friction factors. This was also observed by Jensen and Vlakancic (1999), Webb et al. (2000), and Zdaniuk et al. (2008). A larger number of starts result in higher turbulent friction factors because of higher surface roughness. Increasing the number of starts also delays the end of transition for both inlets (Fig. 7.8, Table 7.4). Therefore, a wider transitional region occurs along with an increase in the number of starts.

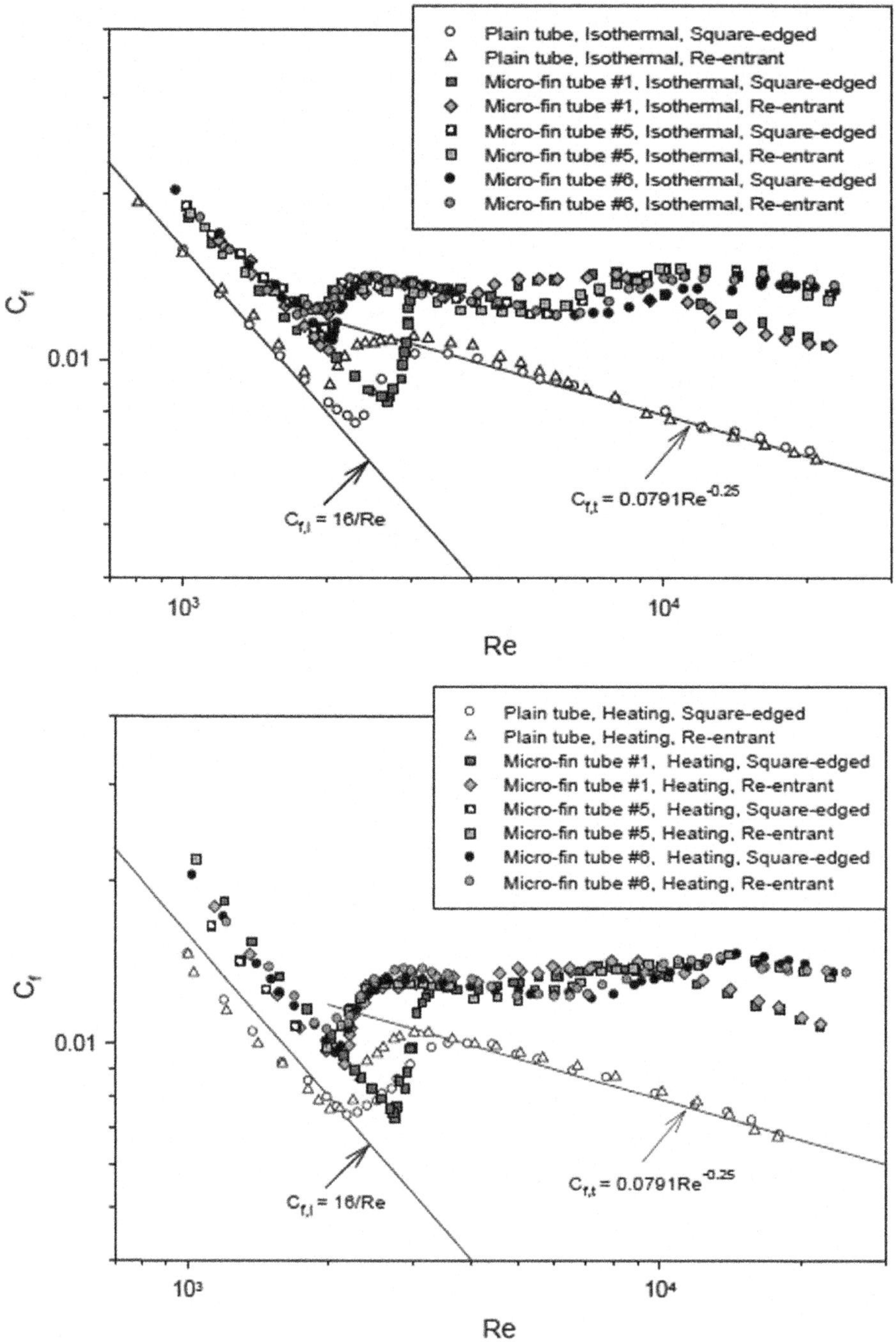

Fig. 7.8 Fully developed friction factor characteristics for the plain and micro-fin tubes #1, #5, and #6 having different numbers of starts and different inlet configurations under isothermal and heating boundary conditions

Table 7.4 Start and end of fully developed friction factor transition in plain and micro-fin tubes #1 ($N_s = 25$), #5 ($N_s = 45$), and #6 ($N_s = 65$) at 200 diameters from the plain and micro-fin tubes' entrance

	Friction factor			
Tube, condition, inlet	Re_{start}	C_f	Re_{end}	C_f
Plain, Isothermal (square-edged)	2222	0.0079	3588	0.0102
Plain, Heating (square-edged)	2316	0.0074	3941	0.0100
Plain, Isothermal (re-entrant)	2032	0.0090	3031	0.0110
Plain, Heating (re-entrant)	2257	0.0079	3250	0.0105
Micro-fin #1, Isothermal (square-edged)	2675	0.0084	8007	0.0144
Micro-fin #1, Heating (square-edged)	2765	0.0073	8925	0.0136
Micro-fin #1, Isothermal (re-entrant)	2021	0.0104	8023	0.0143
Micro-fin #1, Heating (re-entrant)	2141	0.0092	9027	0.0142
Micro-fin #5, Isothermal (square-edged)	2068	0.0112	14,204	0.0142
Micro-fin #5, Heating (square-edged)	2121	0.0108	14,006	0.0146
Micro-fin #5, Isothermal (re-entrant)	1949	0.0123	14,250	0.0142
Micro-fin #5, Heating (re-entrant)	2045	0.0105	14,165	0.0145
Micro-fin #6, Isothermal (square-edged)	2106	0.0109	16,366	0.0140
Micro-fin #6, Heating (square-edged)	2077	0.0096	16,689	0.0144
Micro-fin #6, Isothermal (re-entrant)	1739	0.0120	16,212	0.0143
Micro-fin #6, Heating (re-entrant)	1991	0.0106	16,792	0.0138

Figure 7.8 also represents fully developed friction factor characteristics for plain and micro-fin tubes #1, #5, and #6, which have different inlet configurations under heating (non-isothermal) boundary condition. Parallel shifts from the classic laminar equation are observed in these micro-fin tubes with both inlets. The laminar friction factor for heated micro-fin tubes has the same trend in the laminar region. In other words, the friction factor in that region is also insensitive to the boundary condition and number of starts. For the heated friction factor data, the lower transition Reynolds numbers of micro-fin tube #5 (2121 and 2045) for the square-edged and re-entrant inlets are different from micro-fin tube #1 (2764 and 2141), and an early transition for the micro-fin tube #5 with a larger number of starts ($N_s = 45$) is observed. Furthermore, the early transition for micro-fin tube #6 with a larger number of starts ($N_s = 65$) is also observed when compared with micro-fin tube #1 (Table 7.4). The square-edged inlet delays the start of transition in comparison to the re-entrant inlet. In the transitional region, the heated friction factors for those micro-fin tubes also go through a steep increase followed by a relatively constant C_f section and then a parallel shift from the classic Blasius turbulent friction factor correlation ($C_{f,t} = 0.0791\text{Re}^{-0.25}$). With the larger number of starts in micro-fin tubes #5 and #6, a dip in the friction factor also occurs in the relatively constant C_f section, and a higher friction factor in the turbulent region is observed. Larger number of starts increase surface roughness and consequently friction factors in the turbulent region. Under heated conditions, larger number of starts in the micro-

fin tube delay the end of transition (Fig. 7.8, Table 7.4). With the start and end of transition under heating condition, a wider transitional region occurs with an increase in the number of starts.

Thus, regardless of isothermal or heating boundary condition, the larger number of starts causes a wider transition range, a higher friction factor in the turbulent region, and a dip in the upper transitional region.

Chapter 8
Proposed Correlations for Laminar, Transition, and Turbulent Friction Factors in Micro-fin Tubes with Different Inlet Configurations

Background

For proper design of thermal systems using micro-fin tubes, the experimental data presented in Chap. 7 was used to develop accurate friction factor correlations for the laminar and transitional friction factors in the entrance and fully developed regions of micro-fin tubes under isothermal and non-isothermal (heating) conditions, different inlet configurations (square-edged and re-entrant), and different fin geometries (spiral angle, fin height-to-diameter ratio, and number of starts). For the fully developed turbulent region, Tam et al. (2012) and Tam (2013) did a detailed review of the available fully developed turbulent friction factor correlations for micro-fin tubes and recommended the correlation of Jensen and Vlakancic (1999).

Laminar Region

To develop a correlation for the fully developed laminar friction factor for micro-fin tubes, a total of 439 isothermal friction factor data points from all six micro-fin tubes (Table 2.1) for the two inlet types (square-edged and re-entrant) were used. As was shown in Figs. 7.2, 7.3, and 7.4, a parallel shift from the classic laminar equation ($C_{f,l} = 16/\text{Re}$) occurred for the micro-fin tubes in the laminar region. Furthermore, the laminar isothermal friction factor is not dependent on the inlet configuration (Fig. 7.4) and fin geometry (Figs. 7.6, 7.7, and 7.8). Hence, the proposed isothermal laminar correlation is of the form:

A. J. Ghajar, *Single- and Two-Phase Flow Pressure Drop and Heat Transfer in Tubes*, Mechanical Engineering Series,
https://doi.org/10.1007/978-3-030-87281-6_8

$$C_{f,l} = \frac{20}{\text{Re}} \tag{8.1}$$

for $800 \leq \text{Re} \leq 2675$, $18° \leq \alpha \leq 35°$, $0.0202 \leq e/D \leq 0.0336$, and $25 \leq N_s \leq 65$.

Equation (8.1) gives a representation of the isothermal friction factor experimental data (439 data points) being predicted to within +10.6% to −19.2% with an absolute average deviation of 5.5%, with 83.6% (367 data points) having less than ±10% deviation and 16.4% (72 data points) having ±10 to 20% deviation. Figure 8.1 compares the predicted fully developed isothermal laminar friction factors obtained from the proposed Eq. (8.1) for each inlet with isothermal friction factor measurements.

Since the laminar non-isothermal (heating) friction factors for all six micro-fin tubes for the two inlet types are also not dependent on the inlet configuration (Fig. 7.5) and fin geometry (Figs. 7.6, 7.7, and 7.8), Eq. (8.1) is also applicable to the non-isothermal friction factors in the laminar region. The correlation gives a representation of the non-isothermal (heating) experimental data (328 data points) to within +16.6% to −14.9% with an absolute average deviation of 4.6%, with 92.1% (302 data points) having less than ±10% deviation and 7.9% (26 data points) having ±10 to 20% deviation. Figure 8.1 also compares the predicted fully developed friction factors obtained from the proposed Eq. (8.1) for each inlet with non-isothermal friction factor measurements.

To develop a friction factor correlation for the entrance and fully developed region similar to the correlation proposed for plain tube friction factor (Eq. 4.1), a correction factor in terms of ζ was introduced into the proposed isothermal micro-fin tube fully developed laminar friction factor correlation (Eq. 8.1), $C_{f,l} = 20/\text{Re}$, to account for the variation of friction factors in the entrance region. In development of this correlation, a total of 1271 isothermal experimental data points (686 data points for the square-edged inlet and 585 data points for the re-entrant inlet) were used. For the isothermal condition, the resulting correlation is of the following form:

$$f_{\text{app,lam,iso}} = \frac{1}{\text{Re}} \left(20 + \frac{0.177}{0.0017 + 1.848\,\zeta} \right) \tag{8.2}$$

where $\zeta = \frac{x/D}{\text{Re}}$ and the range of application of Eq. (8.2) is $920 \leq \text{Re} \leq 2675$, $18° \leq \alpha \leq 35°$, $0.0202 \leq e/D \leq 0.0336$, $25 \leq N_s \leq 65$, and $3 \leq x/D \leq 200$.

Equation (8.2) gives a representation of the experimental data to within +39.5% to −38.4%. The average deviation between the results predicted by the correlation and experimental data is 8.7%, with 2.6% (33 data points) having ±30 to 40% deviation, 4.4% (56 data points) having ±20 to 30% deviation, 21.2% (270 data points) having ±10 to 20% deviation, and 71.8% (912 data points) having less than ±10% deviation. Figure 8.2 compares the predicted entrance and fully developed laminar friction factors obtained from the proposed equation (Eq. 8.2), with the actual measurements.

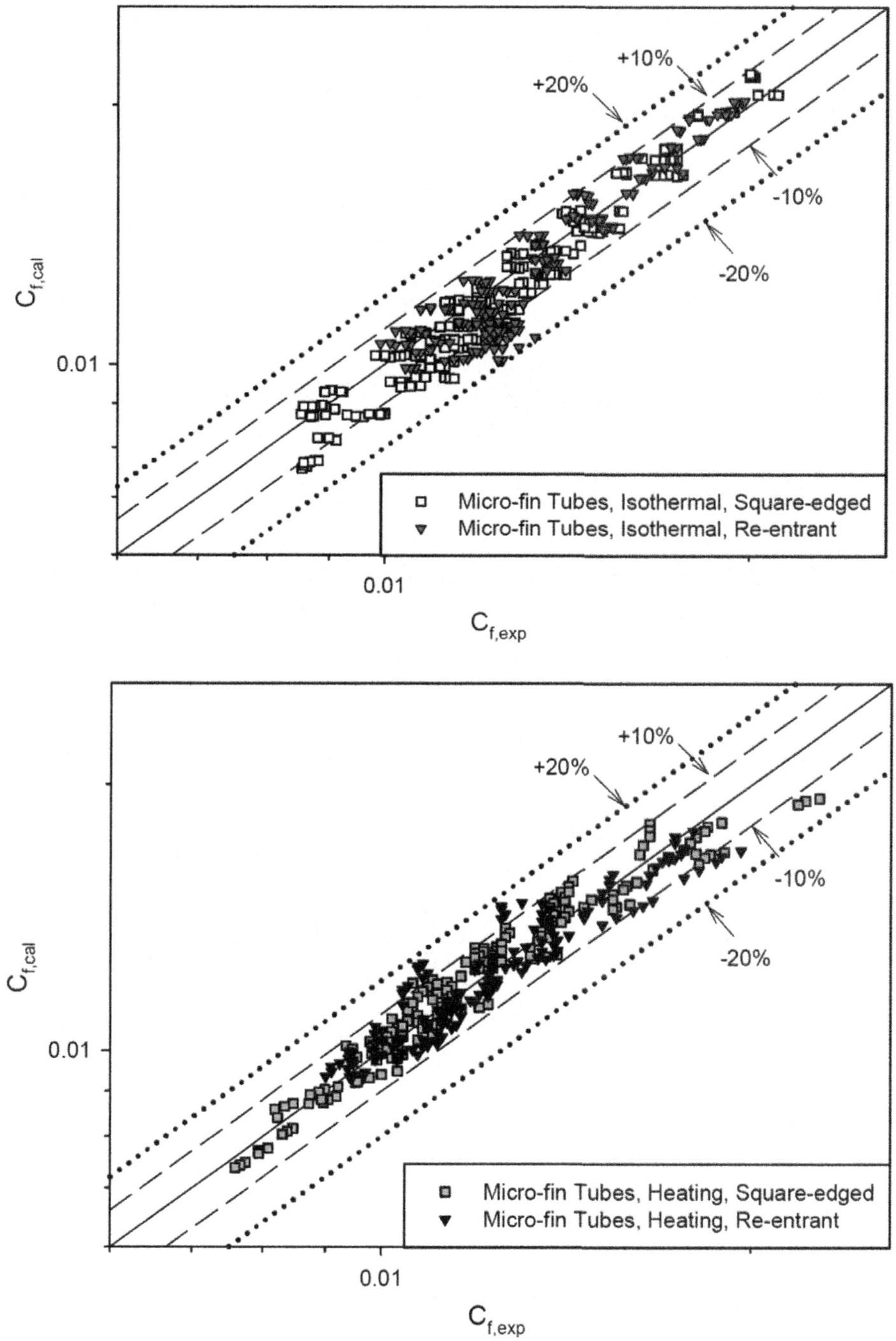

Fig. 8.1 Comparison between experimental fully developed friction factors and the proposed laminar region correlation, Eq. (8.1), for all micro-fin tubes with two different inlet configurations under isothermal and non-isothermal boundary conditions

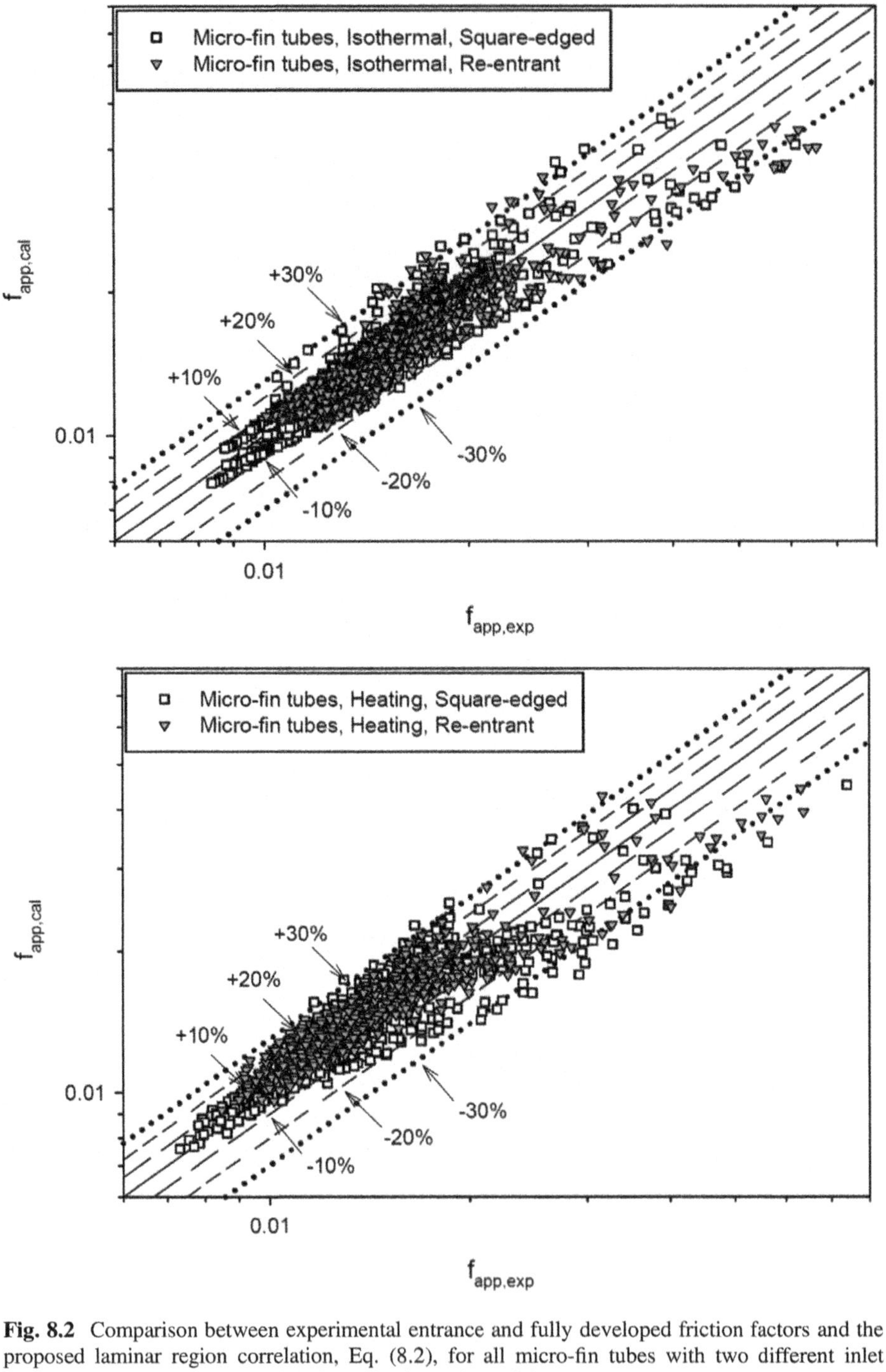

Fig. 8.2 Comparison between experimental entrance and fully developed friction factors and the proposed laminar region correlation, Eq. (8.2), for all micro-fin tubes with two different inlet configurations under isothermal and non-isothermal conditions

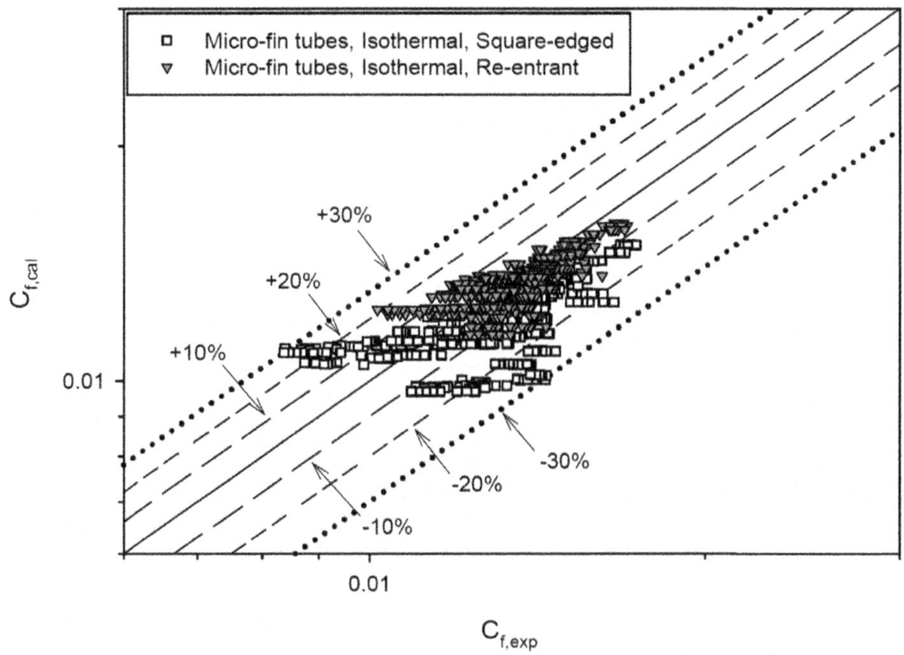

Fig. 8.3 Comparison between experimental fully developed friction factors and the proposed transitional region correlation, Eq. (8.3), for all micro-fin tubes with two different inlet configurations under isothermal condition

Equation (8.2) is also proposed for the non-isothermal (heating) condition because the heating effect on the micro-fin tube entrance flow friction factor is negligible (Fig. 7.3).

Equation (8.2) gives a representation of the experimental data to within +37.2% to −39.8%. To develop the correlation, a total of 1094 experimental data points (585 data points for the square-edged inlet and 509 data points for the re-entrant inlet) were used. The average deviation between the results predicted by the correlation and the experimental data is 10.4%, with 3.7% (40 data points) having ±30 to 40% deviation, 11.2% (122 data points) having ±20 to 30% deviation, 31.5% (345 data points) having ±10 to 20% deviation, and 53.7% (587 data points) having less than ±10% deviation. Figure 8.2 also compares the predicted apparent friction factors obtained from the proposed correlation with measurements for the non-isothermal (heating) condition.

Transitional Region

As shown in Figs. 7.4 and 7.5, the type of inlet configuration influences the start and end of the transition region. Thus, a single correlation cannot predict the data, and a specific correlation for each inlet should be developed. The correlation for the isothermal fully developed friction factor is presented first, and then the correlation for the non-isothermal friction factor is obtained by adding a viscosity correction term. Since the fully developed friction factor in the transition region goes through a steep increase followed by a relatively constant C_f section under isothermal condition, the proposed correlation form is developed based on specific transitional region characteristics. Furthermore, to account for the fin geometry effect, fin geometry parameters (spiral angle, fin height-to-diameter ratio, and number of starts) are also involved in the correlation in the power form. Hence, the proposed correlation is:

$$C_{f,\mathrm{tr,iso}} = \left(\frac{a\ \mathrm{Re}}{b + \mathrm{Re}}\right)^c (\alpha)^d (e/D)^f (N_s)^g \tag{8.3}$$

where the coefficients and exponents, a, b, c, d, f, and g are inlet dependent and are obtained separately for each inlet. The coefficients and exponents for each inlet are as follows:

- Square-edged: $a = 0.02$, $b = 880$, $c = 0.98$, $d = 0.22$, $f = 0.27$, $g = -0.04$ for $1962 \leq \mathrm{Re} \leq 16{,}366$, $80 \leq x/D \leq 200$, $18° \leq \alpha \leq 35°$, $0.0202 \leq e/D \leq 0.0336$, $25 \leq N_s \leq 65$
- Re-entrant: $a = 0.676$, $b = 42$, $c = 10.1$, $d = 0.17$, $f = 0.17$, $g = -0.05$ for $1739 \leq \mathrm{Re} \leq 16{,}212$, $80 \leq x/D \leq 200$, $18° \leq \alpha \leq 35°$, $0.0202 \leq e/D \leq 0.0336$, $25 \leq N_s \leq 65$

Equation (8.3) is applicable to an isothermal and fully developed transitional region and should be used with an appropriate set of constants for each inlet geometry (square-edged and re-entrant). It gives a representation of the experimental data to within +31.0% to −30.3%. A total of 605 data points for the re-entrant inlet and 602 data points for the square-edged inlet were used to develop the correlation. The average deviation between the results predicted by the correlation and the experimental data is 5.8%, with 6.4% (77 data points) having ±20 to 30% deviation, 22.9% (277 data points) having ±10 to 20% deviation, and 70.5% (851 data points) having less than ±10% deviation. Figure 8.3 compares the predicted fully developed transitional region friction factors obtained from the proposed equation, Eq. (8.3), with isothermal measurements.

To account for the entrance effect, a correction factor in terms of length-to-diameter ratio (x/D) was added to Eq. (8.3). Therefore, the correlation for the prediction of entrance and fully developed flows in the transition region under isothermal condition is of the following form:

$$f_{\mathrm{app,tr,iso}} = C_{f,\mathrm{tr,iso}}\left[1 + \left(\frac{m}{x/D}\right)\right] \tag{8.4}$$

where the coefficient m is inlet dependent and obtained separately for each inlet. The coefficient for each inlet and its range of application are as follows:

- Re-entrant: $m = 3.1$; for $1739 \le \text{Re} \le 16{,}212$, $3 \le x/D \le 200$, $18° \le \alpha \le 35°$, $0.0202 \le e/D \le 0.0336$, $25 \le N_s \le 65$
- Square-edged: $m = 2.2$; for $1962 \le \text{Re} \le 16{,}366$, $3 \le x/D \le 200$, $18° \le \alpha \le 35°$, $0.0202 \le e/D \le 0.0336$, $25 \le N_s \le 65$

Equation (8.4) is applicable to an isothermal entrance and fully developed transitional region and should be used with an appropriate constant for each inlet configuration (square-edged and re-entrant). For the development of the transition region correlation for the re-entrant inlet, 1188 experimental data points were used. The correlation gives a representation of the experimental data to within +38.9% to −39.5% with an average deviation of 7.1%, with 2.9% (35 data points) having ±30 to 40% deviation, 4.1% (49 data points) having ±20 to 30% deviation, 16.1% (191 data points) having ±10 to 20% deviation, and 76.9% (913 data points) having less than ±10% deviation. For the square-edged inlet, 1145 experimental data points were used to develop the correlation. The equation correlated the experimental data to within +39.9% to −39.9% with an average deviation of 10.5%, with 5.2% (60 data points) having ±30 to 40% deviation, 12.9% (148 data points) having ±20 to 30% deviation, 26.3% (301 data points) having ±10 to 20% deviation, and 55.5% (636 data points) having less than ±10% deviation. The accuyracy of predition is reasonble since the majority of the re-entrant and square-edged data (around 80% data) were predicted within ±20%. Figure 8.4 compares the predicted entrance and fully developed friction factors obtained from the proposed Eq. (8.4) for each inlet with isothermal measurements.

To account for the effect of heating, the viscosity ratio correction factor proposed by Tam and Ghajar (1997) expressed as a function of the Grashof and Prandtl numbers was applied to the isothermal correlation, Eq. (8.4). The proposed correlation is of the following form:

$$f_{\text{app,tr,heat}} = f_{\text{app,tr,iso}} \times \left(\frac{\mu_b}{\mu_w}\right)^n \tag{8.5}$$

where the exponent n is inlet dependent and obtained separately for each inlet. The exponent of the viscosity ratio term and its range of application for each inlet are as follows:

- Re-entrant: $n = -5.8 + 8.2\text{Gr}^{-0.019}\,\text{Pr}^{-0.07}$
  for $1954 \le \text{Re} \le 16{,}792$, $3 \le x/D \le 200$, $18° \le \alpha \le 35°$, $0.0202 \le e/D \le 0.0336$, $25 \le N_s \le 65$, $4.5 \le \text{Pr} \le 57.3$, $1481 \le \text{Gr} \le 28{,}115$, $1.007 \le \mu_b/\mu_w \le 1.879$
- Square-edged: $n = -6.15 + 9.6\text{Gr}^{-0.027}\,\text{Pr}^{-0.06}$
  for $2077 \le \text{Re} \le 16{,}689$, $3 \le x/D \le 200$, $18° \le \alpha \le 35°$, $0.0202 \le e/D \le 0.0336$, $25 \le N_s \le 65$, $4.5 \le \text{Pr} \le 58.7$, $1933 \le \text{Gr} \le 44{,}839$, $1.007 \le \mu_b/\mu_w \le 1.956$

Equation (8.5) is applicable to the non-isothermal entrance and fully developed transition regions and should be used with an appropriate constant for each inlet

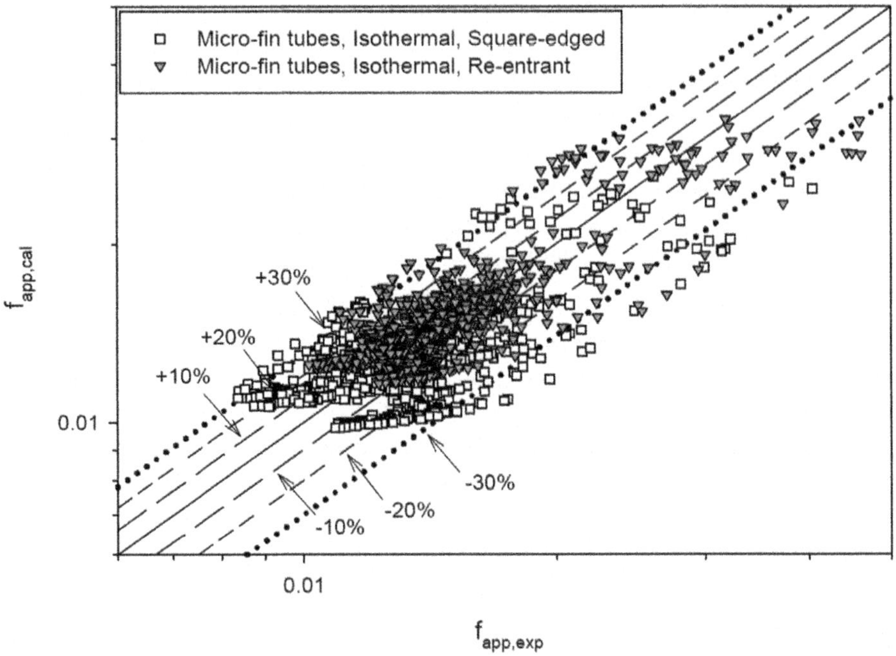

Fig. 8.4 Comparison between experimental entrance and fully developed friction factors and the proposed transitional region correlation, Eq. (8.4), for all micro-fin tubes with two different inlet configurations under isothermal condition

configuration (square-edged and re-entrant). For the development of the transitional region correlation for the re-entrant inlet, 961 experimental data points were used. The correlation gives a representation of the experimental data to within +39.3% to −39.8% and has an average deviation of 11.0%, with 5.9% (57 data points) having ±30 to 40% deviation, 12.5% (120 data points) having ±20 to 30% deviation, 24.8% (238 data points) having ±10 to 20% deviation, and 56.8% (546 data points) having less than ±10% deviation. For the square-edged inlet, 973 experimental data points were used to develop the correlation. The equation correlated the experimental data to within +38.8% to −39.3% and has an average deviation of 9.8%, with 6.8% (66 data points) having ±30 to 40% deviation, 12.4% (121 data points) having ±20 to 30% deviation, 24.3% (236 data points) having ±10 to 20% deviation, and 56.5% (550 data points) having less than ±10% deviation. The accuracy of prediction is reasonable since the majority of the re-entrant and square-edged data (around 80% data) were predicted within ±20%. Figure 8.5 compares the predicted entrance and fully developed friction factors obtained from the proposed Eq. (8.5) for each inlet with isothermal measurements.

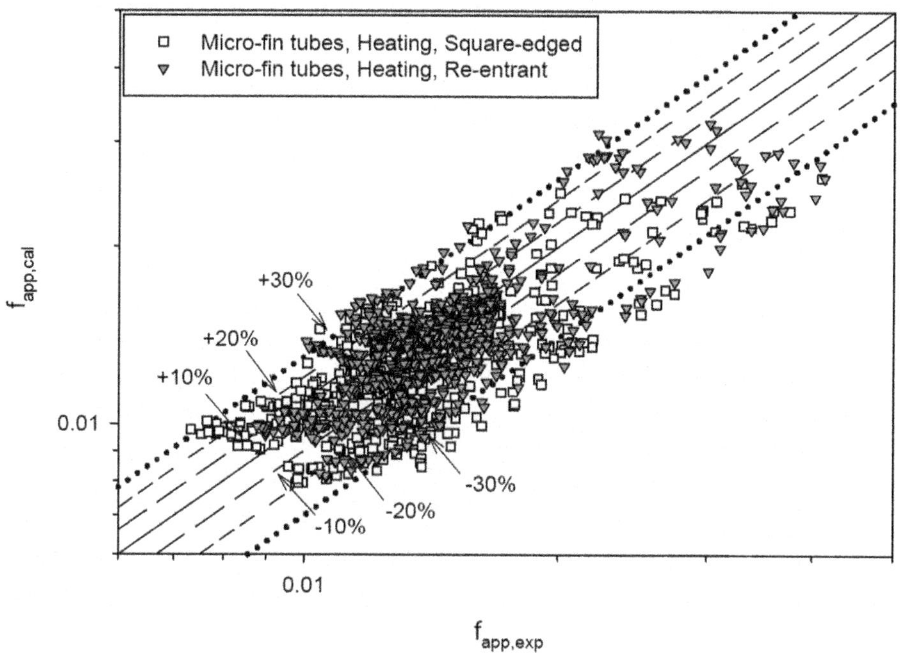

Fig. 8.5 Comparison between experimental entrance and fully developed friction factors and the proposed transitional region correlation, Eq. (8.5), for all micro-fin tubes with two different inlet configurations under non-isothermal (heating) condition

Turbulent Region

Tam et al. (2012) and Tam (2013) compared the predictions of the fully developed turbulent friction factor correlations available in the open literature for micro-fin tubes with the experimental fully developed turbulent friction factor data presented in Chap. 7 for micro-fin tubes. Based on their detailed comparisons, the correlation of Jensen and Vlakancic (1999) performed best in predicting the experimental data presented in Chap. 7, and it is recommended.

The fully developed turbulent friction factor correlation of Jensen and Vlakancic (1999) is given by Eq. (8.6):

$$C_{f,t} = C_{f,p}\left(\frac{l_{\text{csw}}}{D}\right)\left(\frac{A}{A_{\text{xs}}}\right)^{1.75} - 0.0151\left\{\left(\frac{l_{\text{csw}}}{D}\right)^{-1.25}\left(\frac{A_{\text{c}}}{A_{\text{xs}}}\right)^{1.75} - 1\right\}\exp^{\left(-\frac{\text{Re}}{6780}\right)} \quad (8.6)$$

where $\frac{l_{\text{csw}}}{D} = \left[1 - 0.994(\text{SW})^{0.89}\left(\frac{2e}{D}\right)^{0.44}(W)^{0.41}\right]$

$$SW = N_s\left(\frac{\sin\,\alpha}{\pi}\right), W = [(\pi/N_s) - (t/D)](\cos\,\alpha), C_{f,p} = (1.58\ln \mathrm{Re}\ - 3.28)^{-2}$$

and the ranges of variables for the micro-fin tubes friction factor data are $2500 \leq \mathrm{Re} \leq 20{,}000$, $25° \leq \alpha \leq 45°$, $0.02 \leq e/D \leq 0.03$, and $36 \leq N_s \leq 54$.

The correlation of Jensen and Vlakancic (1999), Eq. (8.6), predicted the fully developed turbulent isothermal friction factor data for the square-edged (210 data points) and re-entrant (216 data points) inlets with an absolute deviation of 7.2% (deviation range of −17.5% to 18.1%) and 5.9% (deviation range of −16.4% to 16.0%), respectively. The non-isothermal (heating) fully developed turbulent friction factor data for the squared-edged (197 data points) and re-entrant (209 data points) inlets were predicted with an absolute deviation of 7.8% (deviation range of −22.0% to 19.7%) and 7.1% (deviation range of −19.0% to 15.9%), respectively. Therefore, the correlation of Jensen and Vlakancic (1999) overall does a good job of predicting the fully developed turbulent friction factor data under isothermal and heating (non-isothermal) boundary conditions and is recommended.

Chapter 9
Heat Transfer Results in Micro-fin Tubes

Background

Single-phase liquid flow in internally enhanced tubes is becoming more important in commercial heating, ventilating, and air-conditioning (HVAC) applications, where enhanced tube bundles are used in flooded evaporators and shell-side condensers to increase heat transfer. This enables water chillers to reach high efficiency, which helps mitigate global warming concerns of HVAC systems. One kind of internally enhanced tube is the spiral micro-fin tube. For micro-fin tubes, tube side roughening increases the heat transfer surface area, resulting in an increase in heat transfer for the micro-fin tube over the plain tube. However, there is also an increase in the pressure drop (friction factor) as discussed in Chap. 7. The increase in the surface area causes low flow rates in the heat exchanger tubes, resulting in the flow being at Reynolds numbers that are in the transition region (between laminar and fully turbulent regions). With these high heat transfer rates, it is likely that more HVAC units will operate in the transition region where the understanding of friction factor (pressure drop) and heat transfer is limited. Friction factor characteristics of micro-fin tubes were discussed in Chap. 7. As far as heat transfer characteristics of micro-fin tubes are concerned, there is limited information on the effect of different inlet configurations, buoyancy, and different fin geometries on the heat transfer in the entrance and fully developed regions of the micro-fin tubes. Refer to Tam et al. (2012) and Tam (2013) for a detailed review of related studies on micro-fin tubes.

For this purpose, using the experimental setup detailed in Chap. 2, experiments were conducted to obtain accurate and broad heat transfer database across all flow regimes. This chapter presents heat transfer results in the entrance and fully developed regions of the micro-fin tubes, the effect of inlet configuration and buoyancy on the entrance and fully developed heat transfer in all flow regimes (laminar, transitional, and turbulent), the effect of inlet configuration on the start and end of transition of the fully developed heat transfer, and the effect of fin geometries (spiral

A. J. Ghajar, *Single- and Two-Phase Flow Pressure Drop and Heat Transfer in Tubes*, Mechanical Engineering Series,
https://doi.org/10.1007/978-3-030-87281-6_9

angle, fin height-to-diameter ratio, and number of starts) on heat transfer. Detailed information on these effects may be obtained from Tam et al. (2012) and Tam (2013).

Effect of Inlet Configuration and Buoyancy on Micro-fin Tubes' Heat Transfer

Figure 9.1 clearly shows the influence of inlet configuration on the start and end of the heat transfer transitional region for the plain tube and micro-fin tube #1 (Table 2.1). This figure plots the local average peripheral heat transfer coefficients in terms of the Colburn j factor ($j = St\,Pr^{0.67}$) versus local Reynolds number for all flow regimes at the length-to-inside-diameter ratio of 200. The presented results show how heat transfer in the transitional region varies between the limits of the fully developed turbulent flow correlation [$Nu_t = 0.023Re^{0.8}Pr^{0.33}(\mu_b/\mu_w)^{0.14}$] and fully developed laminar flow equation ($Nu_l = 4.364$) with a uniform wall heat flux boundary condition. The Reynolds number for the start of transition of the micro-fin tube is defined as the Reynolds number corresponding to the first abrupt change from the laminar region (landed on a line parallel to $Nu_l = 4.364$) to the transition region in the Colburn j factor ($St\,Pr^{0.67}$). The Reynolds number for the end of the transition region of the micro-fin tube is defined as the first point where the heat

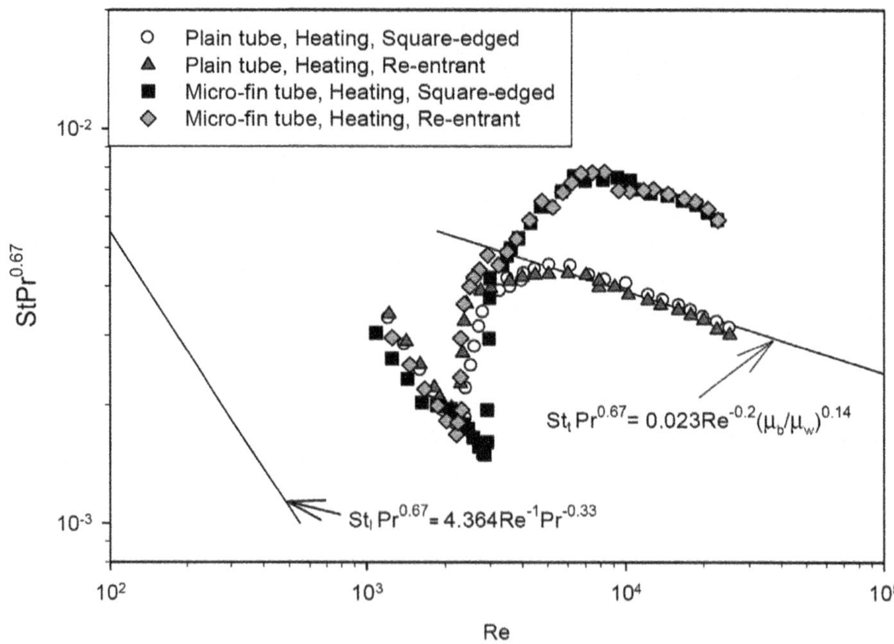

Fig. 9.1 Heat transfer characteristics for the plain tube and micro-fin tube #1

Table 9.1 Start and end of fully developed heat transfer transition for plain and micro-fin tubes at 200 diameters from the plain and micro-fin tubes' entrance

	Heat transfer			
Tube, condition, inlet	Re_{start}	$StPr^{0.7}$	Re_{end}	$StPr^{0.7}$
Plain, heating (square-edged)	2298	0.0017	8357	0.0041
Plain, heating (re-entrant)	2001	0.0020	7919	0.00412
Micro-fin #1, heating (square-edged)	2751	0.0017	8963	0.0094
Micro-fin #1, heating (re-entrant)	2145	0.0019	8014	0.0092

transfer data landed on a line parallel to the fully developed turbulent heat transfer line. As shown in Fig. 9.1, the upper and lower limits of the heat transfer transition Reynolds numbers for the plain and micro-fin tubes are dependent on the inlet configuration. The transition Reynolds numbers for heat transfer for plain and micro-fin tubes are summarized in Table 9.1.

In the laminar flow region of plain and micro-fin tubes, the effect of free convection (or buoyancy) on forced convection can be clearly seen in Fig. 9.1, resulting in a parallel upward shift of Colburn j factors from their fully developed forced convection laminar values. In the transitional region, the effect of mixed convection cannot be easily seen without carefully examining local heat transfer information, which is the ratio of the local peripheral heat transfer coefficient at the top of the tube to the local peripheral heat transfer coefficient at the bottom of the tube (h_t/h_b).

To properly account for the effect of buoyancy on heat transfer, four thermocouples were placed around the periphery of the tube (90° apart) at each of the 16 designated axial locations (Fig. 2.5). According to Ghajar and Tam (1994), h_t/h_b should be close to unity (0.8–1.0) for forced convection and is much less than unity (<0.8) when mixed convection exists. Ghajar and Tam (1994) established that when h_t/h_b is less than 0.8, a strong buoyancy effect is present (Fig. 3.3).

Figure 9.2 shows the effect of buoyancy on the heat transfer coefficient for micro-fin tube #1 with different inlets and flow regimes. For micro-tube #1 with square-edged and re-entrant inlets, when Reynolds numbers were less than 2800 and 2100, respectively, the h_t/h_b ratio started at one and decreased rapidly, finally stabilizing when x/D was larger than 100. The lower transition Reynolds number for the micro-fin tube #1 with the re-entrant inlet was delayed to 2145 from 2001 relative to the plain tube and that for the micro-fin tube #1 with the square-edged inlet was delayed to 2751 from 2298 (Table 9.1). These delays in transition are a direct effect of buoyancy forces in the laminar region and are the main reason for the parallel shift from the classic Nu_l value of 4.364 observed in Fig. 9.1. No buoyancy effect was observed in transition and turbulent regions.

For heat transfer, it can be clearly seen from Table 9.1 and Fig. 9.1 that the lower transition is delayed for the micro-fin tube and the lower transition Reynolds number is 2751 instead of 2298 for the plain tube fitted with the square-edge inlet. The same observation can be made for the re-entrant inlet, but the lower transition Reynolds

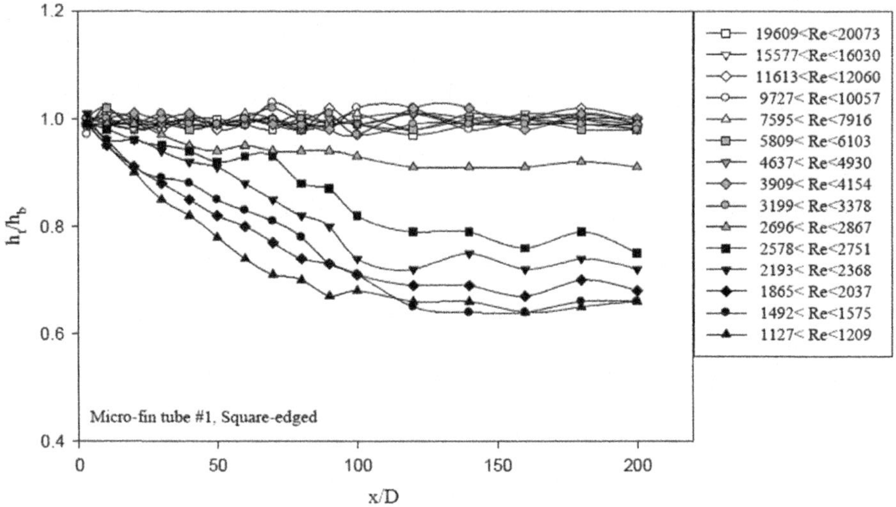

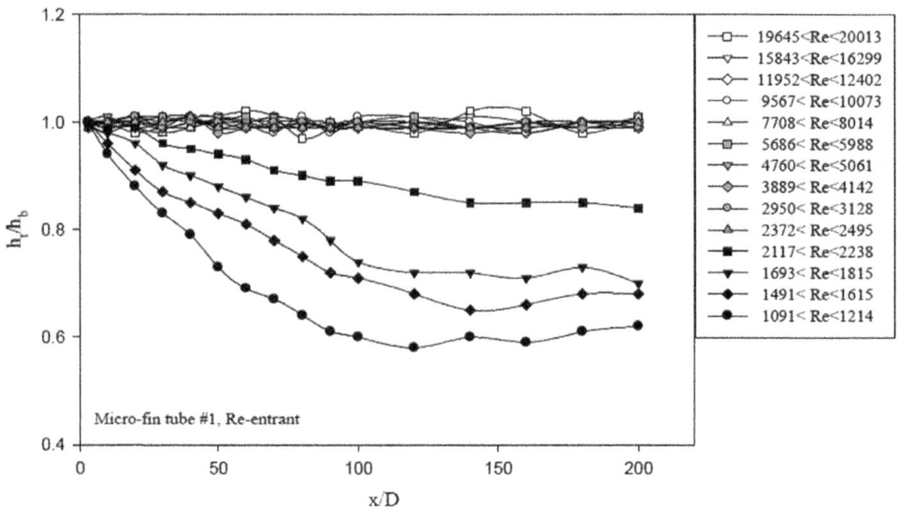

Fig. 9.2 The ratio of heat transfer coefficients at the top and bottom of micro-fin tube #1 with square-edged and re-entrant inlets and all flow regimes (laminar, transition, and turbulent)

number is 2145 instead of 2001. Upper transition heat transfer was also delayed in the micro-fin tube. Furthermore, the onset of the transition was inlet dependent. Heat transfer behavior in the upper transition region of plain and micro-fin tubes is very different (Fig. 9.1). Around a Reynolds number of 4000, plain tube data followed the correlation for the turbulent region. However, micro-fin data continued to increase up to a Reynolds number of about 8000, and then a parallel shift from the turbulent correlation occurred. This increase in the Colburn j factor (St $Pr^{0.67}$) and an upward parallel shift (120% increase for micro-fin tube #1) from the turbulent correlation in the turbulent region was due to the swirling motion induced by the micro-fin.

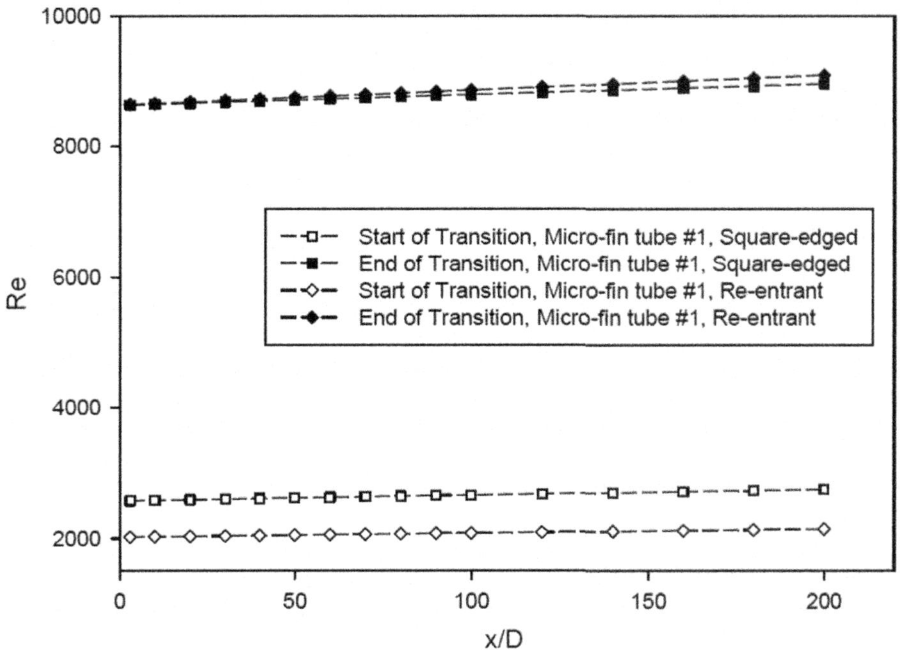

Fig. 9.3 Variation of the lower and upper limits of heat transfer transition Reynolds number along micro-fin tube #1 for square-edged and re-entrant inlets

Figure 9.3 depicts the variation in the lower and upper limits of heat transfer transition Reynolds numbers along the tube for each inlet. To determine the range of heat transfer transition Reynolds numbers along the tube ($3 \leq x/D \leq 200$), graphs similar to Fig. 9.1 were developed for 15 other x/D locations (Ghajar & Tam, 1994). Based on the experimental results, the limits of the ranges of heat transfer transition Reynolds numbers for each inlet over the entire tube ($3 \leq x/D \leq 200$) can be summarized as follows: square-edged ($2578 < Re < 8963$) and re-entrant ($2022 < Re < 9094$).

The above limits for the heat transfer transition Reynolds number indicate that the re-entrant inlet that causes the most disturbance produces the earliest transition from laminar to transition regime (at about 2000), while the square-edged entrance that caused the least disturbance retards this regime change (at about 2500).

Effect of Fin Geometry of Micro-fin Tubes on Heat Transfer

Figure 9.4 shows the heat transfer characteristics of the plain tube and six micro-fin tubes (Table 2.1) used in this study. In general, it can be seen that heat transfer behavior in the upper transition regions of plain and micro-fin tubes are very different. Around a Reynolds number of 4000, plain tube data follows the correlation for the turbulent region. However, the micro-fin data continues to increase up to a Reynolds number of about 8000, and then a parallel shift from the turbulent correlation occurs. This increase in the Colburn j factor (St $Pr^{0.67}$) and an upward parallel shift from the correlation in the turbulent region is due to swirling induced by micro-fins. Each micro-fin tube shape (geometry) has its own particular heat transfer behavior and transition range. It is difficult to observe in Fig. 9.4 the influence of fin geometry on heat transfer characteristics and lower and upper transition limits.

Referring to Jensen and Vlakancic (1999) and Zdaniuk et al. (2008), fin geometry parameters such as fin spiral angle, fin height-to-diameter ratio, and number of starts are the major factors influencing micro-fin tube heat transfer in the turbulent region. In general, turbulent heat transfer increases with an increase of each parameter. However, the effect of those parameters on heat transfer in the laminar and transition regions is not clearly defined in the open literature. Therefore, a qualitative analysis of the influence of the abovementioned parameters on heat transfer is presented next.

Effect of Change in Spiral Angle on Heat Transfer

Micro-fin tubes #1, #2, and #3 have the same fin height-to-diameter ratio ($e/D = 0.0336$) and number of starts ($N_s = 25$) but have different spiral angles ($\alpha = 18°, 25°$, and $35°$) (Table 2.1). Therefore, those micro-fin tubes were selected to analyze the effect of spiral angle on heat transfer.

Figure 9.5 represents the fully developed heat transfer characteristics of plain and micro-fin tubes with different spiral angles (#1, #2, and #3) and different inlet configurations (square-edged and re-entrant). Heat transfer behavior in the upper transition and turbulent regions for plain and micro-fin tubes is very different. Around a Reynolds number of 8000, plain tube data followed the correlation for the turbulent region. However, the micro-fin tube data continued to increase up to a Reynolds number of about 8000 and then made a parallel shift from the turbulent correlation. This increase in the Colburn j factor (St $Pr^{0.67}$) and the parallel shift from the correlation in the turbulent region is due to the swirling motion induced by the micro-fin. In the upper transition and turbulent regions, the larger spiral angle (micro-fin tube #3 with $\alpha = 35°$) led to a higher Colburn j factor. This is because the larger spiral angle increases mixing between the bulk fluid and tube wall, and therefore heat transfer is enhanced. On the other hand, a smaller spiral angle (micro-fin tube #1 with $\alpha = 18°$) leads to a lower Colburn j factor.

The start and end of the transition for the heat transfer (with uniform wall heat flux boundary condition) in plain and micro-fin tubes for the two inlet configurations (square-edged and re-entrant) shown in Fig. 9.5 are summarized in Table 9.2. The plain tube and micro-fin tube #1 results are the same as those presented in Table 9.1. They are repeated here for comparison with the micro-fin tubes #2 and #3 results. The start of the transition is defined as the critical point of sudden change from the laminar region (landing on a line parallel to $Nu_l = 4.364$) to the transition region

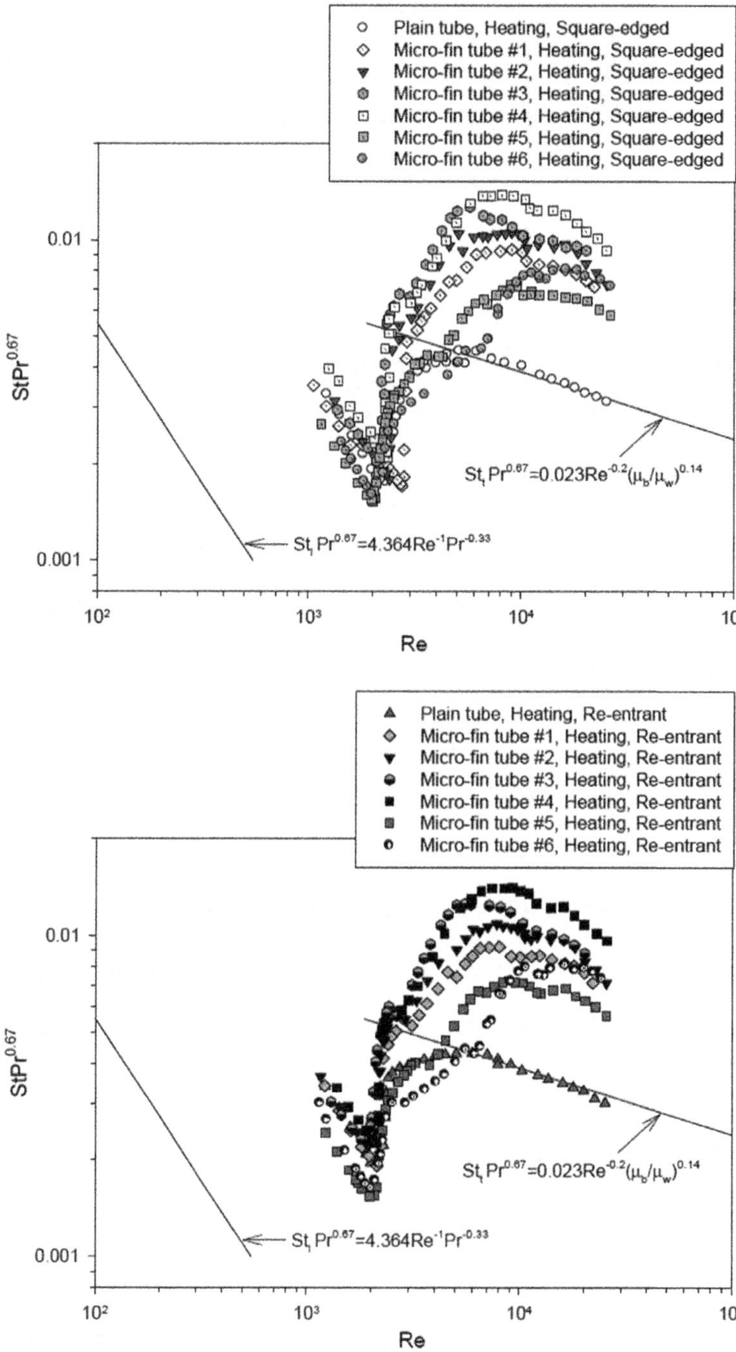

Fig. 9.4 Heat transfer characteristics of the plain tube and micro-fin tubes #1 to #6 with square-edged and re-entrant inlets

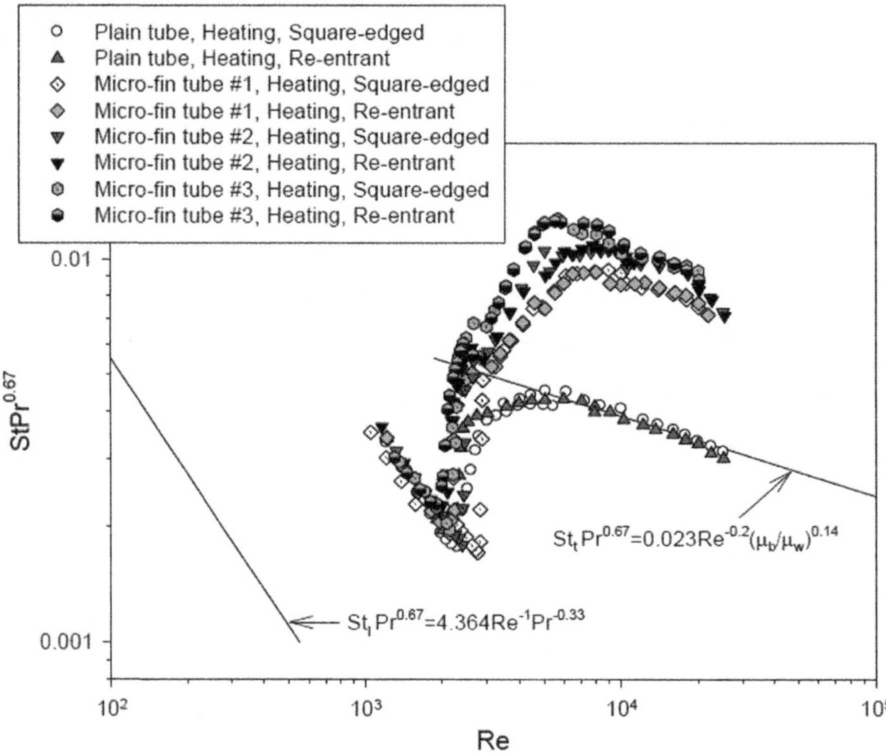

Fig. 9.5 Fully developed heat transfer characteristics for plain tube and micro-fin tubes #1, #2, and #3 with different spiral angles and different inlet configurations

(Fig. 9.5). The end of the transition in micro-fin tubes is defined as the first point where the heat transfer data landed on a line parallel to the turbulent correlation.

In comparison to the plain tube, the heat transfer of micro-fin tube #1 with a spiral angle of 18° in the lower transition was delayed in both square-edged and re-entrant inlets (Table 9.2 and Fig. 9.5). The transition Reynolds numbers for the plain tube with square-edged and re-entrant inlets were 2298 and 2001, respectively. For micro-fin tube #1, the lower transition Reynolds number increased to 2751 for the square-edged inlet and 2145 for the re-entrant inlet. The upper transition Reynolds numbers for micro-fin tube #1 are 8963 for the square-edged and 8014 for re-entrant inlets, and therefore the upper transition was also delayed for both inlet configurations when compared with the plain tube (Re = 8357 for the square-edged and Re = 7919 for the re-entrant inlets).

For heat transfer in micro-fin tube #2 with a larger spiral angle of 25°, the lower transition Reynolds numbers for both square-edged (Re = 2402) and re-entrant (Re = 1946) inlets are advanced when compared with micro-fin tube #1 (Re = 2751 for the square-edged and Re = 2145 for the re-entrant inlets). In

Table 9.2 Start and end of fully developed heat transfer transition in plain and micro-fin tubes #1 ($\alpha = 18°$), #2 ($\alpha = 25°$), and #3 ($\alpha = 35°$) at 200 diameters from the plain and micro-fin tubes' entrance

	Heat transfer			
Tube, condition, inlet	Re_{start}	$StPr^{0.7}$	Re_{end}	$StPr^{0.7}$
Plain, heating (square-edged)	2298	0.0017	8357	0.0041
Plain, heating (re-entrant)	2001	0.0020	7919	0.00412
Micro-fin #1, heating (square-edged)	2751	0.0017	8963	0.0094
Micro-fin #1, heating (re-entrant)	2145	0.0019	8014	0.0092
Micro-fin #2, heating (square-edged)	2402	0.0018	8956	0.0106
Micro-fin #2, heating (re-entrant)	1946	0.0022	7791	0.0109
Micro-fin #3, heating (square-edged)	2144	0.0020	8051	0.0116
Micro-fin #3, heating (re-entrant)	1903	0.0023	7170	0.0124

comparison to the plain tube, for the micro-fin #2 with the square-edged inlet, the lower transition Reynolds number was delayed from 2298 to 2402, but for the re-entrant inlet, the transition Reynolds number was slightly advanced from 2001 to 1946. For the upper transition Reynolds numbers of micro-fin tube #2, the two inlets behaved differently, with the square-edged upper transition Reynolds number being almost the same as that of micro-fin tube #1 (8956 vs 8963). The re-entrant inlet's upper transition Reynolds number for micro-fin tube #2 compared to micro-fin #1 advanced from 8014 to 7791. Compared to the plain tube, the upper transition Reynolds number of micro-fin tube #2 with the square-edged inlet was delayed from 8357 to 8956, but the re-entrant upper transition Reynolds number advanced from 7919 to 7791.

Heat transfer in micro-fin tube #3, with a much larger spiral angle of 35°, had a lower transition Reynolds number for both square-edged ($Re = 2144$) and re-entrant ($Re = 1903$) inlets, in comparison to the micro-fin tube #1 ($Re = 2751$ for the square-edged and $Re = 2145$ for the re-entrant inlets) and micro-fin tube #2 ($Re = 2402$ for the square-edged and $Re = 1946$ for the re-entrant inlets). Comparing the plain tube to micro-fin tube #3, the lower transition Reynolds number of the latter for the square-edged and re-entrant inlets advanced from 2298 to 2144 and from 2001 to 1903, respectively. Also, it was observed that the upper transition Reynolds numbers for the micro-fin #3 ($Re = 8051$ for the square-edged and $Re = 7170$ for the re-entrant inlets) were lower than those of the other two micro-fin tubes, micro-fin tube #1 ($Re = 8963$ for the square-edged and $Re = 8014$ for the re-entrant inlets) and micro-fin tube #2 ($Re = 8956$ for the square-edged and $Re = 7791$ for the re-entrant inlets). Relative to the plain tube, the upper transition Reynolds numbers of micro-fin tube #3 for the square-edged and re-entrant inlets advanced from 8357 to 8051 and from 7919 to 7170, respectively.

Increasing the spiral angle for micro-fin tubes caused an early transition of heat transfer, which is because larger spiral angles increase flow disturbance.

Effect of Change in Fin Height-to-Diameter Ratio on Heat Transfer

As seen in Table 2.1, micro-fin tubes #3 and #4 have the same spiral angle ($\alpha = 35°$) and number of starts ($N_s = 25$) but different fin height-to-diameter ratios ($e/D = 0.0336$ and 0.0202). Therefore, those two micro-fin tubes were selected to analyze the effect of fin height-to-diameter ratio on heat transfer.

Figure 9.6 represents the fully developed heat transfer characteristics of plain and micro-fin tubes #3 and #4, where it is observed that in the laminar region, the heat transfer of micro-fin tube #4 is slightly higher than that of micro-fin tube #3. As can been seen from Fig. 9.6, the heat transfer behavior in the upper transition and turbulent regions of plain and micro-fin tubes is very different. Around a Reynolds number of 8000, plain tube data follows the correlation for the turbulent region. However, micro-fin tube data continued to show an increase up to a Reynolds number of about 8000 and then made a parallel shift from the turbulent correlation. This increase in the Colburn j factor ($St\ Pr^{0.67}$) and the parallel shift from the correlation in the turbulent region is due to the swirling induced by micro-fins. In the turbulent region, the larger fin height-to-diameter ratio of micro-fin tube #3 ($e/D = 0.0336$) led to a lower Colburn j factor. Jensen and Vlakancic (1999) also observed that the turbulent heat transfer of the micro-fin tube with $e/D = 0.04$ was higher than the micro-fin tube with $e/D = 0.06$. They explained that this finding was due to different fin profiles (e.g., fin shape). However, in this study, the fin profiles for all micro-fin tubes were identical. Therefore, the increase of the micro-fin height (from e/D of 0.0202 to 0.0336) does not provide a significant increase in heat transfer. Inversely, the higher fin height can obstruct heat transfer from the heated fin basal surface area to the colder fluid. Therefore, an increase in fin height will reduce turbulent heat transfer.

The start and end of the transition of heat transfer (with uniform wall heat flux boundary condition) in plain and micro-fin tubes for the two inlet configurations (square-edged and re-entrant) shown in Fig. 9.6 are summarized in Table 9.3. The plain tube and micro-fin tube #3 results are the same as those presented in Table 9.2. They are repeated here for comparison with the micro-fin tube #4 results. The start of the transition shown in Fig. 9.6 is defined as the critical point for the sudden change from the laminar region (landing on a line parallel to $Nu_l = 4.364$) to the transition region. The end of the transition is defined as the first point where the heat transfer data landed on a line parallel to the turbulent correlation.

In comparison to the plain tube, the lower transition of heat transfer was advanced in micro-fin tube #3 having a fin height of 0.5 mm with both square-edged and re-entrant inlets (Table 9.3 and Fig. 9.6). The transition Reynolds numbers for the plain tube having square-edged and re-entrant inlets were 2298 and 2001, respectively. For the micro-fin tube #3, the lower transition Reynolds number was lowered to 2144 for the square-edged inlet and 1903 for the re-entrant inlet. The upper transition Reynolds numbers for the plain tube were 8357 for the square-edged

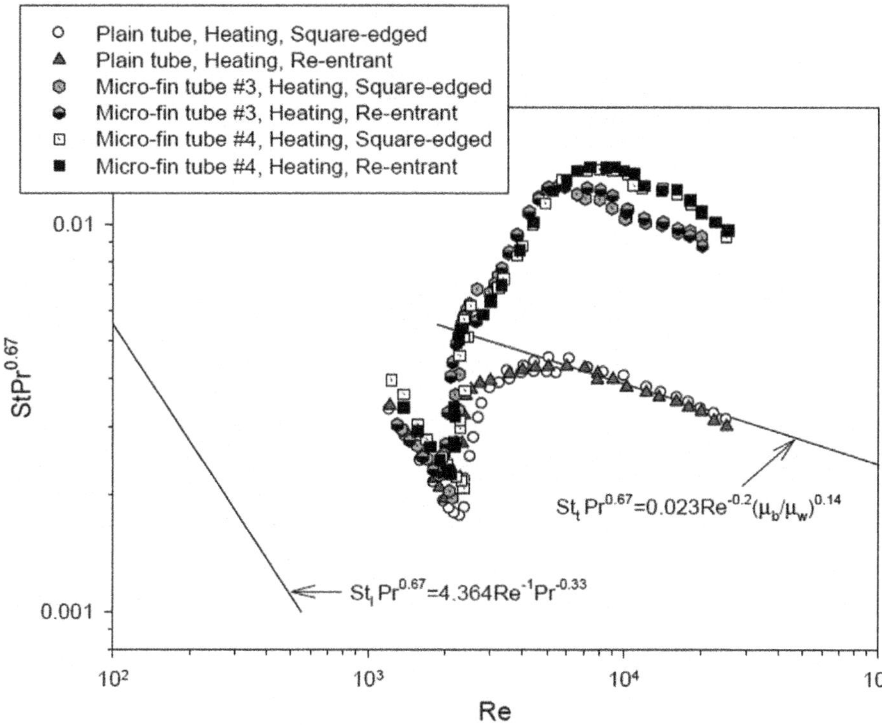

Fig. 9.6 Fully developed heat transfer characteristics for plain tube and micro-fin tubes #3 and #4 with different fin height-to-diameter ratios and different inlet configurations

Table 9.3 Start and end of fully developed heat transfer transition in plain and micro-fin tubes #3 ($e/D = 0.0336$) and #4 ($e/D = 0.0202$) at 200 diameters from the plain and micro-fin tubes' entrance

	Heat transfer			
Tube, condition, inlet	Re_{start}	$StPr^{0.7}$	Re_{end}	$StPr^{0.7}$
Plain, heating (square-edged)	2298	0.0017	8357	0.0041
Plain, heating (re-entrant)	2001	0.0020	7919	0.00412
Micro-fin #3, heating (square-edged)	2144	0.0020	8051	0.0116
Micro-fin #3, heating (re-entrant)	1903	0.0023	7170	0.0124
Micro-fin #4, heating (square-edged)	2364	0.0021	8069	0.0138
Micro-fin #4, heating (re-entrant)	2122	0.0023	8483	0.0140

inlet and 7919 for the re-entrant inlet. For the micro-fin tube #3, the upper transition Reynolds numbers were lowered to 8051 for the square-edged and 7170 for the re-entrant inlets.

For the heat transfer of micro-fin tube #4 with a smaller fin height of 0.3 mm, the lower transition Reynolds numbers for both square-edged ($Re = 2364$) and re-entrant (Re = 2122) inlets were delayed in comparison with micro-fin tube #3 (Re = 2144 for the square-edged and Re = 1903 for the re-entrant inlets). In comparison to the plain tube, the lower transition Reynolds numbers for the micro-fin tube #4 with square-edged inlet were delayed from 2298 to 2364 and for the re-entrant inlet were delayed from 2001 to 2122. For the upper transition Reynolds number, the two inlets behaved differently. For the square-edged inlet, the upper transition Reynolds number of micro-fin tube #4 was almost the same as that of micro-fin tube #3 (8069 vs 8051). For the re-entrant inlet, the upper transition Reynolds number of micro-fin tube #4 was delayed from 7170 to 8483 when compared with micro-fin tube #3. In comparison to the plain tube, the upper transition Reynolds number of micro-fin tube #4 for the square-edged inlet advanced from 8357 to 8069, but for the re-entrant inlet, the upper transition Reynolds number was delayed from 7919 to 8483.

The above results demonstrate that an increase in fin height-to-diameter ratio causes an early transition of heat transfer because of increased flow disturbance.

Effect of Change in Number of Starts on Heat Transfer

Micro-fin tubes #1, #5, and #6 have the same fin height-to-diameter ratio (e/D = 0.0336) and spiral angles ($\alpha = 18°$) but different numbers of starts (N_s = 25, 45, and 65) (Table 2.1). Therefore, those micro-fin tubes were selected to analyze the effect of the number of starts on heat transfer.

Figure 9.7 represents the fully developed heat transfer characteristics of plain and micro-fin tubes #1, #5, and #6 and shows that heat transfer in the laminar region of micro-fin tube #1 is slightly higher than in micro-fin tubes #5 and #6. As can been seen from Fig. 9.7, the heat transfer behavior in the upper transition and turbulent regions of plain and micro-fin tubes is very different. Around a Reynolds number of 8000, the plain tube data follows the correlation for the turbulent region. However, the micro-fin tube data shows a continuous increase up to a Reynolds number of about 8000, and then there is a parallel shift from the turbulent correlation. This increase in the Colburn j factor ($St\,Pr^{0.67}$) and the parallel shift from the correlation in the turbulent region is due to the swirling induced by micro-fins.

Differing numbers of starts have differing heat transfer behaviors within transition and turbulent regions (Fig. 9.7). In the transition region ($3000 < Re < 10{,}000$), the Colburn j factor of micro-fin tube #1 with smaller N_s of 25 is higher than that of micro-fin tubes #5 and #6 with larger N_s of 45 and 65, respectively. An unusual behavior, a dip, is produced for larger N_s tubes in the transition region (Fig. 9.7). In the turbulent region, micro-fin tube #6 exhibits nearly the same behavior as micro-fin tube #1. However, micro-fin tube #5 produces the lowest heat transfer rate in the turbulent region.

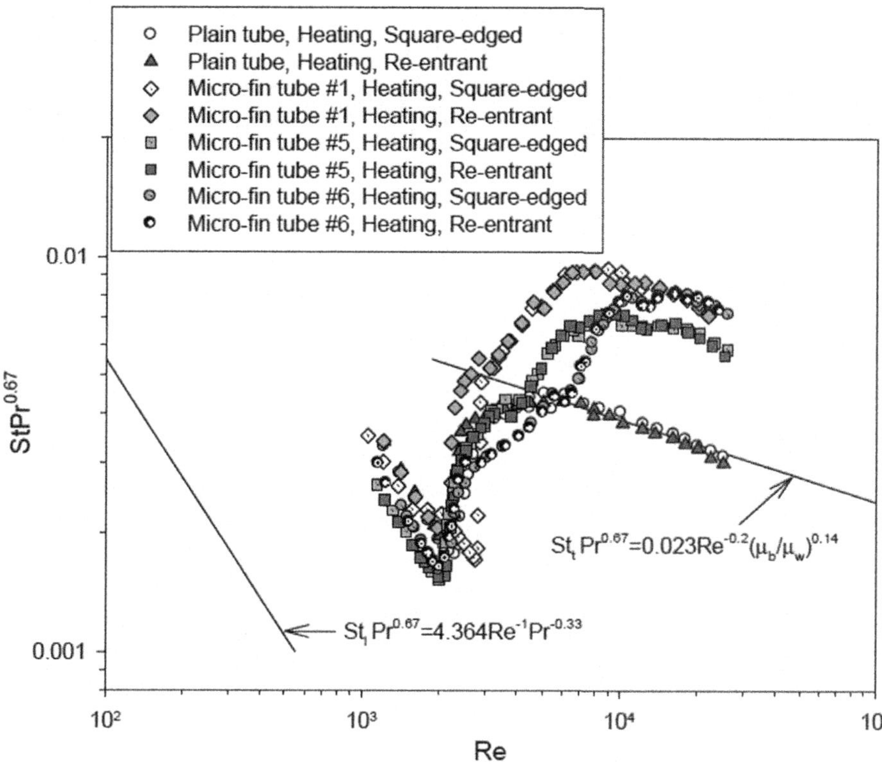

Fig. 9.7 Fully developed heat transfer characteristics for plain tube and micro-fin tubes #1, #5, and #6 with different number of starts and different inlet configurations

It is possible that the unusual dip in the transition region is caused by an increase of N_s reducing space width between the fins. Bulk fluid does not freely contact fin bases when fin space is reduced, therefore decreasing heat transfer in the transition region. When flow is increased and becomes turbulent, fluid mixing is greatly increased, and the influence of the number of starts is reduced. Since the effect of the number of starts on transition flow heat transfer is significant, additional micro-fin tubes with differing N_s are suggested for future studies.

Start and end of transition regions of heat transfer (under uniform wall heat flux boundary condition) in plain and micro-fin tubes for the two inlet configurations (square-edged and re-entrant) shown in Fig. 9.7 are summarized in Table 9.4. The plain tube and micro-fin tube #1 results are the same as those presented in Table 9.2. They are repeated here for comparison with the micro-fin tubes #5 and #6 results. The start of the transition is defined as the critical point for the sudden change from the laminar region (landing on a line parallel to $Nu_l = 4.364$) to the transitional region, and the end of the transition is the first point where heat transfer data landed on a line parallel to the turbulent correlation (Fig. 9.7).

Table 9.4 Start and end of fully developed heat transfer transition in plain and micro-fin tubes #1 ($N_s = 25$), #5 ($N_s = 45$), and #6 ($N_s = 65$) at 200 diameters from the plain and micro-fin tubes' entrance

	Heat Transfer			
Tube, condition, inlet	Re_{start}	$StPr^{0.7}$	Re_{end}	$StPr^{0.7}$
Plain, heating (square-edged)	2298	0.0017	8357	0.0041
Plain, heating (re-entrant)	2001	0.0020	7919	0.00412
Micro-fin #1, heating (square-edged)	2751	0.0017	8963	0.0094
Micro-fin #1, heating (re-entrant)	2145	0.0019	8014	0.0092
Micro-fin #5, heating (square-edged)	2060	0.0016	14,082	0.0067
Micro-fin #5, heating (re-entrant)	1981	0.0015	14,381	0.0067
Micro-fin #6, heating (square-edged)	1990	0.0016	16,145	0.0081
Micro-fin #6, heating (re-entrant)	1885	0.0017	16,243	0.0081

The lower transition regions of heat transfer in micro-fin tube #1 with 25 starts were delayed for square-edged and re-entrant inlets in comparison to the plain tube (Table 9.4 and Fig. 9.7). Plain tube transition Reynolds numbers for square-edged and re-entrant inlets were 2298 and 2001, respectively. Lower transition Reynolds numbers in micro-tube #1 increased to 2751 for the square-edged and 2145 for the re-entrant inlets. Upper transition Reynolds numbers for micro-fin tube #1 are 8963 for the square-edged and 8014 for re-entrant inlets, and therefore the upper transition was also delayed for both inlet configurations relative to the plain tube (Re = 8357 for square-edged and Re = 7919 for re-entrant inlets).

Lower transition Reynolds numbers for square-edged (Re = 2060) and re-entrant (Re = 1981) inlets were advanced in micro-tube #5 with 45 starts when compared to micro-fin tube #1 (Re = 2751 for square-edged and Re = 2145 for re-entrant inlets). Compared with the plain tube, the lower transition Reynolds numbers for the square-edged inlet were slightly advanced from 2298 to 2060 and for the re-entrant inlet from 2001 to 1981. For the upper transition Reynolds numbers, the two inlets behaved similarly. For the micro-fin tube #5 with a square-edged inlet compared to the micro-fin tube #1 with the same inlet, the upper transition Reynolds number was delayed from 8963 to 14,082, and for the micro-tube #5 with a re-entrant inlet compared to the micro-fin tube #1 with the same inlet, the upper transition Reynolds number was delayed from 8014 to 14,381 relative to micro-fin tube #1. Compared to the plain tube, upper transition Reynolds numbers for the micro-tube #5 were delayed from 8357 to 14,082 for the squared-edged inlet and from 7919 to 14,381 for the re-entrant inlet.

For the heat transfer of micro-fin tube #6 with 65 starts, the lower transition Reynolds numbers for square-edged (Re = 1990) and re-entrant (Re = 1885) inlets were the most advanced transition Reynolds numbers when compared with micro-fin tube #1 (Re = 2751 for the square-edged and Re = 2145 for the re-entrant inlets) and micro-fin tube #5 (Re = 2060 for the square-edged and Re = 1981 for the re-entrant inlets). Compared to the plain tube, the lower transition Reynolds numbers of micro-fin tube #6 for the square-edged and re-entrant inlets advanced from 2298 to 1990 and from 2001 to 1885, respectively. The upper transition Reynolds numbers of micro-fin tube #6 (Re = 16,145 for square-edged and Re = 16,243 for re-entrant inlets) were delayed compared to the other two micro-fin tubes: micro-fin tube #1 (Re = 8963 for square-edged and Re = 8014 for re-entrant inlets) and micro-fin tube #5 (Re = 14,082 for square-edged and Re = 14,381 for re-entrant inlets). Compared with the plain tube, the upper transition Reynolds numbers of micro-fin tube #6 with the square-edged and re-entrant inlets advanced from 8357 to 16,145 and from 7919 to 16,243, respectively.

The above observations demonstrate that an increase in the number of starts results in an early initiation of the transition region of heat transfer due to an increase in flow disturbance. Furthermore, increasing the number of starts results in delaying the end of the transition region of heat transfer. Based on the start and ending of transition Reynolds numbers, it can be seen that increasing the number of starts results in a wider transitional range during heat transfer.

Simultaneous Heat Transfer and Friction Factor for Micro-fin Tubes with Different Inlet Configurations

In most studies, heat transfer and friction factor characteristics for micro-fin tubes usually are represented and discussed independently. The simultaneous representation for the heat transfer and friction factor characteristics may not only help in observing the difference of the transition range between the heat transfer and the friction factor but also inspire the researchers to study the relationship between heat transfer and the friction factor in micro-fin tubes. Therefore, the micro-fin tube heat transfer and pressure drop (friction factor) data, which were measured simultaneously in the experiments of Tam et al. (2012) and Tam (2013), were plotted together in a single graph (Fig. 9.8). The figure includes the fully developed heat transfer data (with uniform wall heat flux boundary condition) and the fully developed friction factor data (with isothermal and uniform wall heat flux boundary conditions) for square-edged and re-entrant inlets. On the bases of Fig. 9.8, the transition Reynolds numbers for the heat transfer and friction factors are summarized in Table 9.5.

In the laminar region, it can be observed in Fig. 9.8 that the isothermal and heating friction factors both have the same parallel shift from the fully developed laminar friction factor line ($C_{f,l} = 16/\mathrm{Re}$). The effect of heating is not significant for the micro-fin tube laminar friction factor. In the same plot, the laminar heat transfer data has a similar downward trend as the laminar friction factor data. When Reynolds numbers reach the critical numbers of about 2700 for the square-edged inlet and 2100 for the re-entrant inlet, heat transfer and friction factors start to move away from the laminar region (start of the transition region). As discussed in Chap. 7, the start of transition for the friction factor is inlet dependent. Moreover, the heating

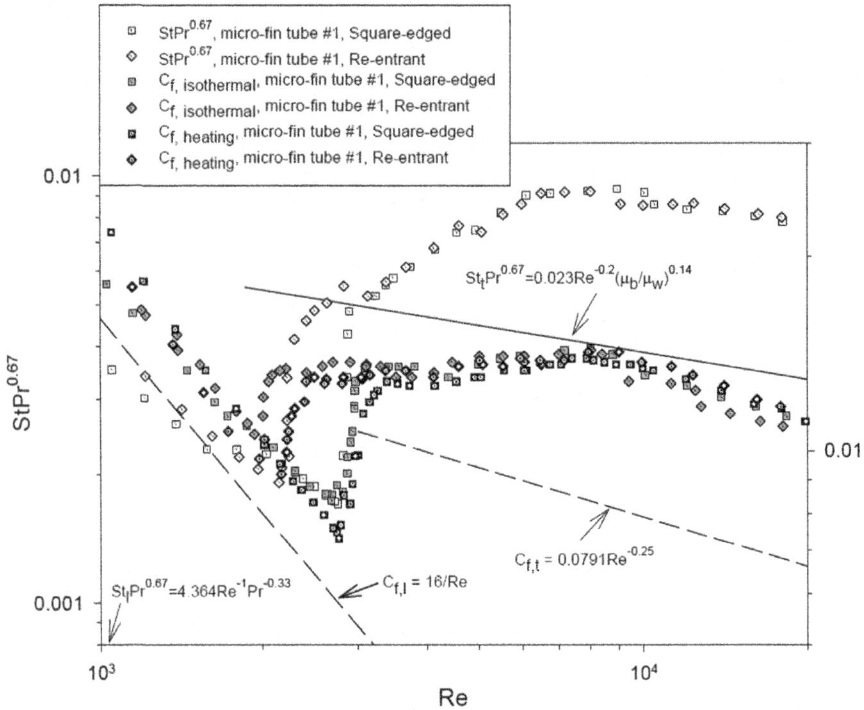

Fig. 9.8 Simultaneous heat transfer and friction factor characteristics of micro-fin tube #1 at 200 diameters from the tube entrance for square-edged and re-entrant inlets under different wall boundary conditions

condition slightly influences the start of transition of the friction factor. As shown in Table 9.5, the start of the transition for heat transfer is also inlet dependent. The square-edged inlet (Re = 2751) will delay the start of transition of heat transfer relative to the re-entrant inlet (Re = 2145). For the heating condition (Table 9.5), it can be seen that the transition Reynolds numbers for the heat transfer (Re = 2751 for square-edged and Re = 2145 for re-entrant inlets) are nearly the same as those for the heating friction factor (Re = 2765 for square-edged and Re = 2141 for re-entrant inlets).

As shown in Fig. 9.8, when the Reynolds number reaches about 3000, none of the friction factor data follow the fully developed turbulent friction factor line ($C_{f,t} = 0.0791Re^{-0.25}$). Instead, it starts to follow a constant C_f line until the Reynolds number is about 9000. Meanwhile, the heat transfer data does not follow the fully developed turbulent heat transfer line [$StPr^{0.67} = 0.023Re^{-0.2}(\mu_b/\mu_w)^{0.14}$]; instead it continues to increase until a Reynolds number of about 9000. At that point, the heat transfer and the friction factor data start to follow a line that is parallel shifted to the fully developed turbulent heat transfer and friction factor lines, respectively. Table 9.5 shows that the end of transition for the heat transfer (Re = 8963 for square-edged and Re = 9094 for re-entrant inlets) is very similar to that of the

Table 9.5 Start and end of the transition of the fully developed heat transfer and friction factor at 200 diameters from the micro-fin tube #1 entrance for square-edged and re-entrant inlets

Tube, condition, inlet	Heat transfer				Friction factor			
	Re_{start}	$StPr^{0.7}$	Re_{end}	$StPr^{0.7}$	Re_{start}	C_f	Re_{end}	C_f
Micro-fin #1, isothermal (square-edged)					2675	0.0084	8007	0.0144
Micro-fin #1, heating (square-edged)	2751	0.0017	8963	0.0092	2765	0.0073	8925	0.0136
Micro-fin #1, isothermal (re-entrant)					2021	0.0104	8023	0.0143
Micro-fin #1, heating (re-entrant)	2145	0.0019	9094	0.0086	2141	0.0092	9027	0.0142

heating friction factor (Re = 8925 for square-edged and Re = 9027 for re-entrant inlets). Based on the start and end of the transition, it can be concluded that for the micro-fin tubes, the transition range is similar for both the heat transfer and the heating friction factor. For the turbulent region, the heat transfer and the friction factor data will continue to decrease along a line parallel shifted to the fully developed turbulent heat transfer and friction factor lines.

Chapter 10
Proposed Correlations for Laminar, Turbulent, and Transition Heat Transfer in Micro-fin Tubes with Different Inlet Configurations

Background

For proper design of thermal systems using micro-fin tubes, the experimental data presented in Chap. 9 was used to develop accurate heat transfer correlations for the laminar, transitional, and turbulent dimensionless heat transfer coefficients (Nusselt number) in the entrance and fully developed regions of micro-fin tubes under uniform wall heat flux boundary condition, different inlet configurations (square-edged and re-entrant), and different fin geometries (spiral angle, fin height-to-diameter ratio, and number of starts).

Laminar Region

The modes of heat transfer in the laminar flow region can be divided into forced and mixed convection (Fig. 9.2). Pure forced convection can be observed near the entrance since secondary flow needs a certain tube length for development. To develop a heat transfer correlation in the entrance and fully developed laminar regions, the mixed and forced convection data for all the micro-fin tubes (Table 2.1) and the two inlets (square-edged and re-entrant) were used (Chap. 9). A correlation similar to the one proposed by Martinelli and Boelter (1942) was curve-fitted to the current data using a least squares curve-fitting program. Furthermore, the effect of spiral angle, fin height-to-diameter ratio, and the number of starts were also involved in the new correlation. The heat transfer correlation for the micro-fin tubes in the laminar region is:

A. J. Ghajar, *Single- and Two-Phase Flow Pressure Drop and Heat Transfer in Tubes*, Mechanical Engineering Series,
https://doi.org/10.1007/978-3-030-87281-6_10

$$\mathrm{Nu_l} = 10.89\left[(\mathrm{RePrD}/x) + 0.025(\mathrm{GrPr})^{0.72}\right]^{0.266}\left(\frac{\mu_b}{\mu_w}\right)^{0.14} N_s^{-0.524}\left(\frac{e}{D}\right)^{-0.015} \tag{10.1}$$

for $918 \le \mathrm{Re} \le 2751, 36.6 \le \mathrm{Pr} \le 56.3, 4913 \le \mathrm{Gr} \le 30{,}168, 1.158 \le \mu_b/\mu_w \le 2.022$, $3 \le x/D \le 200$, $18° \le \alpha \le 35°$, $0.0202 \le e/D \le 0.0336$, and $25 \le N_s \le 65$.

Equation (10.1) is applicable to laminar forced and mixed convection in the entrance and fully developed regions of the micro-fin tubes and can be used for both inlets. The equation gives a representation of experimental data points to within +24.0% to −25.7%. In developing the correlation, 1138 experimental data points (623 data points for the square-edged inlet and 515 data points for the re-entrant inlet) were used. The absolute average deviation between the results predicted by the correlation and the experimental data is 7.2%, with 3.6% (41 data points) having a ± 20 to 30% deviation, 23.0% (262 data points) having a ± 10 to 20% deviation, and 73.4% (835 data points) having less than a ± 10% deviation. Figure 10.1 compares the predicted Nusselt numbers obtained from Eq. (10.1) with the experimental data.

Turbulent Region

In the turbulent flow region (Fig. 9.2), the effect of free convection on the heat transfer data is insignificant, and the flow is considered to be dominated by forced convection effects. Using the turbulent forced convection data for all micro-fin tubes and the two inlets (square-edged and re-entrant), a correlation similar to the one proposed by Webb et al. (2000) was curve-fitted to the experimental data. The proposed correlation is:

$$\mathrm{Nu_t} = 0.025\mathrm{Re}^{0.664}\mathrm{Pr}^{0.298} N_s^{-0.02}\left(\frac{e}{D}\right)^{-0.332}\alpha^{0.361} \tag{10.2}$$

for $7071 \le \mathrm{Re} \le 26{,}180$, $4.5 \le \mathrm{Pr} \le 14.7$, $18° \le \alpha \le 35°$, $0.0202 \le e/D \le 0.0336$, and $25 \le N_s \le 65$.

Equation (10.2) is applicable to turbulent forced convection in the entrance and fully developed regions and can be used for all micro-fin tubes and the two inlets. The equation correlates the experimental data to within +26.9% to −31.8%. In developing the correlation, 1152 experimental data points (572 data points for the square-edged inlet and 580 data points for the re-entrant inlet) were used. The absolute average deviation between the results predicted by the correlation and the experimental data is 6.6%, with 4.3% of the data (49 data points) having more than ±20% deviation, 37.3% (430 data points) having ±10 to 20% deviation, and 58.4% (673 data points) having less than ±10% deviation. Figure 10.2 compares the predicted Nusselt numbers obtained from Eq. (10.2) with the experimental data.

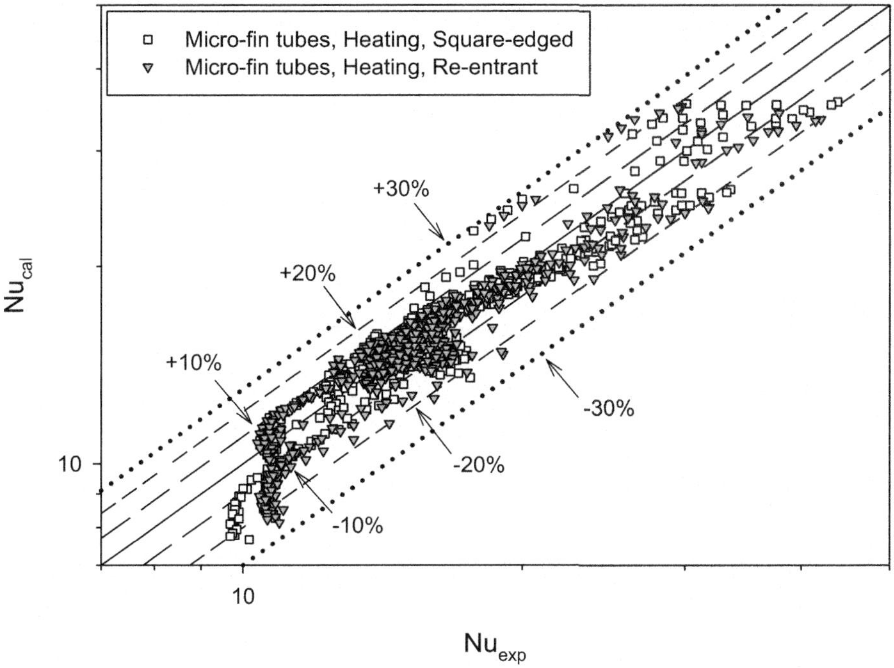

Fig. 10.1 Comparison between micro-fin tube experimental entrance and fully developed Nusselt numbers and those predicted by the proposed laminar region correlation (Eq. 10.1) for the square-edged and the re-entrant inlets

Transition Region

Flow in the micro-fin tube transition region has laminar and turbulent characteristics, and the type of inlet configuration and fin geometry influences the start and end of the transition region. Thus, a single correlation cannot predict the data, and a correlation for each inlet and each fin geometry should be developed. The form of the correlation developed for this region is similar to the one proposed by Churchill (1977a):

$$\mathrm{Nu_{tr}} = \left(\mathrm{Nu_l^{10}} + \left(\frac{\exp\left(\mathrm{a} - \mathrm{Re}\right)/\mathrm{b}}{\mathrm{Nu_l^2}} + \frac{1}{\mathrm{Nu_t^2}}\right)^{-5}\right)^{\frac{1}{10}} \tag{10.3}$$

where the coefficients a and b are listed in Table 10.1 and their ranges of application are $1885 \le \mathrm{Re} \le 16{,}243$, $3 \le x/D \le 200$, $18^\circ \le \alpha \le 35^\circ$, $0.0202 \le e/D \le 0.0336$, and $25 \le N_s \le 65$.

Equation (10.3) is applicable to entrance and fully developed Nusselt numbers in the transition region and should be used with appropriate set of constants

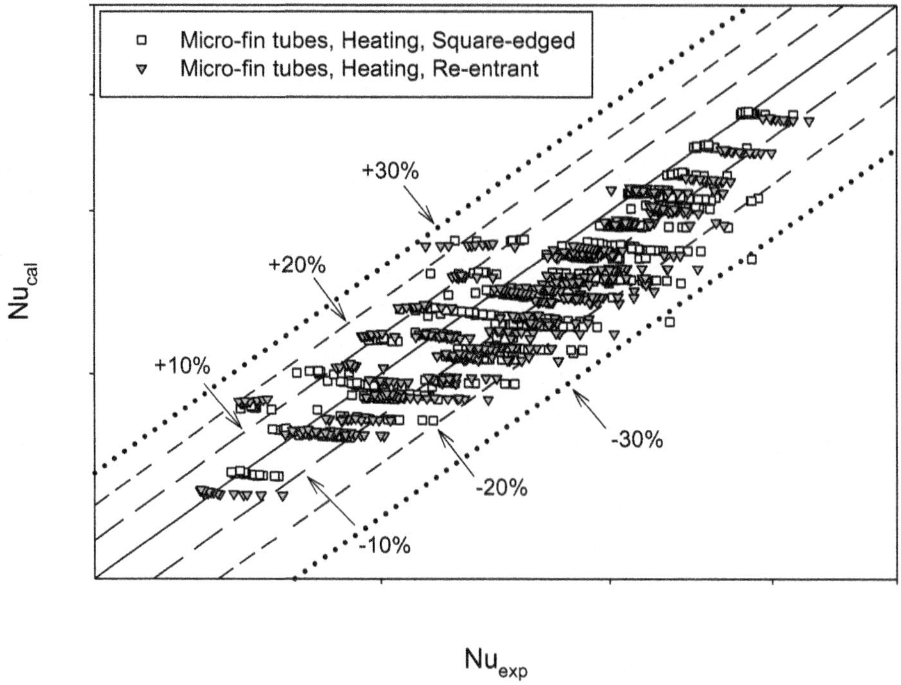

Fig. 10.2 Comparison between micro-fin tube experimental entrance and fully developed Nusselt numbers and those predicted by the proposed turbulent region correlation (Eq. 10.2) for the square-edged and the re-entrant inlets

Table 10.1 Constants for transition heat transfer correlation (Eq. 10.3) for each micro-fin tube and inlet type

Tube	Square-edged		Re-entrant	
	a	b	a	b
Micro-fin tube #1	2550	270	2100	350
Micro-fin tube #2	2100	440	2020	440
Micro-fin tube #3	1880	360	1920	380
Micro-fin tube #4	1960	410	1510	800
Micro-fin tube #5	2120	660	1780	880
Micro-fin tube #6	1990	1120	1840	920

(Table 10.1). To develop the transition region correlation for the re-entrant inlet, 1182 experimental data points were used. The correlation gave a representation of the experimental data to within +51.9% to −56.2% and has an average deviation of 20.7%, with 2.7% (32 data points) having more than ±50% deviation, 12.0% (142 data points) having ±40 to 50% deviation, 14.7% (174 data points) having ±30 to 40% deviation, 16.7% (197 data points) having ±20 to 30% deviation, 24.8% (293 data points) having ±10 to 20% deviation, and 29.1% (344 data points) having less than ±10% deviation. For the square-edged inlet, 1133 experimental

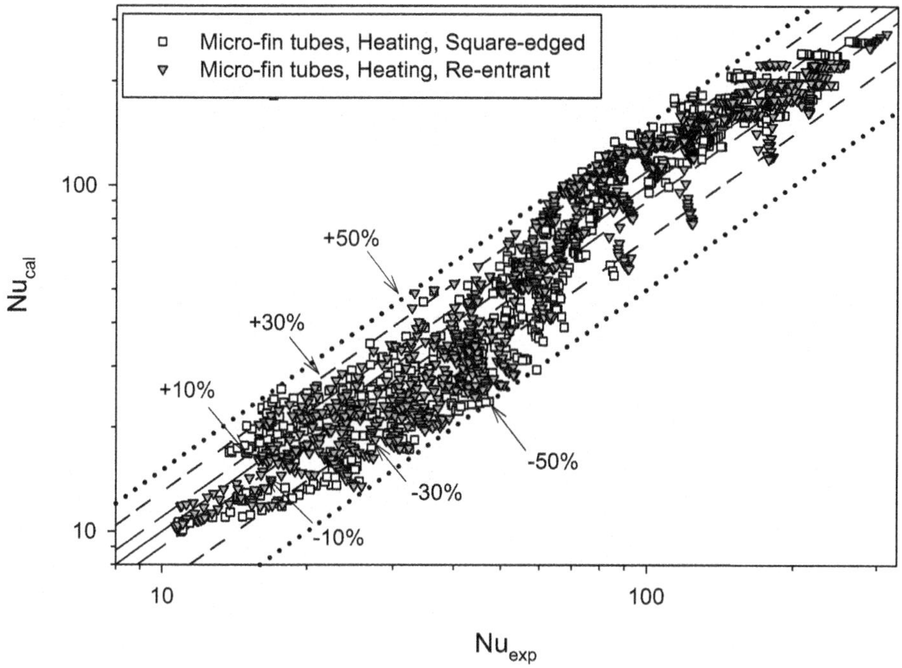

Fig. 10.3 Comparison between micro-fin tube experimental entrance and fully developed Nusselt numbers and those predicted by the proposed transition region correlation (Eq. 10.3) for the square-edged and the re-entrant inlets

data points were used to develop the correlation. The equation correlated the experimental data to within +49.1% to −56.0% and has an average deviation of 20.3%, with 2.4% (27 data points) having more than ±50% deviation, 11.9% (135 data points) having ±40 to 50% deviation, 12.3% (139 data points) having ±30 to 40% deviation, 18.0% (204 data points) having ±20 to 30% deviation, 22.2% (251 data points) having ±10 to 20% deviation, and 33.3% (377 data points) having less than ±10% deviation. Figure 10.3 compares the predicted Nusselt numbers obtained from Eq. (10.3) for each inlet with the experimental data.

Chapter 11
Friction Factor in the Transition Region of Mini- and Micro-tubes

Background

Due to rapid advancement in fabrication techniques, the miniaturization of devices and components is ever increasing in many applications. Whether it is in the application of miniature heat exchangers, fuel cells, pumps, compressors, turbines, sensors, or artificial blood vessels, a sound understanding of fluid flow in micro-scale channels and tubes is required. However, amidst all the investigations in mini- and microchannel flow, there seems to be a lack in the study of the flow in the transition region according to Krishnamoorty and Ghajar (2007). One obvious question is the location of the transition region with respect to the hydraulic diameter of the channel and the roughness of the channel. To better understand the flow behavior in small tubes, Ghajar et al. (2010) extended their isothermal transitional pressure drop (friction factor) studies in standard-size plain tubes presented in Chap. 3 to mini- and micro-tubes. They performed a systematic and careful experimental investigation of friction factor in the transition region for single-phase water flow in 12 stainless steel tubes with diameters ranging from 2083 to 337 μm (Fig. 11.1). For the study, mini- and micro-tubes were chosen over other noncircular channels to negate the effect of aspect ratio, which may serve to alter flow characteristics at these small scales. The pressure drop (friction factor) measurements were carefully performed by paying particular attention to the sensitivity of the pressure-sensing diaphragms used in the pressure transducer.

This chapter presents the details of the experimental setup used for the study of the pressure drop measurements in the mini- and micro-tubes over a wide range of Reynolds numbers from laminar to the turbulent region and the start and end of the transition region in these small sized tubes. For more details on these results, refer to Ghajar et al. (2010).

At the conclusion of this chapter, a brief summary of several studies that extends the work of Ghajar et al. (2010) is presented.

A. J. Ghajar, *Single- and Two-Phase Flow Pressure Drop and Heat Transfer in Tubes*, Mechanical Engineering Series,
https://doi.org/10.1007/978-3-030-87281-6_11

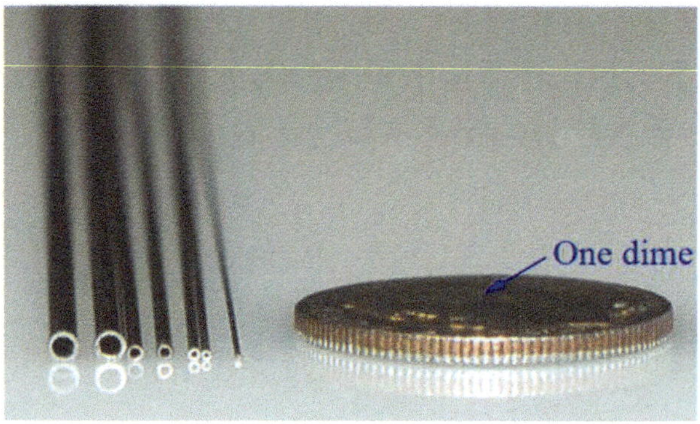

Fig. 11.1 Representative mini/micro stainless steel tubes used in the experiments of Ghajar et al. (2010)

Details of the Single-Phase Flow Pressure Drop Experimental Setup for Mini- and Micro-tubes

The experimentation for this study was performed using a relatively simple but highly effective apparatus. The apparatus used was designed with the intention of conducting highly accurate pressure drop measurements. In addition to accurate measurements, the apparatus was also designed to be versatile, accommodating the use of multiple diameters and lengths of test sections. The apparatus consists of four major components. These are the fluid delivery system, the flow meter array, the test section assembly, and the data acquisition system. Each of these different components is discussed independently. An overall schematic for the experimental test apparatus is shown in Fig. 11.2.

The fluid delivery system is a pneumatic and hydraulic combination, consisting of a high-pressure cylinder filled with ultra-high purity nitrogen in combination with a stainless steel pressure vessel. The system is an open loop. Thus, after the working fluid passes through the apparatus, it is passed into a sealed collection container and recycled manually. Nitrogen in the high-pressure cylinder is pressurized to 17.2 MPa by the distributor. This pressurized nitrogen is then fed to the stainless steel pressure vessel via a two-phase regulator and line. The working fluid, distilled water for the purposes of this research, is stored in the stainless steel pressure vessel. As the pressurized nitrogen is fed into the stainless steel pressure vessel, the working fluid is forced up a stem extending to the bottom of the vessel, out of the pressure vessel, and through the flow meter array and test section. An Airgas regulator is used for the purposes of controlling the pressure of the nitrogen inlet to the stainless steel pressure vessel. This dual stage regulator is capable of providing pressures ranging from 0 to 1.72 MPa. The stainless steel pressure vessel used is an Alloy Products model 72–05, providing a maximum working pressure of 1.37 MPa and a capacity of 19.0 L.

After exiting the pressure vessel, the distilled water travels to the flow meter array. Flow rate of the water entering the array is further regulated using a Parker N-Series Model 6A-NLL-NE-SS-V metering valve, which allows fine tuning of the

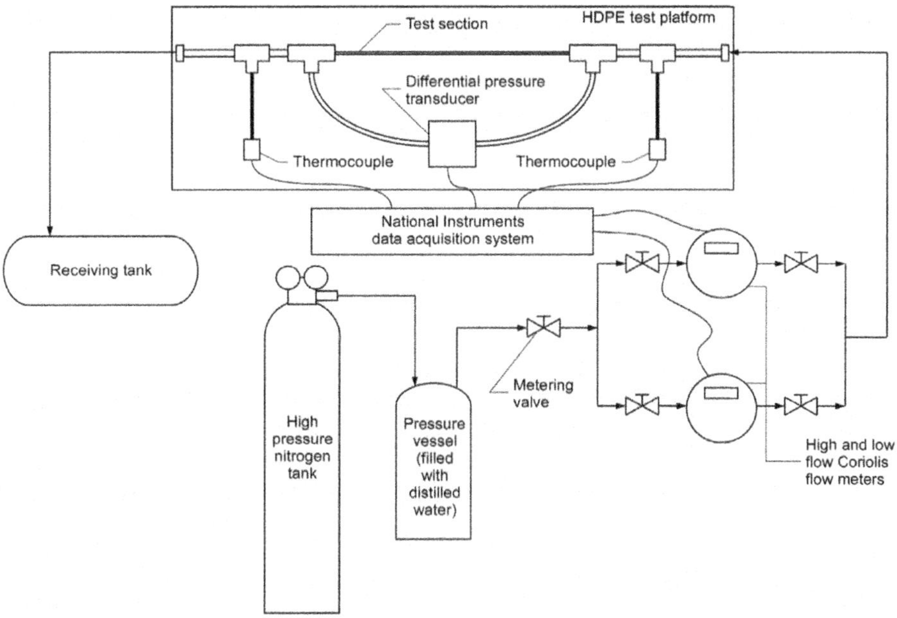

Fig. 11.2 Schematic of the experimental setup used for the measurement of friction factor in mini- and micro-tubes

flow rate. Fluid passes through the metering valve and into one of the two Micro Motion Coriolis flow meters. Two flow meters are necessary in order to accommodate the large range of flow rates that are studied using the experimental apparatus. The larger of the two meters used is a CMF025 coupled with a Model 1700 transmitter. This meter is designed to measure mass flows ranging from 54 to 2180 kg/h for liquids. Within this range of mass flows, this meter is accurate to 0.05%. However, much smaller flow rates can be measured with very little loss in accuracy. The smaller of the two meters is a Micro Motion Model LMF3M, coupled with a LFT transmitter. This second meter is designed to measure mass flows ranging from 0.001 to 1.5 kg/h.

After passing through the flow meter array, fluid enters the test section assembly, via a second section of PFA tubing to the test section assembly. The test section assembly contains the test section as well as the equipment necessary for measurement of inlet and outlet fluid temperatures and pressure drop. This test section was constructed for the incorporation of a very broad range of test section diameters, encompassing both mini- and micro-tube sizes. For the experiments, 12 different tube sizes were used. The inner diameters of these tubes varied from 2083 to 337 μm. All of the tubes used are available from Small Parts Inc. The tubes used are stainless steel-type 304 hypodermic tubes with factory cut lengths of 30.5 cm (for $2083 \leq D \leq 667$ μm) and 15.2 cm (for $559 \leq D \leq 337$ μm). Since the friction factor measurements are conducted for fully developed flow, the length of the tube bears no effect on the results. As pointed out by Krishnamoorty and Ghajar (2007), the effect of tube length on the friction factor is negligible as long as the flow is fully developed.

The pressure transducer used for pressure drop measurements was a Validyne model DP15. This pressure transducer utilized a series of interchangeable diaphragms to provide the ability to measure a very large range of differential pressures. The research facilities used for experimentation have different pressure transducer diaphragms to encompass a range of differential pressures from 1.38 to 1380 kPa. The Validyne pressure transducer had an accuracy of ±0.25% full scale of the diaphragm used. Careful attention was given to ensure that the range of the diaphragm used is conducive to the pressure being measured. The use of the numerous interchangeable diaphragms is an important factor in ensuring the accuracy of the pressure drop measurements. All data from the thermocouples and pressure transducer was acquired using a National Instruments data acquisition system and recorded with the laboratory PC and LabView software.

Nine different pressure transducer diaphragms were used to cover differential pressures ranging from 1.38 to 1380 kPa. Calibrations were performed at the beginning of each experiment with the appropriate diaphragms. During calibration, the voltage output of the differential pressure transducer at numerous applied pressures was compared against the reading of one of the four research grade test gauges. Of these four gauges, the highest rated in terms of pressure is a Perma-Cal 2070 kPa test gauge with an accuracy of ±0.25% full scale. For lower pressures, two Wika test gauges, also with accuracy of ±0.25% full scale, were used. The first of these has a pressure rating of 1100 kPa, and the second is rated up to 103 kPa. For low-pressure diaphragm calibration, a Cole-Parmer digital manometer was used. This instrument has a pressure rating of 103 kPa, an accuracy of ±0.3% full scale, and a resolution of 69 Pa.

The Micro Motion Coriolis flow meters are factory calibrated as well. For the CMF-025, the manufacturer's specified tolerance for calibration error is ±0.1%. For the LMF3M, the manufacturer's specified calibration tolerance is ±1.0%. For both of these meters, in lab calibration consisted of checking the manufacturer's calibrations over a range of flow rates via timed collection of fluid passing through the meters. In addition, the maximum and minimum milliamp outputs of the CMF-025 were tuned to improve the resolution of the meter.

Developing an understanding of the experimental uncertainty in the calculated friction factor is absolutely necessary. From the measured pressure drop data, the friction factor can be calculated with:

$$f = \frac{2\Delta p D}{\rho L V^2} = \frac{\Delta p D^5 \pi^2 \rho}{8 L \dot{m}^2} \tag{11.1}$$

The uncertainty in the pressure drop measurement can be readily obtained from the manufacturer's specifications on the Validyne pressure transducer. As has been previously stated, this specification is given as ±0.25% of the full-scale reading of each diaphragm used. Diaphragms were carefully selected during experimentation in order to obtain the highest accuracies possible. The worst-case scenario occurs with small tube size and low Reynolds number. In this region, the uncertainty in the pressure drop measurement can be estimated at ±1.0%. It is more representative to look at intermediately sized tubes and/or flow rates through the transition and turbulent regions. Uncertainty of the pressure drop measurement in these areas drops to ±0.4%.

The uncertainty in mass flow rate measurement given by the Micro Motion flow meter specifications for the CMF-025 meter is ±0.05%. For the LMF3M meter, an uncertainty of reading of ±0.50% is given. However, it has to be taken into consideration that the larger meter is being utilized at flow rates lower than its range in order to cover the entire range of flow rates for all of the tubes under research. Based upon uncertainty equations given in the Micro Motion specifications, the worst-case scenario accuracy between the two flow meters is ±1.8%. Both the use of the LMF3M and the under-ranging of the CMF-025 occur at smaller tube sizes and lower Reynolds numbers. Thus, it is necessary to calculate uncertainty for either of the meters running within their specified mass flow ranges and to estimate uncertainty when the CMF-025 is pushed to its lowest range of measurement.

Uncertainty in tube length is determined by the accuracy of the cutting of the high-density polyethylene tube cradles. The cradles serve as a reference point for the mounting of the different tube sections in order to ensure consistency. Measured uncertainty in the cradle lengths is ±0.26% of length.

Due to the fact that uncertainty in both the Validyne pressure transducer and the Micro Motion meters is dependent upon tube size and Reynolds number, three different uncertainty values have been established. In order to quantify the overall uncertainty, analysis was conducted using the method described by Kline and McClintock (1953). In the case of larger tube sizes and higher Reynolds numbers, the CMF-025 meter is used and is functioning at the manufacturer's specified uncertainty level. The pressure transducer is operating at the better of its two calculated uncertainty levels. Taking this into consideration, the overall uncertainty is calculated at ±0.83%. As the tube size and Reynolds number decrease, the CMF-025 meter begins to operate under range. In this area, the pressure transducer is considered to be operating at the lesser of its two uncertainty levels. In this range, the overall uncertainty associated with the experimental apparatus is calculated to be ±2.78%. Finally, at the lowest ranges of tube size and Reynolds number, the LMF3M meter is used. In this area, the pressure transducer is still operating at the lesser of its two uncertainty levels. For this lowest Reynolds number and smallest tube size situation, the overall uncertainty decreases to ±1.51%. The uncertainties discussed are for the experimental results when the pressure-sensing diaphragms were used appropriately for the measured pressure drop ranges.

Since the mini- and micro-tubes under research were purchased from an outside source, data obtained from these tubes is only as accurate as the manufacturer's specifications. The diameters of the tubes as well as the roughness of the inner walls of the tubes are of particular concern due to the type of research being undertaken. In order to ensure that the data recorded was of the highest quality possible, it was deemed necessary to determine the degree of accuracy of the manufacturer's specifications. In order to do this, both the scanning electron microscope (SEM) and the scanning probe microscope (SPM) at the Oklahoma State University Microscopy Laboratory were utilized. Diameter measurements were taken using the SEM, while roughness measurements were taken using the SPM.

Two different tube sizes were examined using the SEM in order to check the accuracy of the manufacturer's tolerances. The first of these two tubes had an inner diameter and tolerance of 5330 ± 76 μm. The second tube examined had an inner diameter and tolerance of 584 ± 38 μm. Thus, a broad range of tube size was covered between the two tubes examined. Imaging was done using JEOL JXM 6400 Scanning Electron Microscope System in combination with a digital camera system. The resolution of the microscope ranged from 30 to 50 nm. Once images had been captured, it was possible to determine image pixel size in terms of length scale. With a known pixel-to-length scale, the inner diameter of the tubes could be estimated from the SEM images.

For the first tube with the manufacturer specified inner diameter of 5330 ± 76 μm, the average inner diameter was estimated to be 5280 μm from the SEM image. For the second tube with the manufacturer specified inner diameter of 584 ± 38 μm, the average inner diameter was estimated to be 574 μm from the SEM image (Fig. 11.3). The SEM imaging of these two tubes verified that the manufacturer's specifications of the tube diameters and tolerances were verifiable and reasonably dependable.

Roughness measurements were conducted using a SPM station in combination with Digital Instruments Multimode V electronics and an optical microscope for tip positioning. The system used is capable of three-dimensional spatial mapping and an ultimate resolution of 0.1 nm laterally and 0.01 nm vertically. Scans were taken of multiple sections of two stainless steel tubes with different inner diameters: 5330 μm and 2390 μm. Roughness data were taken from three different sections of each of these tubes. In order to negate the effect of the curvature of the tubes upon the roughness

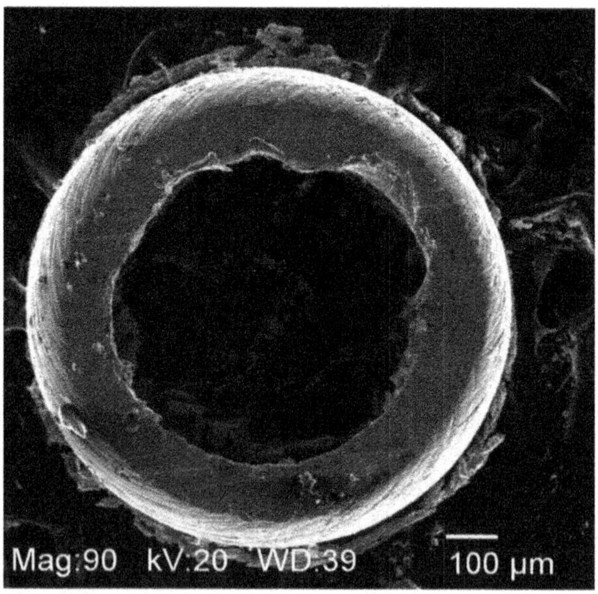

Fig. 11.3 SEM image of a 584 ± 38 μm (manufacturer's specification) diameter stainless steel tube; based on this SEM image, the tube diameter was found to be 574 μm

measurement generated by the SPM, a flattening feature was utilized. From the SPM, information such as average roughness (R_a), maximum roughness profile peak height (R_{max}), and root mean square roughness (R_q) of the samples were obtained. The topographic image for a section of the 5330 μm diameter tube is shown in Fig. 11.4.

From the SPM measurements, the inner surface of the 5330 μm diameter stainless steel tube has average roughness (R_a), maximum roughness profile peak height (R_{max}), and root mean square roughness (R_q) of 240 nm, 2628 nm, and 292 nm, respectively. In similar manner, the inner surface of the 2390 μm diameter stainless steel tube has R_a, R_{max}, and R_q of 150 nm, 1710 nm, and 194 nm, respectively. Some variability was found between the two tubes, though this was to be expected. The manufacturer specified an inner wall root mean square roughness of 410 nm. Thus,

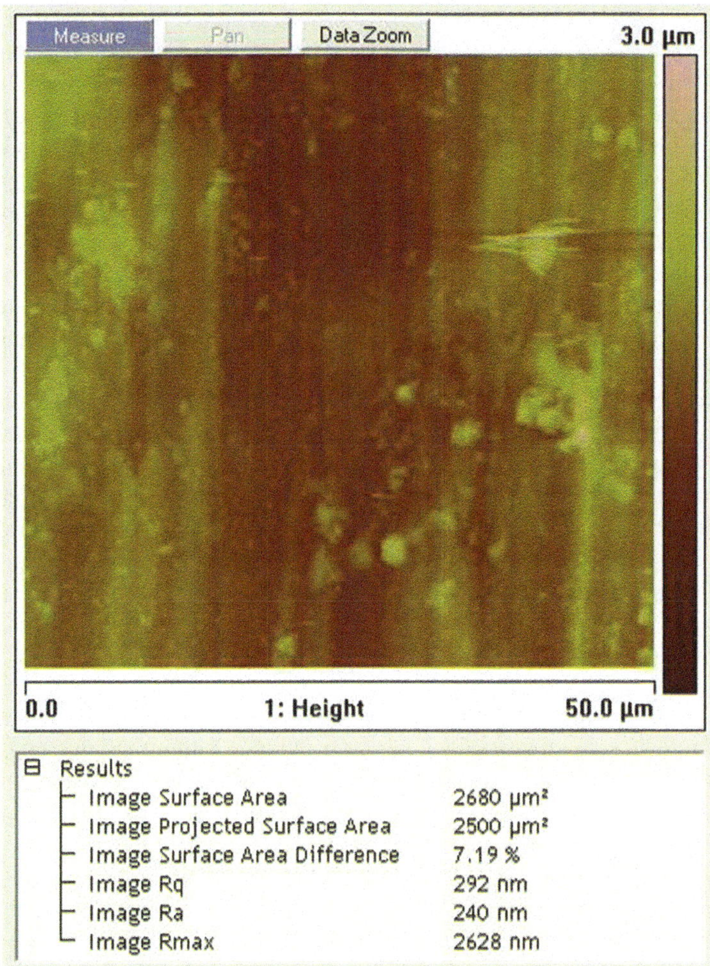

Fig. 11.4 SPM topographic image for a section of a 5330 μm inner diameter stainless steel tube ($R_a = 240$ nm, $R_{max} = 2628$ nm, $R_q = 292$ nm)

the root mean square roughness measured by the SPM for the two tubes was within the manufacturer's specifications.

Effect of Mini- and Micro-tube Diameter on the Start and End of the Transition Region of the Friction Factor: The Work of Ghajar et al. (2010)

The review documented by Krishnamoorty and Ghajar (2007) pointed out the need of further experiments to confirm the start and end of transition region in mini- and micro-tubes. In large part, this need for further experimentation is exemplified by the highly contradictory observations that have been reported by various investigators. The disparity found in the literature may be attributed to factors such as tube diameter, surface roughness, experimental facilities, and instrumentation. Without any doubt, the sensitivity of the instruments used in the measurement of pressure drop plays one of the most crucial roles in collecting accurate data. In addition, in properly addressing the effect of tube diameter on pressure drop for flow in mini- and micro-tubes, the importance of systematically investigating various tube sizes cannot be overlooked.

In order to be able to clearly pick up the transition region along with the laminar and turbulent regions, the sensitivity of the pressure-sensing diaphragms used in the Validyne DP15 pressure transducer had to be given meticulous consideration. Even the numerous studies that covered the laminar and turbulent regions fail to explicitly capture the transition region. In many cases, this failure may be attributed to the questionable sensitivity of the instrumentation used. The dilemma that arises from this is that if one is not confident with the results for the transition region, then the confidence in the results for laminar and turbulent regions is also questionable.

To properly recognize sensitivity for each pressure-sensing diaphragm, it is necessary to collect pressure drop data for the entire pressure-sensing range of each diaphragm before changing to the next diaphragm. Collecting pressure drop data through the entire pressure-sensing range of each diaphragm further enhances collection of accurate data. The overall uncertainty associated to each pressure-sensing diaphragm was estimated to be roughly $\pm 1.5\%$ full scale of the diaphragm's pressure-sensing range. This $\pm 1.5\%$ uncertainty includes the uncertainties of the pressure transducer, the pressure gauges used for calibrating the pressure diaphragms, and the standard deviation of the pressure drop data. An uncertainty of $\pm 1.5\%$ full scale of the diaphragm's pressure-sensing range implies that a 345 kPa pressure diaphragm would have an uncertainty of ± 5.18 kPa.

Figure 11.5 illustrates the comparison of the friction factor data points measured using various pressure-sensing diaphragms for 1600 and 1067 μm diameter stainless steel tubes. Figure 11.5a brings out the obvious scenario that using both 55.2 and 138 kPa pressure diaphragms for $700 < Re < 3500$ would easily bring one to conclude that the higher friction factor values are due to surface roughness. The appropriate error bars, based on the uncertainty of ±1.5% full scale of the diaphragm's pressure-sensing range, are attached on two selected data points obtained by 55.2 and 138 kPa pressure diaphragms to illustrate their uncertainties. The data point measured by the 55.2 kPa pressure diaphragm at Re = 1100 carried an uncertainty of ±28%, while the data point measured by the 138 kPa pressure diaphragm at Re = 1400 carried an uncertainty of ±46%. In addition, the data point measured by the 138 kPa pressure diaphragm at Re = 1400 shows the error bar with a 14% extension below the f = 64/Re line. Such scenario implies the possibility of having a wrong conclusion that the value of f·Re is 55 rather than the conventional value of 64. For comparison purposes, in Fig. 11.5a, error bars are also attached on two selected data points obtained by the 3.45 and 13.8 kPa pressure diaphragms which show significantly lower uncertainties than the error bars on the data points obtained by the 55.2 and 138 kPa pressure diaphragms. As shown in Fig. 11.5a, pressure diaphragms with ratings of 13.8 kPa or lower would be appropriate for Re < 3500, and pressure diaphragms with ratings of 55.2 kPa or higher would be appropriate for Re > 3500. For 500 < Re < 1700, data points measured by the 13.8 kPa pressure diaphragm were also verified by the 3.45 kPa pressure diaphragm to be within experimental uncertainties. Similarly, data points measured by the 55.2 kPa pressure diaphragm were also verified by the 138 kPa pressure diaphragm to be within experimental uncertainties for 3500 < Re < 5500.

Figure 11.5b illustrates that the proper selection of pressure-sensing diaphragms is essential to accurately measure friction factor in the transition region. The appropriate error bars, based on the uncertainty of ±1.5% full scale of the diaphragm's pressure-sensing range, are attached on a data point measured by a 138 kPa pressure diaphragm at Re = 1700 and another data point measured by a 345 kPa pressure diaphragm at Re = 1900 (Fig. 11.5b). The data point measured by the 138 kPa pressure diaphragm at Re = 1700 carried an uncertainty of ±15%, while the data point measured by the 345 kPa pressure diaphragm at Re = 1900 carried an uncertainty of ±31%. In addition, the data point measured by the 345 kPa pressure diaphragm at Re = 1900 shows the error bar with a 14% extension below the f = 64/Re line. As seen previously in Fig. 11.5a, this scenario shown in Fig. 11.5b also implies the possibility of having a wrong conclusion that the value of f·Re is 55 rather than the conventional value of 64. For comparison purposes, in Fig. 11.5b, error bars are also attached on two selected data points obtained by the 34.5 and 55.2 kPa pressure diaphragms which show significantly lower uncertainties than the error bars on the data points obtained by the 138 and 345 kPa pressure diaphragms. The discrepancies between the data points measured by the 138 and 345 kPa pressure diaphragms show that these diaphragms could not accurately capture the transition region. The actual friction factor values were accurately measured by the 55.2 kPa diaphragm and verified by the 34.5 kPa diaphragm to be within experimental uncertainties. At the trough of the transition region (Re ≈ 2750), the data measured by the 345 kPa diaphragm was 15% higher than the data measured by the 55.2 kPa diaphragm. The data measured by the 138 kPa diaphragm at Re ≈ 2750 was 8% higher than the data measured by the 55.2 kPa diaphragm.

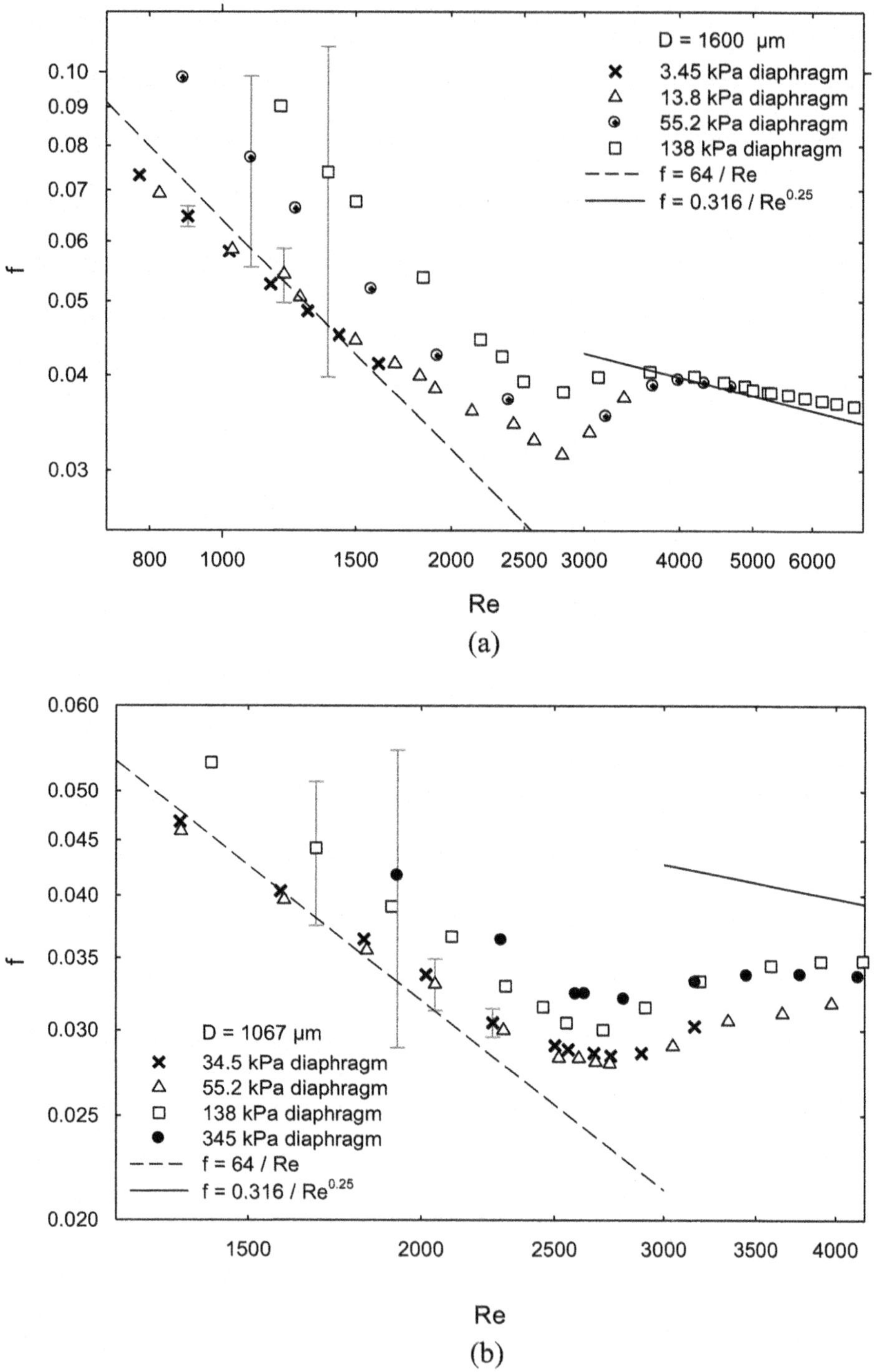

Fig. 11.5 Comparison of results measured by various pressure-sensing diaphragms for two different tubes: (**a**) 1600 μm diameter tube and (**b**) 1067 μm diameter tube

Based on the illustrations of Fig. 11.5, improper use of pressure-sensing diaphragm could easily lead to erroneous conclusions about the flow phenomena in the micro-tubes tested. It should be noted that the ability to capture transition decays quite rapidly when diaphragms inappropriate to the range of pressure drop under investigation are utilized. Thus, extreme care in diaphragm selection is imperative in order to capture the transition region with the greatest possible accuracy. Even small failures in terms of accuracy can lead to flatter transition regions, leaving the actual physics of the flow unobserved. With the effect of surface roughness and diameter in micro-tubes still largely unexplored, these types of failures are unacceptable. Although it seems trivial to discuss the sensitivity of pressure diaphragm, Figs. 11.5a, b clearly illustrate the consequences of ignoring it. It should be noted that all the friction factor data reported in Figs. 11.6, 11.7, 11.8, and 11.9 have an uncertainty of $\pm 1.5\%$.

The experimental results of 12 different stainless steel tubes with diameters ranging from 2083 to 337 μm were investigated in detail with regard to the laminar, transition, and turbulent regions over Reynolds numbers ranging from 500 to 10,000. Experimentally determined friction factor in the laminar region was compared with the conventional friction factor equation for fully developed laminar flow, $f = 64/Re$. For turbulent region, the experimental friction factor was compared with the Blasius (1913) friction factor equation for turbulent flow, $f = 0.316/Re^{0.25}$. A

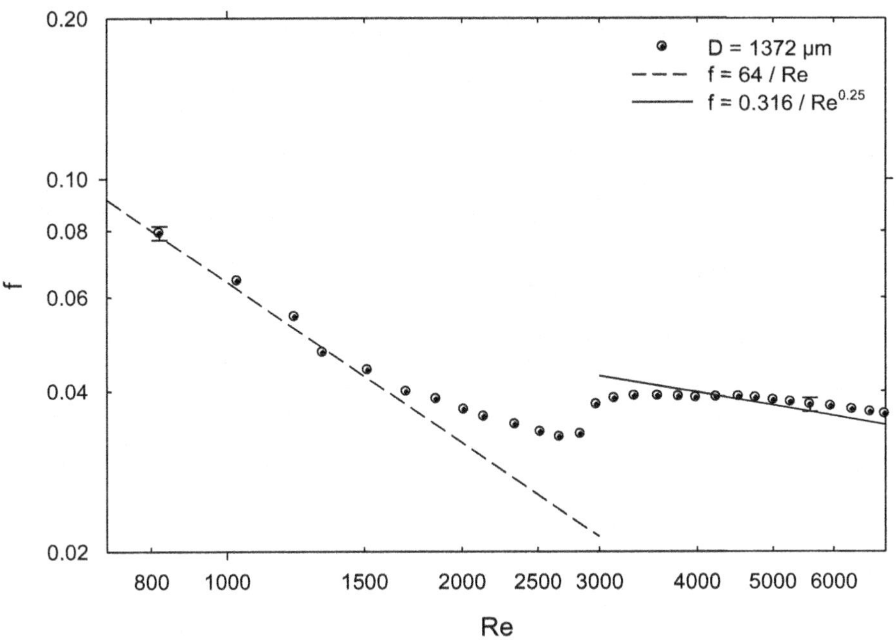

Fig. 11.6 Experimental friction factor of 1372 μm diameter stainless steel tube

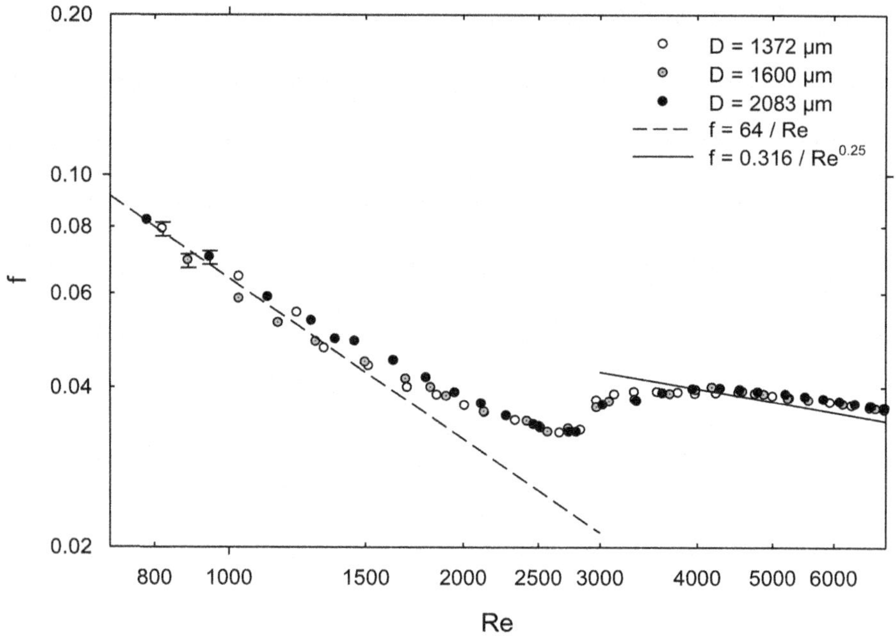

Fig. 11.7 Transition region of stainless steel tubes with diameters from 2083 to 1372 μm

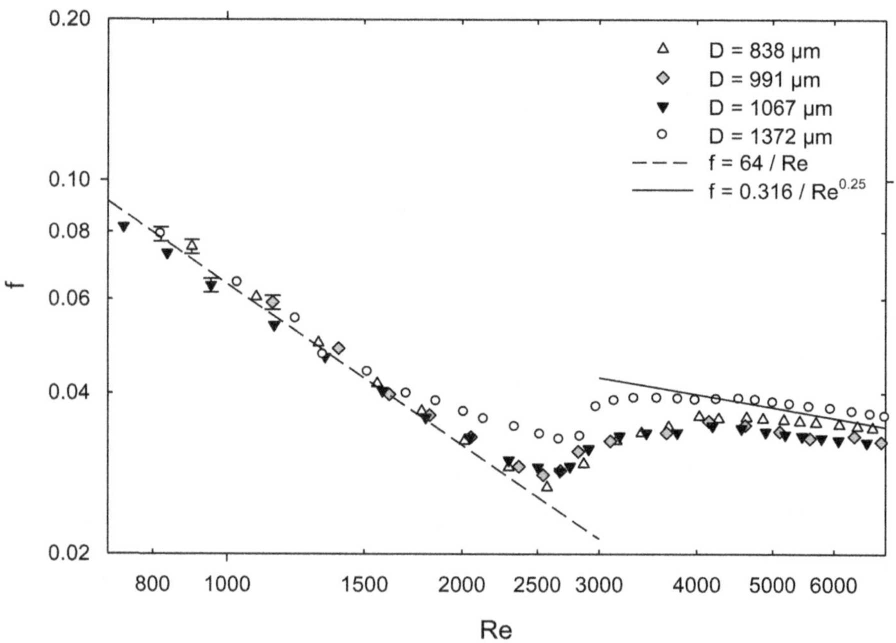

Fig. 11.8 Transition region of stainless steel tubes with diameters from 1372 to 838 μm

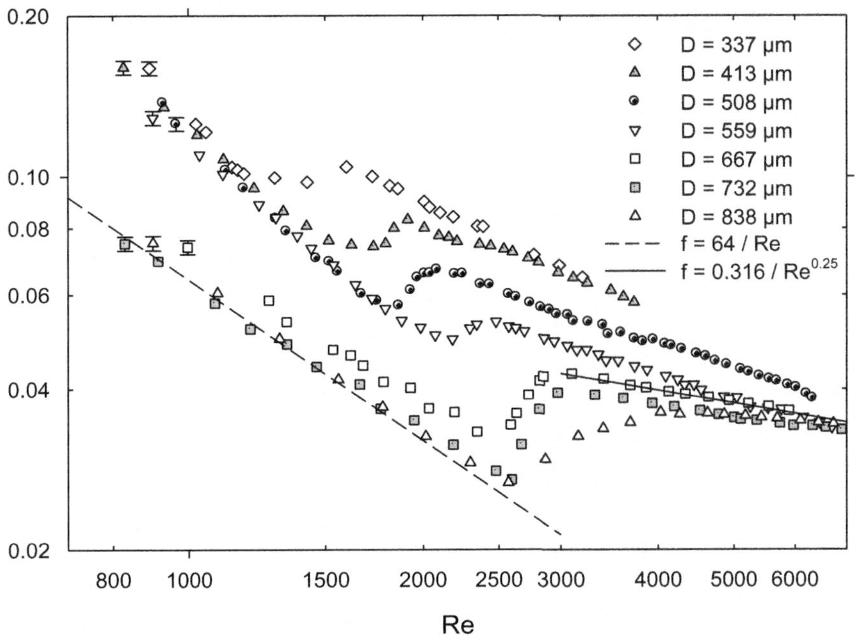

Fig. 11.9 Transition region of stainless steel tubes with diameters from 838 to 337 μm

representative of the experimental results for the friction factor of the 1372 μm diameter stainless steel tube is shown in Fig. 11.6. As illustrated in Fig. 11.6, the onset of transition for the 1372 μm diameter tube is at Reynolds number of 1900, and end of transition is at Reynolds number of 4000.

Based on the experimental friction factors, the transition region can be estimated by locating the Reynolds numbers where the friction factor departs from the laminar line and merges with the turbulent line. The transition Reynolds number ranges summarized in Table 11.1 were determined by using a 5% deviation criterion from the laminar and turbulent lines. According to our experimental uncertainty analysis, the maximum error in friction factor measurements was estimated to be no more than 3%, and the 5% deviation criterion was used to encompass the experimental uncertainty. In the estimation of the transition Reynolds number range for each tube diameter, the transition region begins with the first data point that is 5% higher than the laminar line and ends with the data point that is 5% lower than the perceived turbulent line. The perceived turbulent line is a straight line connecting the data points in the turbulent region (Re > 4000) on the base-10 logarithms fiction factor versus Reynolds number plot.

The experimental friction factor results provided an interesting observation. The decrease in tube diameter from 2083 to 667 μm actually delayed the onset of transition region. For the 2083 μm diameter tube, the transition region began at Reynolds number of 1500. As the tube diameter decreased to 667 μm, the Reynolds

Table 11.1 Summary of transition Reynolds number ranges for various stainless steel tube sizes

Tube ID (μm)	Transition range	Tube ID (μm)	Transition range
2083	1500 < Re < 4000	732	2200 < Re < 3000
1600	1700 < Re < 4000	667	2200 < Re < 3000
1372	1900 < Re < 4000	559	1900 < Re < 2500
1067	2000 < Re < 4000	508	1700 < Re < 2100
991	2000 < Re < 4000	413	1500 < Re < 1900
838	2200 < Re < 4000	337	1300 < Re < 1700

number for the onset of transition region was shifted to approximately 2200 (see Table 11.1). The end of transition region for tubes with diameters from 2083 to 838 μm is consistently located at Reynolds number of 4000. However, for tube sizes of 732 and 667 μm, the end of transition region shifted forward to Reynolds number of 3000. Further decrease in the tube diameter from 667 to 337 μm caused the onset of transition region to shift from Reynolds number of 2200 to 1300, while the end of transition region shifted from Reynolds number of 3000 to 1700. The experimental results indicated that the Reynolds number range for transition flow becomes narrower with the decrease in tube diameter.

By focusing on the transition region, the effect of tube diameter on the friction factor profile can be clearly seen. The friction factor profiles of the 12 stainless steel tubes in the transition region are shown in Figs. 11.7, 11.8, and 11.9. Figure 11.7 shows that the decrease in tube diameter from 2083 to 1372 μm did not significantly affect the profile of the friction factor, with the exception of the onset of transition region. The decrease in the tube diameter from 2083 to 1600 to 1372 μm showed the onset of transition region shifted from Reynolds number of 1500 to 1700 to 1900, respectively.

As the tube diameter is further decreased, the friction factor profiles also shifted (see Fig. 11.8). When the tube diameter is decreased from 1372 to 1067 μm, another group of similar friction factor profiles in the transition region is seen. Figure 11.8 shows that the decrease in tube diameter from 1067 to 838 μm did not significantly affect the profile of the friction factor. Since the 991 μm diameter tube is only 7% smaller than the 1067 μm diameter tube, it is expected that they have the same friction factor profile and transition range (2000 < Re < 4000). However, as the diameter is decreased from 991 to 838 μm, there is a slight noticeable change in the friction factor profile, with the onset of transition region shifted from Reynolds number of 2000 to 2200 and the increase in the depth of the trough in the transition region.

Figure 11.9 shows that further decrease in the tube diameter from 838 to 337 μm caused the transition region to become significantly narrower. The onset of transition region for 732 and 667 μm tubes is the same as that of the 838 μm tube, at Reynolds number of 2200. However, the end of transition region is shifted forward to Reynolds number of 3000, making the transition region of the 732 and 667 μm tubes narrower than that of the 838 μm tube. Further decrease in the tube diameter from 667 to 337 μm caused the onset and end of transition region to shift to lower

Reynolds numbers, while the transition range became narrower. The decrease in the tube diameter from 732 to 337 μm also caused the friction factor profile to shift higher. This suggests that the effect of surface roughness may be beginning to influence the friction factor. Also, the irregularities in the stainless steel tube at such small scale, as illustrated in Fig. 11.3, could have contributed to the shift in the friction factor profile.

A Brief Summary of the Extensions to the Work of Ghajar et al. (2010)

The studies of Ghajar et al. (2010) on the effect of mini- and micro-tube diameter on the transition region of friction factor were extended by Ghajar and co-workers to account for several different effects on both friction factor and heat transfer in mini- and micro-tubes. The following is a brief summary of those studies.

Tam et al. (2011) This study deals with the simultaneous heat transfer and pressure drop measurements in a stainless steel micro-tube with a diameter of 1000 μm. Distilled water was used as the test fluid. For this purpose, the experimental setup presented in Fig. 11.2 was modified to handle simultaneous measurement of heat transfer and non-isothermal pressure drop (friction factor) under uniform wall heat flux boundary condition similar to the results presented in Fig. 5.5.

For heat transfer, the results indicated that the micro-tube had an earlier start and end of transition compared to the macro-tube and, in the turbulent region, an increase in heat transfer due to the surface roughness was observed. For friction factor under isothermal condition, the micro-tube had a narrower transition range due to the roughness compared to the macro-tube. For friction factor under heating condition (non-isothermal), the laminar data and the start of transition were different from the isothermal case, and the effect of heating was not seen on the end of transition.

Tam et al. (2014) The objective of this study was to address the effect of inner surface roughness and heating on the friction factor of mini- to micro-tubes. The modified experimental setup of Tam et al. (2011) was used in this study. Distilled water was used as the test fluid. The experiments were conducted in seven stainless steel mini- to micro-tubes (2000–508 μm in inner diameter) with various surface roughness values (1.94–5.30 μm).

The experimental results clearly showed the effect of heating and roughness on friction factors in the laminar, transition, and turbulent regions. For the friction factors under isothermal condition, the micro-tubes had a narrower transition range due to the roughness when compared to the mini-tubes. For the friction factor under heating condition, the laminar data and the start of transition were different from the isothermal case. Under heating condition, the narrower transition range due to the roughness was also observed in the micro-tubes. However, the effect of heating was not seen on the end of transition and in the turbulent region.

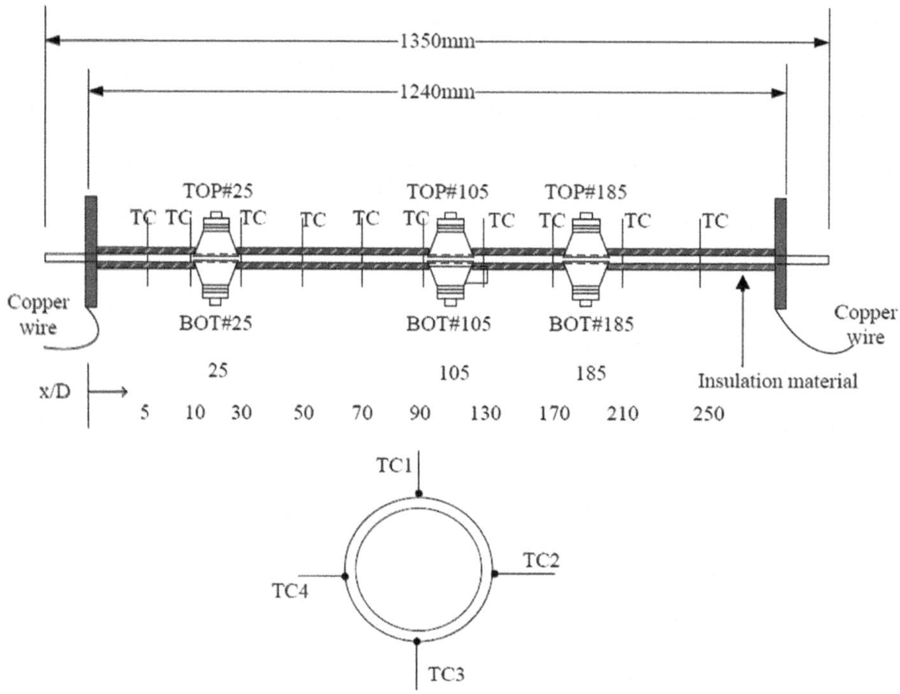

Fig. 11.10 Arrangement of the thermocouples and the ultrasonic heads on the test section

Tam et al. (2017) The objective of this study was to investigate the heat transfer enhancement in a 4-mm stainless steel mini-tube in the laminar region by using ultrasonic effect. The experimental setup of Tam et al. (2011, 2014) was further modified for this purpose. Figure 11.10 shows the arrangement of the thermocouples (TC) and ultrasonic heads on the test section of Fig. 11.2. From Fig. 11.10, it can be seen at each station that the thermocouples denoted as TC1, TC2, TC3, and TC4 are placed 90° apart around the periphery of the test tube. The figure also shows the total of six ultrasonic heads placed on the test section (three at the top and three at the bottom). The ultrasonic heads are arranged at x/D of 25, 105, and 185 and labeled as TOP#25, TOP#105, TOP#185, BOT#25, BOT#105, and BOT#185. Figure 11.11 shows the arrangement of the ultrasonic heads on the top and bottom of the test tube. As shown in Fig. 11.12, each ultrasonic head is a piezoelectric transducer which has a base diameter of 45 mm. According to Steinke and Kandlikar (2004), vibration induced by piezoelectric materials being attached to the periphery of the conventional macro-channel and mini- and microchannels is potentially an active technique of heat transfer enhancement.

The experimental results showed that a substantial enhancement in heat transfer by ultrasound was observed in the laminar region. Heat transfer enhancement by ultrasound in the entrance region was more significant than that in the fully developed region. To increase the fully developed region heat transfer, the installation of

Fig. 11.11 Photograph of the test section showing the arrangement of the ultrasonic heads on the top and bottom of the test tube

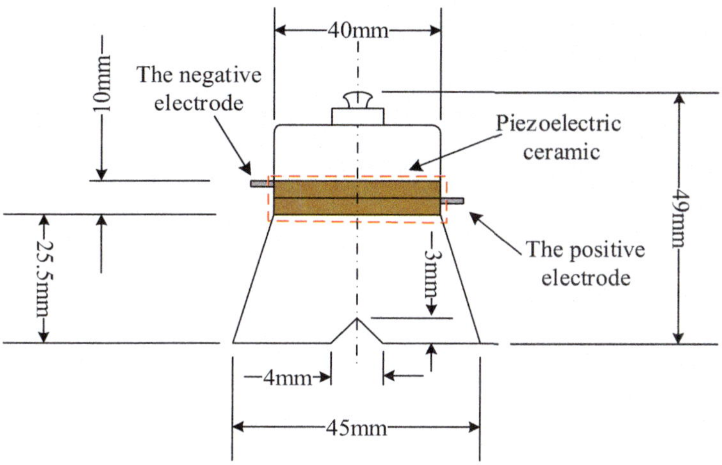

Fig. 11.12 Specifications of the ultrasonic heads used in the experiments

ultrasonic head at the earlier entrance location (i.e., x/D = 25) is not recommended. Based on the combinations of different numbers and positions of ultrasonic heads, two ultrasonic heads were observed to give a better heat transfer enhancement in the entrance and fully developed regions.

Tam et al. (2017, 2019) The reviews documented by Krishnamoorthy et al. (2007) and Celeta (2008) pointed out the need to investigate the effect of inlet on heat transfer inside the mini- and micro-tubes. As documented in Chap. 5, such studies were done for plain macro-tubes by Ghajar and co-workers in all flow regimes (laminar, transitional, and turbulent) for the typical inlet configurations (square-edged and re-entrant) used in heat exchangers. The focus of the studies by Tam et al. (2017, 2019) was to investigate the effect of bending inlets with different angles (bending entrance before a straight tube) that are also used in miniature heat exchangers. For this purpose, the experimental setup of Tam et al. (2011, 2014) was further modified. Kandlikar and Grande (2002) defined the channel with the hydraulic diameter of larger than 3 mm as the "conventional channel" and the channel with the diameter range between 3 and 0.2 mm as the "mini-channel." In the studies of Tam et al. (2017, 2019), heat transfer was measured in two horizontal mini-tubes of 0.8 and 2 mm inner diameters under the uniform wall heat flux boundary condition. Distilled water was used as the test fluid, and the Reynolds number range for the entire experiments was approximately between 800 and 11,000.

Figures 11.13a–c show the three inlet configurations (straight tube length, 90° bend, and 180° bend) of two different diameters (0.8 and 2 mm) installed before the horizontal test tube and the arrangement of the thermocouples on the test section. As shown in Figs. 11.13a–c, for the 2 and 0.8 mm mini-tubes, the total length of the heating section (L_t) was 680 mm and 300 mm, respectively. In Fig. 11.3a, there is a straight tube length arranged horizontally before the heating section. In Figs. 11.13b, c, the 90° and 180° bends, which were arranged horizontally and vertically in different tests, were formed by bending a straight tube. For the vertical bending inlet, the fluid flowed downward from the inlet to the horizontal test tube. For both bend orientations, the heating section was placed horizontally on the test table. The curvature ratio (R/r_i) of the bends is 12.5, where R is bend radius and r_i is the inner radius. To ensure a hydrodynamically developed flow condition, a certain length of straight tube is arranged before each bend.

From the heat transfer results, for the two mini-tubes with different inner diameters, regardless of the inlet type, the influence of buoyancy on the laminar heat transfer was present in the mini-tubes. However, the start of transition was inlet dependent, and the bending inlet delayed the start of transition when compared with the hydrodynamically fully developed length. From the entrance to the fully developed region, the laminar region entry length was almost at the length-to-inside diameter ratio of 150, and the laminar heat transfer of all the horizontal and the vertical 90° and 180° bends was observed to be higher than that of the straight tube inlet, especially in the entrance region.

Tam et al. (2021) The objective of this study was to address the effect of heating on the fully developed friction factors in mini-tubes in the complicated transition region, similar to the work presented in Chap. 3 for plain macro-tubes. For this purpose, experimental setup of Tam et al. (2014) was used to measure the mini-tube pressure drop (friction factor) under the uniform wall heat flux boundary condition. The fully developed pressure drops were measured in a test section fitted with horizontal mini-tubes of 4 and 2 mm inner diameters. A straight tube length was used as the inlet configuration. The entire experiments covered the Reynolds number range between 1000 and 5000 (laminar-transition-turbulent regions).

For the heating (non-isothermal) friction factors, heat fluxes of 9 and 15 kW/m^2 were applied to the 4 mm mini-tube, and heat fluxes of 8 and 16 kW/m^2 were applied to the 8 mm mini-tube. The friction factor results showed that the higher heat fluxes applied to the mini-tubes caused a decrease in the laminar and transitional friction factors and delayed the start and end of the transition region. However, the effect of heating was not obvious in the turbulent region.

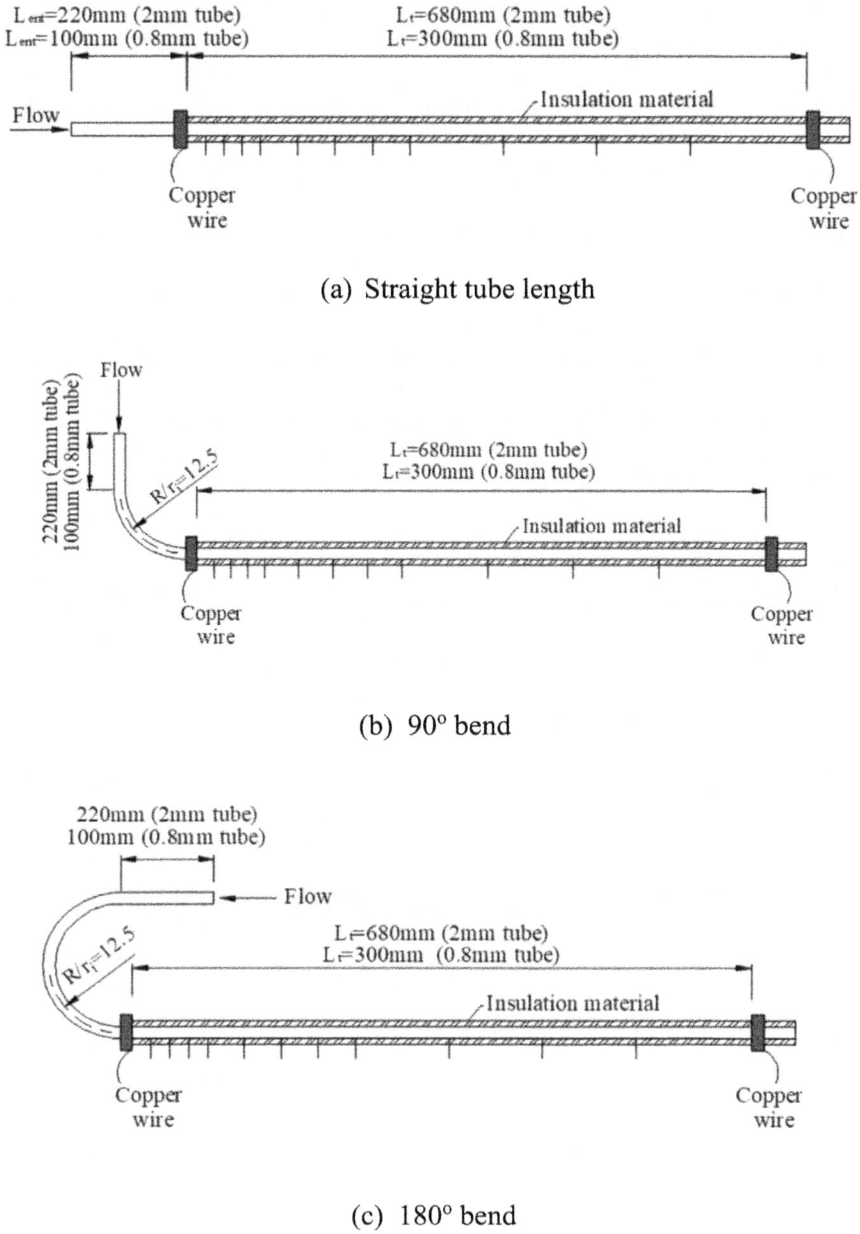

Fig. 11.13 Test sections with different inlets and thermocouples

Yang et al. (2021) The objective of this study was to numerically study the forced and mixed convection heat transfer in the laminar region for eight horizontal macro- and mini-tubes (4.0, 3.0, 2.5, 2.0, 1.9, 1.8, 1.6, and 1.2 mm) under the uniform wall heat flux boundary condition. The ultimate objective was to determine the critical diameter at which the mode of heat transfer changes from mixed convection to forced convection. For each tube diameter studied, the Reynolds number was set at 1000, and the uniform wall heat flux boundary condition was set at 10 kW/m^2. To establish the presence of the mixed convection (buoyancy-driven flow), the work of Ghajar and Tam (1994) was referred to as presented in Chaps. 3 and 5. According to Ghajar and Tam (1994), when the heat transfer coefficient ratio (h_t/h_b) was below 0.8, the flow was dominated by the mixed convection. Inversely, for h_t/h_b values above 0.8, the flow was dominated by the forced convection. Based on the calculated heat transfer coefficient ratios (h_t/h_b) and the temperature contour plots of the cross section of the tubes with different diameters, the existence of the buoyant flow inside the 2.5 mm and larger diameter tubes was confirmed. For 2.0 mm and smaller diameter tubes, the mode of heat transfer was pure forced convection. Based on these observations, the 2.0 mm diameter tube was defined as the critical diameter in which heat transfer changes from mixed convection to forced convection at the length to diameter ratio of 250. Also, a critical natural-to-forced convection ratio ($Gr/Re_{critical}$) of 0.16 was calculated accordingly. Based on their results, it can be summarized that, for horizontal mini-tubes at x/D of 250, forced convection is dominant when $Gr/Re \leq 0.16$ and mixed convection is dominant when $Gr/Re > 0.16$.

Part II
Two-Phase Flow Pressure Drop and Heat Transfer in Tubes

Chapter 12
Introduction

Background

The term two-phase flow designates a flow situation where two distinct phases flow simultaneously usually through a pipe. The two phases can be a combination of any two phases (gas-liquid, liquid-solid, gas-solid, or even two immiscible liquids). However, the main focus of the research work presented here is the simultaneous flow of gas-liquid two-phase flow in round pipes. The two-phase flow of gas and liquid is classified as one-component and two-component two-phase flow. One-component two-phase flow consists of the two phases of a single chemical species such as steam-water and refrigerant vapor-liquid flow used in nuclear and refrigeration applications, whereas two-component two-phase flow consists of the two phases of two chemically different species such as air-water and air-oil. The two-component two-phase flow is often referred to as non-boiling two-phase flow and is usually encountered in industrial applications such as artificial lift systems and simultaneous transportation of oil and natural gas from remote extraction locations to the processing units. Chemical operations requiring flow of two chemical species together for enhanced mass transfer, as in the case of ozone treatment of water, also rely on the two-component two-phase flow phenomenon. For the purpose of flow assurance and designing, sizing, and optimization of industrial processes and equipment handling two-phase flow, determination of the void fraction, total two-phase pressure drop, and heat transfer is crucial. In particular, for non-boiling two-phase flow applications, the total two-phase pressure drop based on flow patterns, void fraction, and pipe geometry influences the design of a two-phase flow system. Knowledge of non-boiling heat transfer in two-component two-phase flows is equally important to estimate and prevent advent of wax deposition in subsea pipeline carrying hydrocarbons.

A. J. Ghajar, *Single- and Two-Phase Flow Pressure Drop and Heat Transfer in Tubes*, Mechanical Engineering Series,
https://doi.org/10.1007/978-3-030-87281-6_12

Depending on the mass flow rates of individual phases, pipe orientation, and pipe diameter, the two phases align themselves in a particular fashion across the pipe cross section identified as a flow pattern. The principal problem associated with the gas-liquid two-phase flow is its flow pattern-dependent complex structure due to turbulent mixing, compressibility nature of the gas phase, significantly different thermophysical properties of the two phases, phase flow rates, and the pipe orientation that renders the analytical relationships for void fraction, two-phase pressure drop, and heat transfer difficult. Thus, it is necessary to study the gas-liquid two-phase flow phenomenon experimentally and device a systematic method to correctly predict the two-phase flow parameters such as void fraction, pressure drop, and heat transfer.

Two-phase flow literature reports a plethora of correlations for determination of flow patterns, void fraction, two-phase pressure drop, and non-boiling heat transfer correlations; however, the validity of a majority of these correlations is restricted over a narrow range of two-phase flow conditions. Consequently, it is quite a challenging task for the end user to select an appropriate correlation/model for the type of two-phase flow under consideration. Selection of a correct correlation also requires some fundamental understanding of the two-phase flow physics and the underlying principles/assumptions/limitations associated with these correlations. Thus, it is of significant interest for a design engineer to have knowledge of the flow patterns and their transitions and their influence on two-phase flow variables.

Need of Two-Phase Flow Study in Inclined Systems

The research in the field of gas-liquid two-phase flow can be classified in terms of its relevance to flow patterns, void fraction, two-phase pressure drop, and heat transfer. The accurate prediction of void fraction is very important in correct estimation of two-phase mixture density and hence the two-phase hydrostatic pressure drop and the convective heat transfer coefficient. The two-phase pressure drop essentially varies from that of its single-phase counterpart due to the significant density difference and hence the slippage between the two phases. Although the two-phase flow literature reports a number of void fraction and pressure drop correlations, they are found to be inadequate in their range of application in terms of flow patterns, pipe orientation, and the fluid physical properties. Most of these correlations are of empirical form and are developed based on limited experimental data and hence perform well only over the certain range of two-phase flow conditions.

In comparison to the horizontal and vertical pipe orientations, the two-phase flow literature is devoid of flow patterns, void fraction, two-phase pressure drop, and heat transfer data in upward and downward inclined pipes. Consequently, the accuracy and performance of the existing two-phase flow correlations for the case of upward and downward inclined two-phase flow are unknown and uncertain. Moreover, due to lack of experimental initiatives, the two-phase flow literature lacks sound understanding of the effect of pipe orientation on two-phase flow parameters. This uncertainty and lack of sound understanding of the inclined two-phase flow phenomenon possess a challenge in the selection procedure of the most appropriate correlation to be used for upward or downward inclined two-phase flow situations.

Thus, this creates a room for further experimental investigation of two-phase flow parameters for upward and downward inclined pipe orientations and contribution to our fundamental understanding of the two-phase flow mechanism. Additionally, due to wide industrial applications of two-phase flow and high uncertainty in correct prediction of flow patterns, it is strongly desired to have unified two-phase flow models that can predict two-phase flow parameters such as void fraction, two-phase pressure drop, and heat transfer independent of flow patterns, pipe orientation, and thermophysical properties of gas and liquid phases.

Basic Definitions in Gas-Liquid Two-Phase Flow

The gas-liquid two-phase flow is realistically a three-dimensional flow with flow conditions and fluid properties varying with respect to the pipe cross section, length, and time. Moreover, the simultaneous gas-liquid two-phase flow is marked by the continuous interaction between the two phases with a dynamic interface along the pipe length and cross section, thus making it difficult to model two-phase flow phenomenon mathematically. Thus, this complexity limits the application of conventional equations designed for single-phase flow to the two-phase flow situations. In order to understand the two-phase flow parameters and its deviation from single-phase flow, basic definitions of two-phase flow are presented in this chapter. A valid assumption of one-dimensional two-phase flow to a great extent simplifies several two-phase flow variables and modeling techniques.

Table 12.1 introduces some of the primary variables that have been adopted by different two-phase flow models. The parameters listed in Table 12.1 are defined in context to single-phase flow of gas and liquid phase and their simple relationships that represent some two-phase flow variables. The mass flow rate of the gas ($\dot{m}_G$) and liquid ($\dot{m}_L$) phase through the pipe can be measured directly by mass flow meters. Based on these mass flow rates, the total or two-phase mixture mass flow rate ($\dot{m}$) is defined as the sum of the mass flow rates of individual phases. The two-phase flow quality (x) is defined as the ratio of mass flow rate of the gas phase ($\dot{m}_G$) to the two-phase mixture mass flow rate ($\dot{m}$). One of the important two-phase flow variables is the void fraction (α_G) and can be defined in different ways based upon its measurement technique such as cross-sectional or volumetric method. For example, if the void fraction is measured using cross-sectional (conductance/capacitance probe) technique, then it is defined as the ratio of pipe cross-sectional area occupied by the gas phase (A_G) to the total pipe cross-sectional area ($A = A_G + A_L$), whereas in the case of volumetric void fraction measurement using quick closing valves, the void fraction is defined as the ratio of volume of the pipe occupied by the gas phase (V_G) to the total volume of test section (V). Another important two-phase flow parameter that is usually used in the analysis of two-phase flow is the gas volumetric

Table 12.1 Basic definitions of interest in two-phase flow

Parameter	Definition	Description
$\dot{m}$	$\dot{m} = \dot{m}_G + \dot{m}_L$	Mixture mass flow rate (kg/s)
x	$\frac{\dot{m}_G}{\dot{m}_G+\dot{m}_L}$	Two-phase flow quality
α_G	$\frac{A_G}{A}$ or $1-\frac{A_L}{A}$, $\frac{V_G}{V}$ or $1-\frac{V_L}{V}$	Void fraction
λ	$\frac{U_{SG}}{U_{SG}+U_{SL}}$ or $\left[1+\frac{1-x}{x}\left(\frac{\rho_G}{\rho_L}\right)\right]^{-1}$	Gas volumetric flow fraction
U_{SG}	$\frac{Gx}{\rho_G}$ or $\frac{\dot{m}_G}{A\rho_G}$	Superficial gas velocity (m/s)
U_{SL}	$\frac{G(1-x)}{\rho_L}$ or $\frac{\dot{m}_L}{A\rho_L}$	Superficial liquid velocity (m/s)
U_G	$\frac{U_{SG}}{\alpha_G}$ or $\frac{\dot{m}_G}{A_G\rho_G}$	Actual gas velocity (m/s)
U_L	$\frac{U_{SL}}{1-\alpha_G}$ or $\frac{\dot{m}_L}{A_L\rho_L}$	Actual liquid velocity (m/s)
U_M	$U_{SL}+U_{SG}$ or $U_L(1-\alpha_G)+U_G\alpha_G$	Two-phase mixture velocity (m/s)
S	$\frac{U_G}{U_L}$ or $\left(\frac{x}{1-x}\right)\left(\frac{1-\alpha_G}{\alpha_G}\right)\left(\frac{\rho_L}{\rho_G}\right)$	Slip ratio
Re_{SL}	$\frac{\rho_L U_{SL} D}{\mu_L}$ or $\frac{G(1-x)D}{\mu_L}$	Superficial liquid Reynolds number
Re_{SG}	$\frac{\rho_G U_{SG} D}{\mu_G}$ or $\frac{GxD}{\mu_G}$	Superficial gas Reynolds number

flow fraction (λ) which is essentially the homogeneous void fraction under the assumption of no slippage between the two phases. It is defined as the ratio of the superficial gas velocity (U_{SG}) to the mixture velocity ($U_{SG}+U_{SL}$). Velocity of liquid and gas phase assuming that phase is flowing alone through the pipe occupying the entire cross section is called as phase superficial velocity. The superficial gas or liquid velocity (U_{SG} or U_{SL}) is defined assuming the phase mass flow rate/mass flux (G) through the entire pipe cross section (A). Actual phase velocity (U_G or U_L) is quite higher than that of the superficial phase velocity due to the reduced pipe cross-sectional area available for each phase. The actual phase velocity is based on the phase mass flow rate through the pipe cross-sectional area (A_G or A_L) occupied by that phase. The two-phase mixture velocity (U_M) is simply the summation of the superficial velocity of each phase. The slip ratio (S) is the ratio of actual gas phase velocity (U_G) to actual liquid phase velocity (U_L). Finally, the superficial liquid or gas Reynolds number (Re_{SL} or Re_{SG}) is based on the phase superficial velocity (U_{SG} or U_{SL}), phase density (ρ_L or ρ_G), and phase dynamic viscosity (μ_L or μ_G).

Chapter 13
Two-Phase Flow Experimental Setup for Inclined Systems

Background

As mentioned earlier in Chap. 12, two-phase flow literature reports a plethora of correlations for determination of flow patterns, void fraction, two-phase pressure drop, and non-boiling heat transfer correlations. However, majority of these correlations were developed based on experimental data that covered a limited range of pipe orientations, e.g., vertical upward and horizontal, and certain flow patterns, e.g., the annular flow regime. In order to understand the phenomenon and physics of void fraction, pressure drop, and non-boiling heat transfer in two-phase flow and study its dependency on various flow parameters, it is necessary to carry out systematic experimental measurements. The experimental work presented in this study intends to complement the two-phase flow data available in the literature and aid to analyze the behavior of void fraction, two-phase pressure drop, and non-boiling heat transfer with varying flow patterns and pipe orientations. Once the necessary experimental data has been collected and the nexus and interdependencies between flow patterns, void fraction, two-phase pressure drop, and non-boiling heat transfer have been established, the next major move to accurately quantify the void fraction, two-phase pressure drop, and non-boiling heat transfer is the development of the two-phase flow models.

In order to have a sound understanding of the effect of pipe orientation on two-phase flow phenomenon, the collected experimental data should help in addressing the following issues:

1. What is the qualitative and quantitative effect of pipe orientation on the physical structure of flow patterns and their transition boundaries?
2. What is the relationship between flow patterns and void fraction as a function of pipe orientation?
3. How sensitive is the two-phase pressure drop to the void fraction for different flow patterns at any given pipe orientation?

A. J. Ghajar, *Single- and Two-Phase Flow Pressure Drop and Heat Transfer in Tubes*, Mechanical Engineering Series,
https://doi.org/10.1007/978-3-030-87281-6_13

4. What is the effect of flow pattern structure and their transition from one to the other on the frictional component of the two-phase pressure drop?
5. What is the effect of pipe surface roughness on the frictional component of two-phase pressure drop for different flow patterns at any given pipe orientation?
6. What is the effect of change in gas and liquid flow rates in different pipe inclinations on the two-phase heat transfer coefficient?
7. What is the effect of pipe inclination on the two-phase heat transfer coefficient?
8. What is the most reliable and accurate method to predict void fraction, pressure drop, and heat transfer in upward and downward inclined gas-liquid two-phase flow?

Details of the Two-Phase Flow Experimental Setup

The versatile experimental setup described here is located at the School of Mechanical and Aerospace Engineering, Two-Phase Flow Laboratory, at Oklahoma State University (OSU), Stillwater, Oklahoma. The setup can be used for flow visualization and measurements of void fraction, pressure drop, and non-boiling heat transfer for different pipe orientations (+90° to −90°), all using a single experimental setup.

The experimental setup used in the present study consists of two test sections: a 12.7 mm I.D. polycarbonate transparent pipe section and a 12.5 mm I.D. stainless-steel pipe. The experimental setup is mounted on a variable inclination frame (made up of rectangular steel tubing) with the help of pulleys and bolts, thus making it flexible to be rotated to any inclination between +90° and −90° including the horizontal orientation. The test section of 12.7 mm I.D. and 890-mm-long transparent polycarbonate pipe is used for flow visualization, void fraction, and pressure drop measurements, while 12.5 mm I.D. and 1030-mm-long schedule 40S stainless-steel pipe having a surface roughness of 0.02 mm is used for pressure drop and heat transfer measurements. An overview of the experimental setup and the variable inclination frame is shown in Figs. 13.1 and 13.2, respectively.

The fluid combination used for generating two-phase flow is compressed air and distilled water. The air supplied by Ingersoll Rand T-30 Model 2545 compressor first passes through a regulator, filter, and lubricator circuit and then through a submerged helical coil heat exchanger. Next, the air is again passed through a filter and then fetched to Micro Motion Elite Series Model LMF 3M and CMF 025 Coriolis gas mass flow meters where the mass flow rate of air is controlled precisely using a Parker (24NS 82(A)-8LN-SS) needle valve. The compressed air is then allowed to enter the test section through a spiral static mixer. The liquid phase in the form of distilled water stored in a 50-gallon tank is circulated in the system using a Bell and Gossett (series 1535, model number 3445 D10) centrifugal pump. The distilled water is first passed through Aqua-Pure AP12-T purifier and then through ITT model BCF 4063 shell and tube heat exchanger. The distilled water is then passed through an Emerson Coriolis mass flow meter (Micro Motion Elite Series model number CMF 100) where the mass flow rate of the liquid phase entering the test

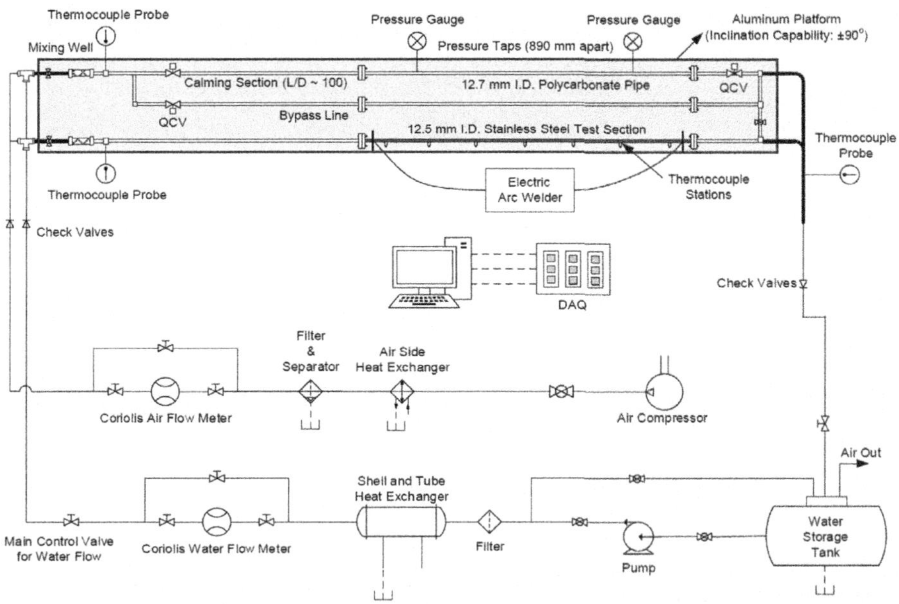

Fig. 13.1 Schematic of the two-phase flow experimental setup used for flow visualization and void fraction, pressure drop, and heat transfer measurements

section is controlled. Later, the water is allowed to mix with air in Koflo model 3/8-40C-4-3V-23/8 static mixer. The mixer is mounted right before the entrance to the test section. Two-phase flow measurements are carried out for all major flow patterns by systematically varying the gas flow rates for fixed values of liquid flow rates. The gas and liquid flow rates are varied in a range of 0.001–0.2 kg/min and 1–10 kg/min, respectively, to generate all possible flow patterns observed in horizontal and upward and downward inclined two-phase flow. For isothermal measurements (void fraction and pressure drop), system temperature is maintained between 20 °C and 25 °C, and the system pressure is found to vary between 1 and 3 bar. For these flow conditions, the Reynolds number for gas and liquid phase was found to vary between 100–19,000 and 2000–18,000, respectively. During heat transfer measurements, although the pipe wall temperature varied between 40 °C and 60 °C, the bulk temperature (T_b) of two-phase mixture was regulated between 17 °C and 25 °C using the shell and tube heat exchanger. Thus, the effect of system temperature on the superficial liquid phase Reynolds number (Re_{SL}) during heat transfer measurements was not considered significant. Consequently, Re_{SL} measured during heat transfer experiments was comparable to that for void fraction and pressure drop measurements. For each combination of gas and liquid flow rate, void fraction is measured using quick closing solenoid valves (SV) with a closing time of 0.1 s. When the valves are triggered, the two normally open valves close, and the normally closed valve opens simultaneously. In this manner, a two-phase sample is trapped in the void fraction section, while the air-water mixture is allowed to continue flowing

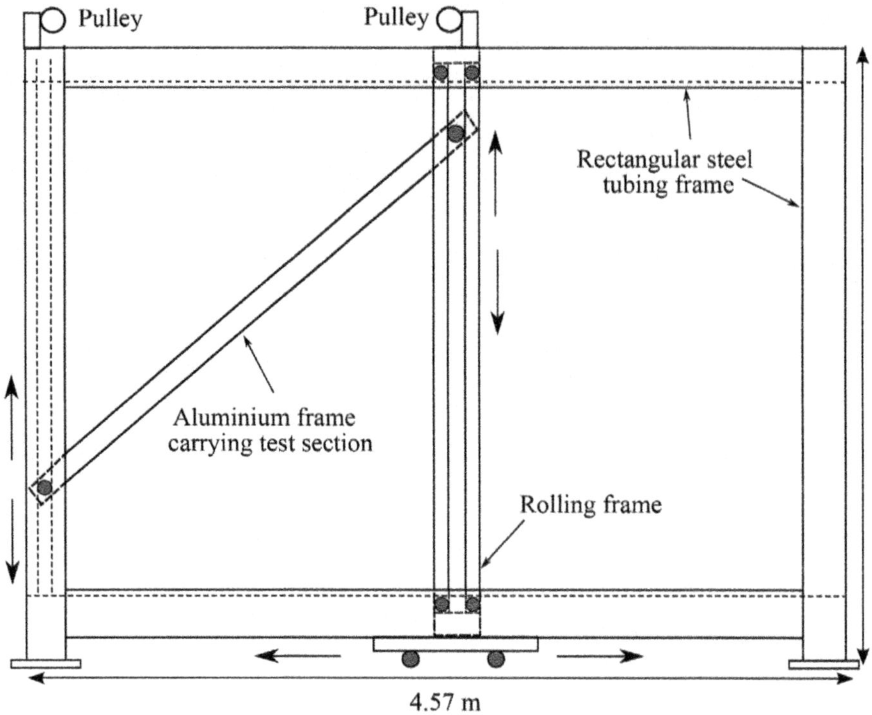

Fig. 13.2 Schematic of variable inclination frame carrying test section

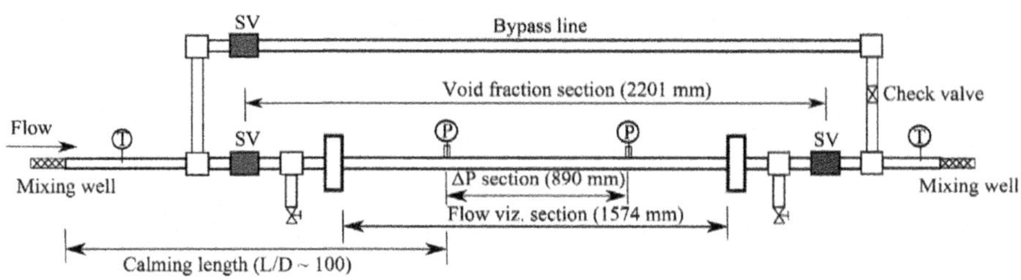

Fig. 13.3 Schematic of the details of the void fraction, pressure drop, and flow visualization test sections

through the bypass line (see Fig. 13.3). Air-water mixture trapped in the void fraction section is drained, and the volume of the liquid is measured. With the measured volume of the liquid phase (V_L), the void fraction (α_G) can be determined from the total volume of the test section (V_t) using $\alpha_G = 1 - V_L/V_t$. The uncertainty associated with the volume of the liquid is $\pm 0.5\ cm^3$, and the uncertainty associated with the total volume of the test section is $\pm 2\ cm^3$. Using the uncertainty propagation method prescribed by Kline and McClintock (1953), the uncertainty associated with measurements of void fraction is estimated to be between $\pm 1.25\%$ and $\pm$ 10%. The high percentage value of uncertainty is due to the small values of void fraction. Note that, in the transient region of the two-phase flow, void fraction measurements were repeated eight to ten times and the average value was used as a representative void fraction.

The total two-phase pressure drop is measured using a DP15 variable reluctance Validyne pressure transducer having an accuracy of 0.25% of full-scale range of the diaphragm. The two pressure diaphragms with a range of 0.5 psi (DP-26) and 2 psi (DP-32) are used to measure total two-phase pressure drop, and hence the associated worst-case uncertainty with total two-phase pressure drop is $\approx \pm 35$ Pa/m. The upstream and downstream pressure taps are placed 890 mm apart with the upstream pressure tap placed at $\approx 100D$ from the entrance of the test section (see Fig. 13.3). Since the pressure taps are connected to the differential pressure transducer through flexible connecting lines, the measured value of total two-phase pressure drop given by Eq. (13.1) is corrected for the amount of liquid phase trapped in these lines at any given pipe orientation. For the non-boiling two-phase flow conditions at near atmospheric system pressures, the accelerational component is usually very small since there is no expansion of the two-phase mixture and/or change in two-phase flow quality due to phase change process. Based on the absolute pressure measured at upstream and downstream pressure taps, the maximum contribution of the accelerational component is found to be $\leq 1\%$ of the total two-phase pressure drop, and hence it is neglected in the calculation of frictional pressure drop. The hydrostatic pressure drop, as a function of void fraction and phase densities, is as expressed by Eq. (13.2). Assuming negligible contribution from the accelerational component of two-phase pressure drop, the frictional component of two-phase pressure drop is obtained by subtracting the hydrostatic component from total component of two-phase pressure drop as given by Eq. (13.3):

$$\left(\frac{dp}{dz}\right)_t = \left(\frac{dp}{dz}\right)_h + \left(\frac{dp}{dz}\right)_f + \left(\frac{dp}{dz}\right)_a \tag{13.1}$$

$$\left(\frac{dp}{dz}\right)_h = [\rho_L(1-\alpha_G) + \rho_G\alpha_G]g\sin\theta = \rho_M g\sin\theta \tag{13.2}$$

$$\left(\frac{dp}{dz}\right)_f = \left(\frac{dp}{dz}\right)_t - \left(\frac{dp}{dz}\right)_h \tag{13.3}$$

In order to accomplish heat transfer measurements, a constant heat flux in a range of 7 kW/m^2 to 60 kW/m^2 is provided to the test section using Miller Maxtron 750 electric arc welder that has a capacity to produce steady output of currents up to 450 Amps. The electric input is delivered to the test section using 6.35-mm-thick copper plates soldered to the ends of test section. These plates completely encircled the test section to ensure uniform distribution of electric input. Also, 1.27-cm-thick phenolic resin boards are placed on outer sides of these copper plates to prevent loss of heat from the heated test section to the non-heated portions of the pipe. A thick Micro-Lok fiberglass insulator is used to provide insulation to ensure minimal heat loss from the setup to the surrounding.

The two-phase inlet and outlet temperature is measured using Omega TMQSS-06U-6 thermocouple probes. These thermocouple probes are inserted inside through the pipe wall till they almost touch the other end of the pipe wall. This ensures that the probes are always in contact with the two-phase mixture. CO1-T-type thermocouples with an accuracy of ±0.5 °C are used to measure wall temperatures at seven different stations spaced 12.7 cm apart along the 1016 mm ($\approx 80D$) long heated test section. At each axial location, four thermocouples equally spaced over the pipe circumference (top, bottom, and two side walls of the pipe) are cemented to the pipe outer wall. Thus, the average values of pipe outer and inner wall temperature are based on 28 temperature measurements along the test section length. See Figs. 13.4 and 13.5 for the details of the heat transfer test section and the placement of thermocouples on the heat transfer test section.

The average values of inside wall temperature ($\overline{T}_{w,i}$), heat flux, and convective heat transfer coefficient ($\overline{h}$) based on pipe outer wall temperature ($\overline{T}_{w,o}$) are calculated using a finite difference formulation-based data reduction program developed by Ghajar and Kim (2006). The local (circumferential average at each axial station) inner wall temperature ($\overline{T}_{w,i}$) and the associated heat transfer coefficient are calculated using Eqs. (13.4) and (13.5), respectively. The two-phase convective heat

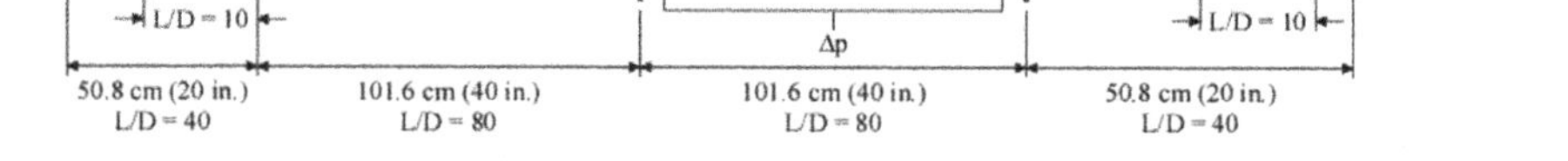

Fig. 13.4 Schematic of the details of the heat transfer test section

Fig. 13.5 Placement of the thermocouples on the heat transfer test section

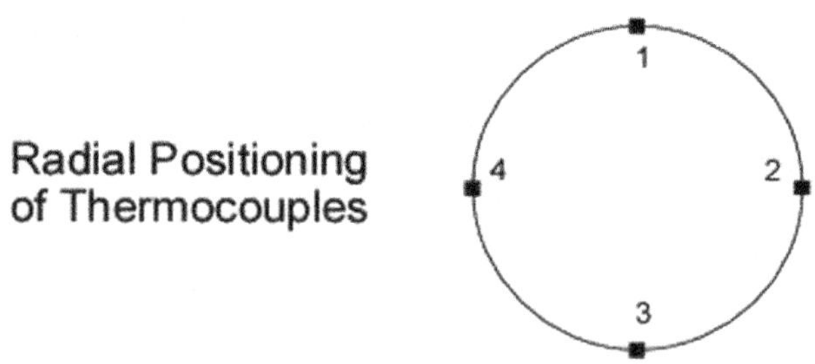

transfer coefficient (h_{TP}) is represented by the average of the measured local values of $\overline{h}$ at each thermocouple station (N_{ST}) given by Eq. (13.6):

$$\overline{T}_{w,i} = \overline{T}_{w,o} + \frac{\dot{q}}{2\pi k\Delta z} \ ln\left(\frac{D_o}{D_i}\right) \tag{13.4}$$

$$\overline{h} = \frac{\dot{q}}{2\pi D\Delta z\left(\overline{T}_{w,i} - T_b\right)} \tag{13.5}$$

$$h_{TP} = \frac{1}{L}\int \overline{h}dz = \frac{1}{L}\sum\nolimits_{j=1}^{N_{ST}} \overline{h}_j\Delta z_j \tag{13.6}$$

The uncertainty associated with measurements of two-phase heat transfer coefficient is determined using Kline and McClintock (1953) method. Higher experimental uncertainties (~25%) are found to be associated with the annular, intermittent, and stratified flow patterns. For slug and bubbly flow patterns, the average uncertainty associated with two-phase heat transfer coefficient is less than 10%. The higher values of uncertainty in measurement of h_{TP} are essentially due to two factors, namely, high heat balance error (difference between actual and enthalpy-based heat transfer rates) and low temperature difference $\left(\overline{T}_{w,i} - T_b\right)$, i.e., the difference between average inner wall temperature and bulk temperature. The bulk temperature is the average of the temperatures measured at the inlet and exit of the test section. The heat balance error results from losses due to convection to the surrounding and/or storage within the test section itself. It is difficult to control the heat balance error; however, it is found that uncertainty in measurement of h_{TP} could be reduced by maintaining at least a 4 °C temperature difference between pipe inner wall and mixture bulk temperature. Thus, the electrical input to the test section is adjusted to obtain $\left(\overline{T}_{w,i} - T_b\right) \geq 4$ °C such that the uncertainty in calculation of this temperature difference is not significantly propagated in calculation of two-phase heat transfer coefficient. The validity of pressure drop and heat transfer measurements carried out using this setup is checked by comparing the single-phase friction factor and heat transfer coefficient against the correlations of Churchill (1977a) and Sieder and Tate (1936). Measured values of single-phase data are found to be within ±10% of the values predicted by these correlations. For detailed information regarding experimental setup and uncertainties in measurement of different two-phase flow variables, please refer to the work of Cook (2008) and Bhagwat (2015).

Chapter 14
Flow Patterns, Flow Pattern Maps, and Flow Pattern Transition Models

Background

The physical structure of two-phase flow known as "flow pattern" is most fundamental among all two-phase flow variables and also forms the basis of two-phase flow modeling methods. Unlike single-phase flow theory, two-phase flow cannot be distinguished as laminar, transitional, or turbulent but rather is classified on the basis of several user-defined flow patterns.

Flow Patterns

The flow patterns observed in gas-liquid two-phase flow are quite intriguing in nature and depend on the alignment of one phase with respect to the other across the pipe cross section. The morphological variations in the structure of flow patterns are essentially due to the significantly different physical properties of the two phases such as viscosity and density, compressible nature of the gas phase, and the interaction between gravity, inertia, and buoyant forces. Correct understanding of the physical structure of flow patterns and their transition is instrumental in the general understanding of the mechanism of mass, momentum, and energy transfer in two-phase flow.

This preliminary section of the chapter is aimed at describing the physical structure of some key flow patterns observed in horizontal, upward, and downward vertical pipes as shown in Figs. 14.1, 14.2, and 14.3, respectively, and their physical variations as a function of varying two-phase flow conditions. The different flow patterns were generated by systematically varying gas and liquid flow rates in a range of 0.001–0.2 kg/min ($Re_{SG} \sim 100$ to 18,500) and 1–10 kg/min ($Re_{SL} \sim 2000$ to 18,000), respectively. Existence of a certain flow pattern was confirmed by visual

A. J. Ghajar, *Single- and Two-Phase Flow Pressure Drop and Heat Transfer in Tubes*, Mechanical Engineering Series,
https://doi.org/10.1007/978-3-030-87281-6_14

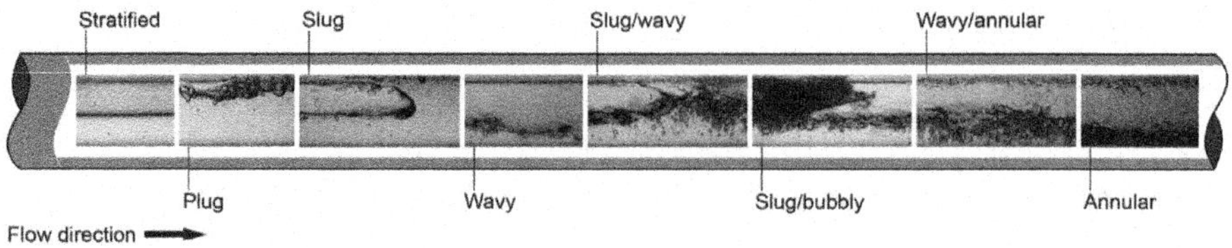

Fig. 14.1 Flow patterns in horizontal air-water two-phase flow. (Flow visualization done at the Two-Phase Flow Lab, OSU, Stillwater, OK)

observation as well as through still photographs and videos recorded over a certain period of time using Nikon D3100 camera and 200 mm f/5.6 lens with a shutter speed of 1/4000 s. What follows is a brief description of two-phase flow patterns and their subtle morphological variations as a function of phase flow rates and pipe inclinations.

Bubbly Flow Bubbly flow pattern is observed at high liquid and low gas flow rates. Bubbly flow often called as dispersed bubbly flow is characterized by the flow of small gas bubbles dispersed in continuous liquid medium. The size (much smaller than the pipe diameter), shape (typically spherical or oblong), and distribution (uniform, center peaked, or wall peaked) of bubble density depend upon the pipe geometry and inclination, fluid properties, and phase flow rates. For horizontal and inclined two-phase flow (both upward and downward), the gas bubbles are always located near the pipe upper wall region (asymmetric flow) as shown in Fig. 14.4, while for vertical flows (both upward and downward), the gas bubbles are evenly distributed across the pipe cross section (symmetric flow); refer to Fig. 14.4. It is observed that the gas and liquid flow rates significantly affect the size, shape, and distribution of the gas bubbles dispersed in the liquid medium. In the case of vertical downward flow, at low gas flow rates, the gas bubbles appear to be distributed only in the central region of the pipe with single-phase liquid occupying the near wall region. This appearance of bubbly flow in vertical downward flow is essentially due to the "coring phenomenon" characterized by the repulsive force exerted by the pipe wall on the liquid phase. This type of flow behavior has been observed and reported by other investigators (Oshinowo 1971, Usui and Sato 1989, Yijun and Rezkallah 1993, Terekhov and Pakhomov 2008) for pipe diameters in a range of 9–45 mm. This coring flow behavior in vertical downward flow is evident from Fig. 14.4.

Slug Flow Slug flow is characterized by the alternate flow of gas and liquid slugs. The size (elongated and comparable to pipe diameter), shape (usually bullet shaped with hemispherical nose and a blunt tail), and frequency of slugs depend upon the pipe geometry, pipe orientation, and fluid properties. For horizontal and non-vertical upward pipe inclinations, the gas slug is in the vicinity of the pipe upper wall, while it is symmetrically distributed around the pipe axis in vertical flow orientation. The translational velocity of the gas slug along the pipe axis is a function of pipe orientation such that the upward pipe inclinations aid the slug motion, while the downward pipe inclinations oppose the slug motion. The length of the gas slugs

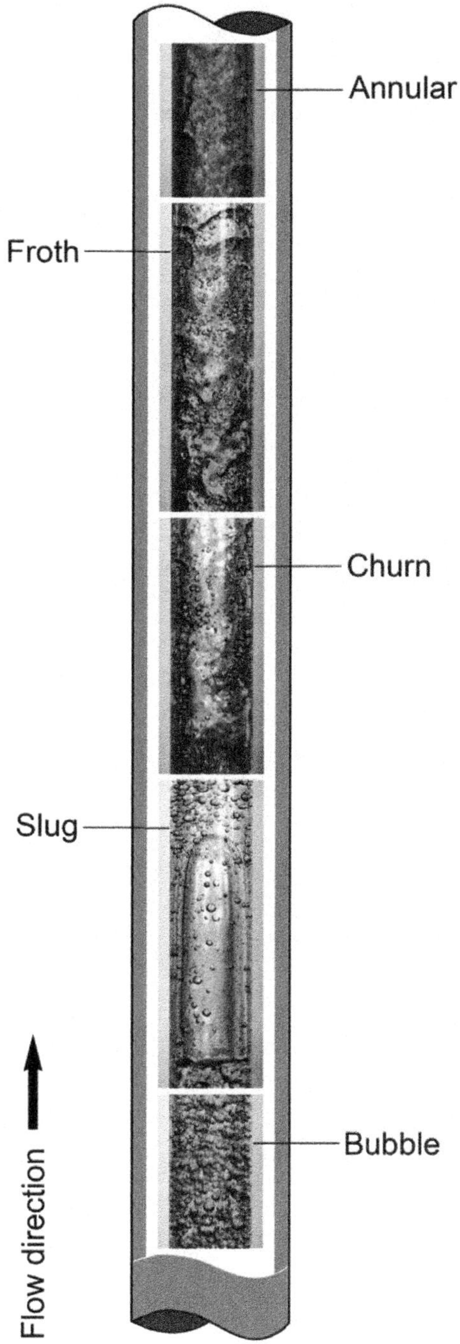

Fig. 14.2 Flow patterns in upward vertical air-water two-phase flow. (Flow visualization done at the Two-Phase Flow Lab, OSU, Stillwater, OK)

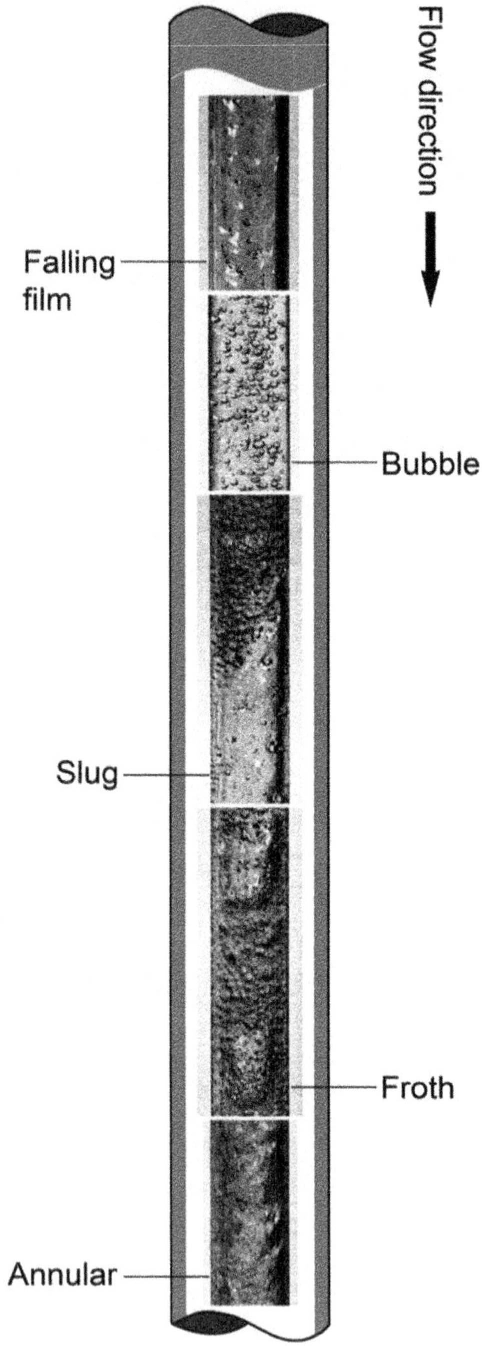

Fig. 14.3 Flow patterns in downward vertical air-water two-phase flow. (Flow visualization done at the Two-Phase Flow Lab, OSU, Stillwater, OK)

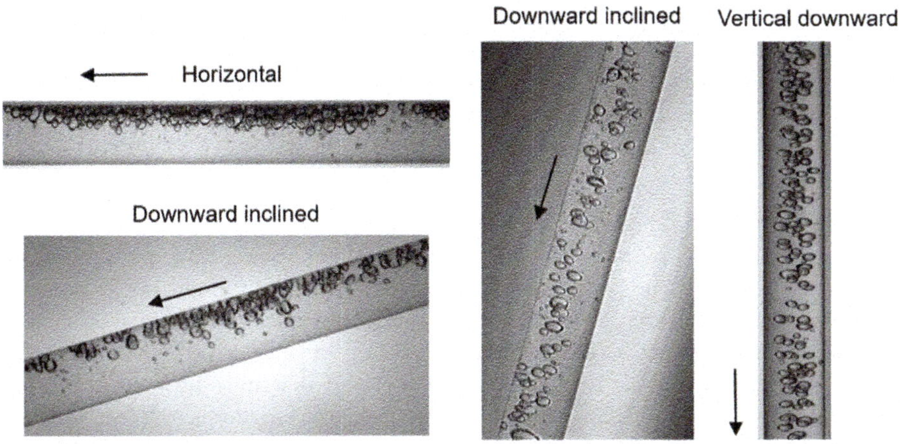

Fig. 14.4 Varying bubbly flow structures in horizontal, downward inclined, and vertical downward two-phase flow. (Flow visualization done at the Two-Phase Flow Lab, OSU, Stillwater, OK)

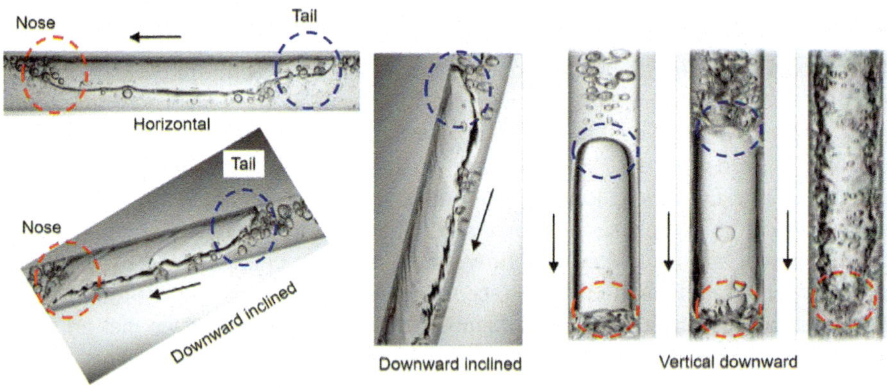

Fig. 14.5 Physical structure of slug flow pattern in downward inclined two-phase flow. (Flow visualization done at the Two-Phase Flow Lab, OSU, Stillwater, OK)

reduces, while their frequency increases with increase in the liquid flow rate. The opposite is true for a fixed liquid flow rate and decreasing gas flow rates.

In downward inclined flow (see Fig. 14.5), the buoyancy acting on the gas phase is in the direction opposite to that of the mean flow that causes distortion of the slug nose, and a tail pointing in the direction opposite to that of the flow direction is observed. It is also worthwhile to mention that the shape of the slug nose in downward inclinations is found to change with change in the phase flow rates depending upon the interaction between buoyant and inertia forces. In the case of vertical downward flow, at low liquid and gas flow rates, the elongated slug bubble appears very similar to that of the vertical upward flow, i.e., gas bubble nose appears to be pointed in the upward direction and remains undistorted. As shown in Fig. 14.5 with increase in the gas and liquid flow rates, the slug nose and tail (both ends of slug) become flat, and with further increase in phase flow rates, the slug nose points in the direction of the mean flow. These different types of slug nose in vertical downward flow are an indication of the different residence times of the gas phase in the test section (different magnitudes of interaction between buoyant and inertial forces) that influence the void fraction, pressure drop, and heat transfer characteristics of two-phase flow.

Intermittent Flow The intermittent two-phase flow that typically exists at moderate gas and liquid flow rates is a representation of two-phase flow having chaotic, pulsating, and indefinite phase alignment characteristics. Different subcategories of key flow patterns such as slug/wavy, slug/bubbly, and wavy/annular shown in Fig. 14.1 for horizontal two-phase flow and churn and froth flow (see Figs. 14.2 and 14.3) exhibit disorderliness, intermittency, and indefinite flow structure, and hence it is appropriate to club all these different sub-flow patterns together to call as intermittent flow. This approach of calling certain flow patterns as intermittent flow has been adopted by many researchers to reduce the ambiguity in identifying the transition boundaries between key flow patterns. Previous works reported in the literature by Beggs and Brill (1973), Mukherjee (1979), and Barnea et al. (1982) have adopted similar flow pattern nomenclature for this type of flow behavior.

Stratified Flow Stratified flow exists in the form of gas and liquid layers flowing parallel to each other such that the pipe bottom wall is in contact with the liquid phase, while the upper wall is in contact with the gas phase. Stratified flow in general can be classified as *smooth stratified* and *wavy stratified* flow. Smooth stratified flow appears at low liquid and low gas flow rates and is characterized by the complete separation of the two phases sharing a smooth and stable interface, whereas, at high gas flow rates, the wavy stratified flow exists, and it is characterized by the wavy, dynamic, and rough gas-liquid interface. At relatively high gas and liquid flow rates, the interfacial instability generates disturbance waves at the interface that grow and tend to touch the pipe wall giving an appearance of rolling wave. This type of stratified flow is recognized as *rolling wave* flow; however, the general flow structure still resembles the stratified flow pattern. The flow visualization of the three different forms of stratified flow described above is presented in Fig. 14.6.

A specific case of stratified flow at low liquid and low to moderate gas flow rates is identified as "falling film flow." This is observed in vertical downward and near vertical downward pipe inclinations of two-phase flow. The falling film flow is characterized by the downward flow of thin liquid film flowing quiescently along the pipe wall, while the gas phase flows through the central region. Falling film flow at low liquid flow rates has a strong tendency of creating dry spots on the pipe wall and hence may be regarded as undesirable flow pattern for the non-boiling two-phase flow involving high heat flux at the pipe wall. The physical form of falling film flow for a fixed liquid flow rate and increasing gas flow rates is depicted in Fig. 14.6. The dry spots on the pipe wall surface at low gas flow rates are quite evident, and with increase in the gas flow rate, liquid droplets are observed to entrain the gas core. Due to the physical resemblance of falling film with the annular flow, some researchers have classified this flow pattern as a sub-region of annular flow. However, there is a

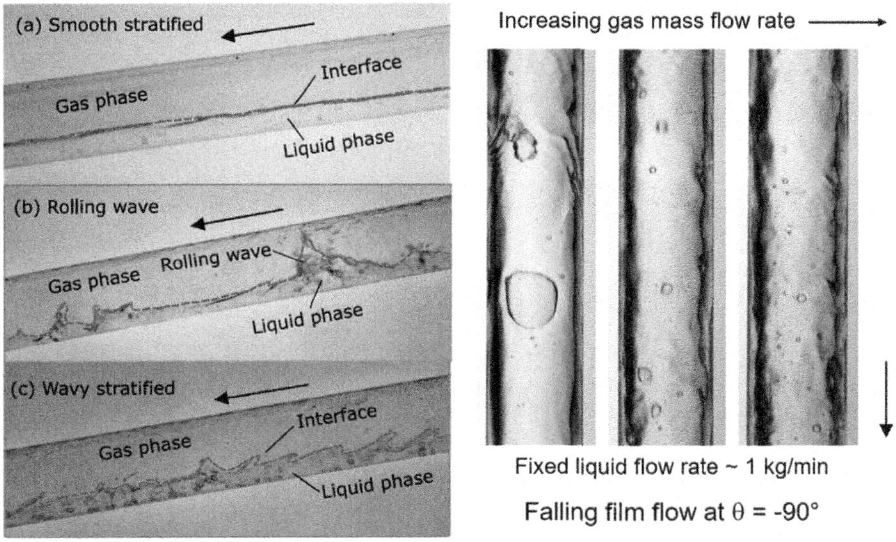

Fig. 14.6 Physical structure of stratified flow pattern in downward pipe inclinations and falling film flow in vertical downward flow. (Flow visualization done at the Two-Phase Flow Lab, OSU, Stillwater, OK)

significant difference in these two flow patterns in terms of the intensity of gas core turbulence, interfacial shear, thickness of liquid film, and the amount of liquid entrainment.

Annular Flow Annular flow appears in the form of a turbulent gas core surrounded by the thin/thick (depending upon liquid flow rates) and wavy liquid film. The liquid film distribution in annular flow is affected by the pipe orientation such that for horizontal flow, the asymmetric liquid film is thick at the pipe bottom, while it is evenly distributed around the circumference in the case of vertical two-phase flow; see Fig. 14.7. Based on the visual observations in Fig. 14.7, the distribution of film thickness around the pipe circumference (or film thickness at the pipe top and bottom wall for a 2D flow visualization) is found to depend on the gas and liquid phase flow rates and the pipe orientation. For a fixed liquid and gas flow rate, the effect of change in the pipe orientation from horizontal is to make the liquid film distribution more symmetric, whereas for a fixed pipe orientation and liquid flow rate, the effect of increase in the gas flow rate is to reduce the liquid film thickness in contact with the pipe wall. This effect of pipe inclination and gas flow rate on the circumferential distribution of liquid film thickness plays a pivotal role in the determination of the local as well as overall two-phase heat transfer coefficient.

Depending upon the gas and liquid flow rates, the gas-liquid interface may be stable or perturbed by small amplitude waves causing the liquid droplets to detach from liquid film and enter the turbulent gas core known as the liquid entrainment phenomenon. The amplitude and frequency of interfacial disturbance waves and the

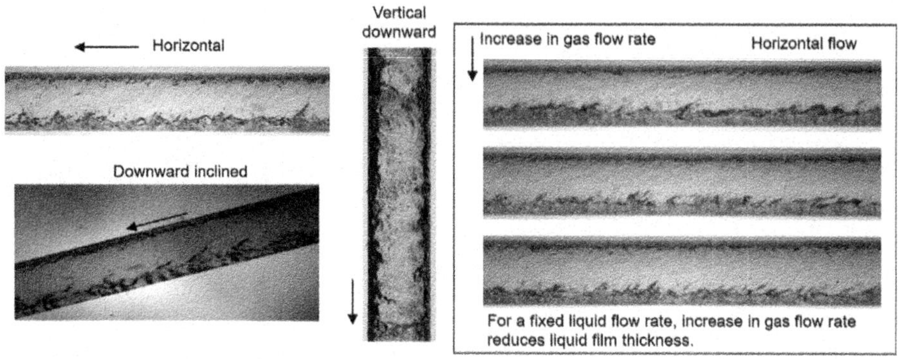

Fig. 14.7 Annular flow in horizontal, downward inclined, and vertical downward two-phase flow. (Flow visualization done at the Two-Phase Flow Lab, OSU, Stillwater, OK)

amount of liquid entrainment depend on the phase flow rates, fluid properties, pipe geometry, and pipe inclination. A special form of annular flow also known as annular-mist flow exists when most of the liquid phase initially flowing in the form of film is entrained into the central gas core by the shearing action of the fast-moving gas core. This type of flow pattern is mostly encountered in applications involving high system pressure and high heat flux conditions.

Flow Pattern Maps

The flow pattern maps or flow maps serve as a tool to estimate the span and sequence of the appearance of different flow patterns with change in gas and liquid flow rates for a given set of flow conditions. The flow patterns and their transitions are defined qualitatively based on visual observations, and hence the accurate mapping of the transition between different flow patterns is highly subjective and totally depends upon the observer's perception. Moreover, the transition between different flow patterns is gradual and sensitive to several parameters such as pipe diameter, pipe orientation, and fluid properties making it extremely difficult to have a universal flow pattern map that can correctly predict the existence of different flow patterns covering a wide range of two-phase flow conditions. Although some researchers have attempted to develop quantitative methods such as the probabilistic flow regime mapping and power spectral density analysis of the acquired pressure signal to predict the existence of certain flow patterns, these methods are not universal and are mostly limited to certain flow conditions. Moreover, these methods also rely upon prior information of visual inspection of flow patterns.

Despite the lack of the universal flow pattern map, some of the existing most referred to flow pattern maps are those of Mandhane et al. (1974) and Taitel and Dukler (1976) for horizontal two-phase flow and Hewitt and Roberts (1969) for vertical upward flow. These flow pattern maps provide an idea about the sequence, extent, and transitions between different flow patterns and serve as a guideline in understanding the appearance of flow patterns as a function of gas and liquid flow rates. In general, keeping in mind the uncertainty, sensitivity, and qualitative identification of flow patterns, the following facts and limitations about the flow pattern maps must be perceived: (a) it is difficult to have a universal flow pattern map that can accurately predict the transition of one flow pattern to another for a wide range of two-phase flow conditions; (b) unlike represented in most flow pattern maps, the transition from one flow pattern to another is always gradual and cannot be presented in the form of a thin continuous line; and (c) correct identification of a flow pattern in the vicinity of the transition line depends on the judgment of an individual. For more discussion on different existing flow pattern maps, refer to Ghajar and Bhagwat (2017).

The purpose of this section is to present an overview of the effect of pipe orientation on the transition between different two-phase flow patterns. The flow pattern maps for different pipe orientations are presented with reference to the horizontal flow pattern map. It should be mentioned that the flow pattern maps presented in this section were developed for two-phase flow of air-water through a 12.7 mm I.D. pipe at different orientations and are applicable for similar flow conditions. The flow pattern maps for each orientation in upward and downward inclinations are presented in Figs. 14.8, 14.9, 14.10, and 14.11. These flow pattern maps are based on the experimental observations and data collected at the Two-Phase Flow Lab, Oklahoma State University (OSU).

In the case of both upward (Fig. 14.8) and downward (Fig. 14.10) inclined two-phase flow, regardless of the pipe orientation, the bubbly flow occurs for low values of gas and high values of liquid mass flow rates, whereas the annular flow exists for low liquid and high values of gas mass flow rates. In the case of upward inclined two-phase flow, slug flow exists at low to moderate liquid flow rates and low gas flow rates. The boundary between slug and intermittent flow is influenced by the change in pipe orientation. It is found that the increase in pipe orientation from horizontal causes this transition boundary to shift toward lower gas flow rates. This is possibly because increase in the pipe orientation assists buoyancy and facilitates formation (agglomeration of gas bubbles) and motion of gas slugs.

The transition between intermittent and annular flow regime is virtually unaffected by the pipe orientation due to the fact that this is a shear-driven flow regime where the pipe orientation has a little effect on the physical structure of this flow pattern. However, during intermittent flow regime, an early transition from wavy slug to wavy annular flow could be observed with increase in the upward pipe inclination. The transition boundary between bubbly and slug flow is also found to be affected by the change in pipe orientation. The increase in pipe orientation from horizontal to upward direction causes early transition from slug to bubbly flow. The transition between bubbly-slug, slug-intermittent, and intermittent-annular is evident from Fig. 14.9. In the case of horizontal flow at low liquid flow rates, a short region of intermittent flow exists in the transition from stratified to annular. The intermittent flow pattern identified in this narrow range of gas flow rates is mostly wavy annular

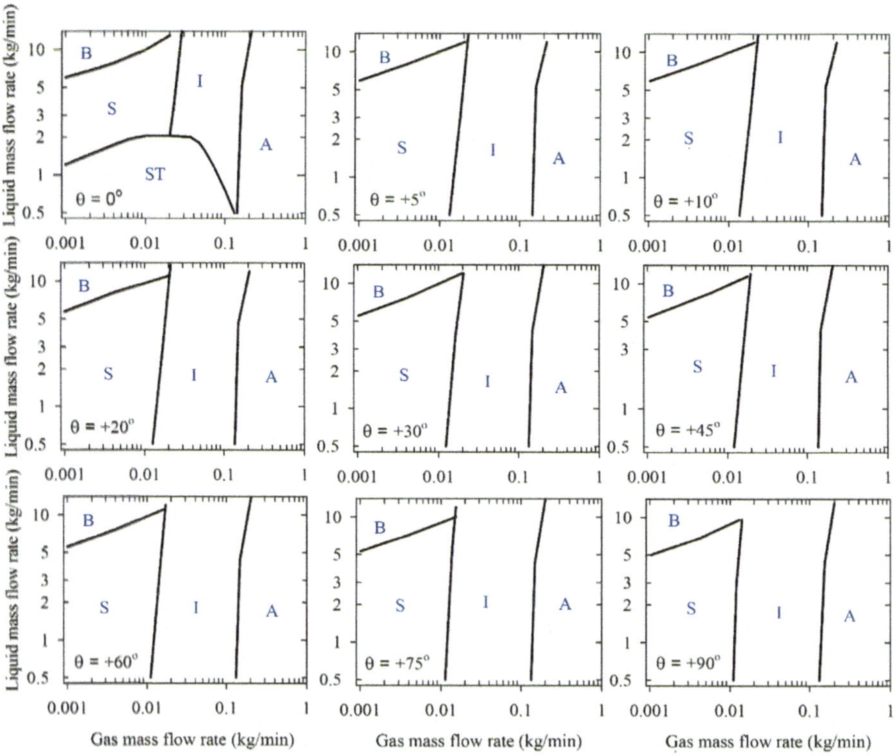

Fig. 14.8 Flow pattern map for upward inclined two-phase flow. (Based on flow visualization done at the Two-Phase Flow Lab, OSU, Stillwater, OK) (B bubbly, S slug, I intermittent, A annular)

in nature. At low liquid flow rates approximately 0.5 kg/min, stratified flow directly transits to annular flow without entering into intermittent flow regime.

In the case of downward inclined two-phase flow (Fig. 14.10), the transition between stratified-slug and slug-intermittent flow regimes is seen to be significantly affected by the change in pipe orientation. As mentioned earlier, the stratified flow appears for low liquid and low to moderate gas flow rates and exists for horizontal and all downward inclinations considered in this study. It is found that the increase in downward pipe inclination expands the region occupied by the stratified flow on the flow pattern map toward higher liquid flow rates and at the expense of slug and intermittent flow regimes.

It is seen from Fig. 14.11 that for low gas flow rates ($\dot{m}_G < 0.01$ kg/min), the transition boundary between stratified and slug flow consistently shifts toward higher liquid flow rates with increase in the gas flow rate up to $-45°$ from horizontal. With further increase in downward pipe inclination, the transition line between stratified and slug flow shifts back to lower liquid flow rates. For lower gas flow rates, it is found that the stratified flow exists for liquid flow rates as high as 5.5 kg/min for -45° pipe orientation. With further increase in pipe orientation at

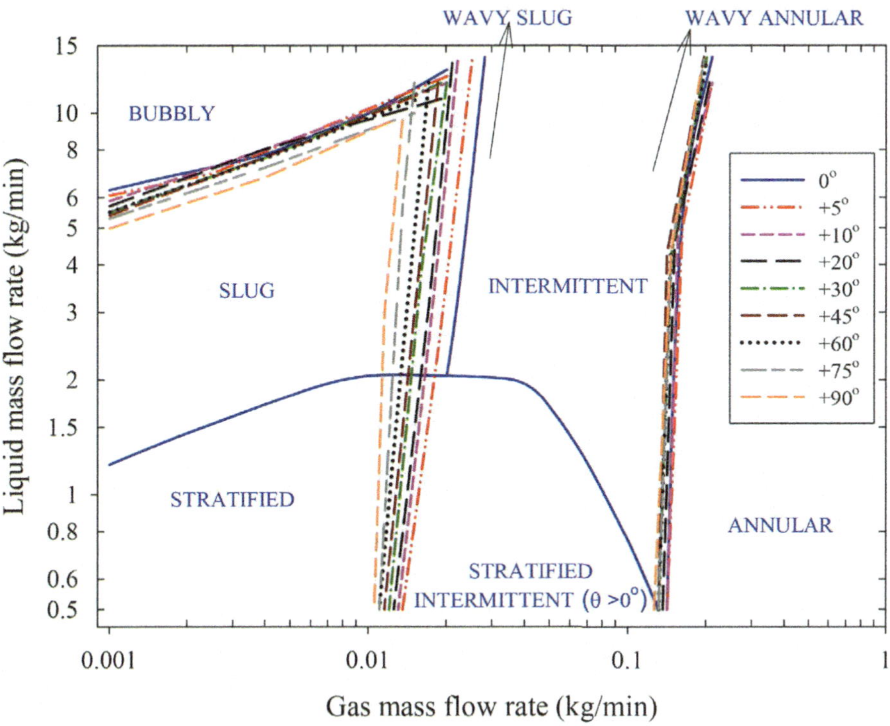

Fig. 14.9 Combined flow pattern map for upward inclined pipe orientations. (Based on flow visualization done at the Two-Phase Flow Lab, OSU, Stillwater, OK)

-60°, the stratified flow cannot sustain beyond liquid flow rate of 4 kg/min. For moderate gas flow rates ($0.01 \leq \dot{m}_G \leq 0.05$ kg/min), the transition between stratified and intermittent flow regime is very gradual (as a function of liquid flow rate and relatively independent of gas flow rate) until a threshold value of the gas flow rate is attained where the gas-liquid surface becomes significantly unstable and the liquid phase is splashed frequently on the pipe top wall. This is the point where the flow pattern gradually shifts to intermittent (wavy annular) flow and the transition line between stratified and intermittent flow patterns is steep and very sensitive to the change in gas and liquid flow rates. Figure 14.11 shows that this threshold value of gas mass flow rate is a function of pipe orientation, i.e., it increases until $\theta = -45^{\circ}$ and then decreases thereafter. Note that for $\theta = -75^{\circ}$, falling film flow coexists with stratified flow at low liquid and low to moderate gas flow rates. At this near vertical pipe orientation, the edges of gas-liquid interface climb up the tube periphery such that it occupies most of the pipe circumference. With increase in liquid flow rate, a thin film of liquid is observed at the pipe upper wall, and falling film flow is said to exist. As mentioned earlier, falling film flow is a special case of stratified flow for vertical and near vertical downward pipe orientations, and hence no distinction is made between these two-phase flow patterns on the flow map in Fig. 14.10. For

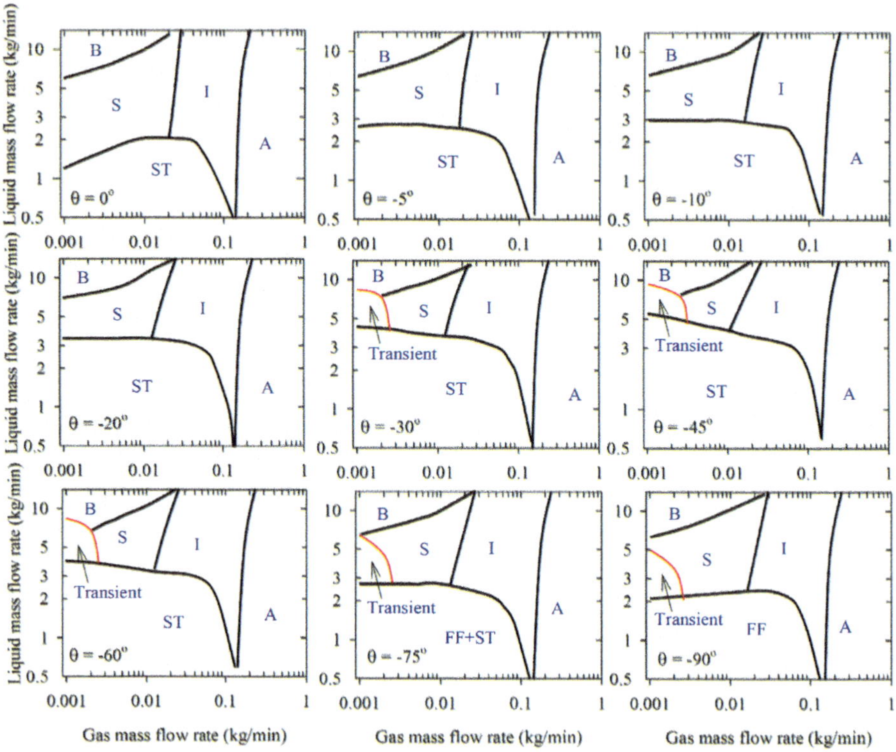

Fig. 14.10 Flow pattern map for downward inclined two-phase flow. (Based on flow visualization done at the Two-Phase Flow Lab, OSU, Stillwater, OK) (B bubbly, S slug, I intermittent, ST stratified, A annular, FF falling film)

vertical downward pipe inclination ($\theta = -90^{\circ}$), stratified flow is always in the form of falling film due to flow symmetry.

In the case of transition between slug and intermittent flows, with increase in the downward pipe inclination, the transition boundary is found to shift toward lower gas flow rates up to $\theta = -45^{\circ}$ and then slightly moves toward higher gas flow rates thereafter. Similar to upward inclined flows, the transition between intermittent and annular flows is virtually not affected by the change in the pipe orientation. For very low liquid flow rates $\dot{m}_L < 1$ kg/min, it is found that the flow pattern shifts directly from stratified to annular without entering the intermittent flow regime. Typically for $\dot{m}_L < 2$ kg/min, the intermittent region is very narrow and is wavy annular in nature.

Another interesting characteristic of gas-liquid two-phase flow observed in steeper downward pipe inclinations is the transient behavior of the flow patterns. As graphically illustrated in Fig. 14.12, this transient behavior results in simultaneous or sequential existence of multiple flow patterns (usually stratified, slug, and bubbly flows) in the test section. This phenomenon found in steeper downward pipe inclinations is verified visually as well as through the two-phase pressure drop signal

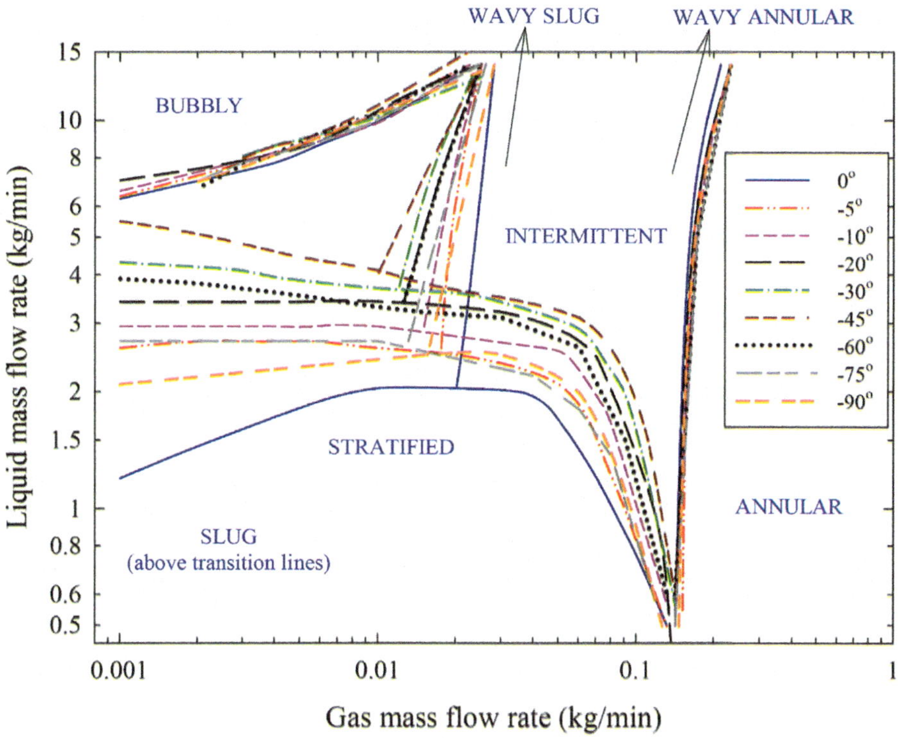

Fig. 14.11 Combined flow pattern map for downward inclined pipe orientations. (Based on flow visualization done at the Two-Phase Flow Lab, OSU, Stillwater, OK)

recorded over a period of time as shown in Fig. 14.13. Comparison of visual observations with the pressure drop signal shows that the flat portion of the signal corresponds to the stratified flow, while the spiked portion of the signal toward negative values indicates the presence of bubbly and/or slug flow patterns. Note that the total pressure drop is negative because of higher negative hydrostatic head compared to positive frictional head in downward pipe inclinations. The pressure drop signal over a period of time confirms that these flow patterns (stratified/bubbly/slug) repeat one after another continuously over a period of time without establishing one single flow pattern. However, with increase in the gas flow rate, the time period between the occurrences of multiple flow patterns decreases. The flow pattern map reported in Fig. 14.10 shows that with increase in the gas flow rate, the transient behavior gradually diminishes, and the slug/intermittent flow behavior of the two-phase flow is restored. A plausible explanation for transient behavior of flow patterns can be given considering the two-phase flow physics in downward pipe inclinations. Near the stratified/non-stratified transition line, the downward moving gas-liquid interface is unstable. The unstable interface frequently splashes liquid on the top wall of the pipe causing frequent bridging of pipe cross section. Thus, the elongated gas slug gets trapped in the test section that appears to remain virtually

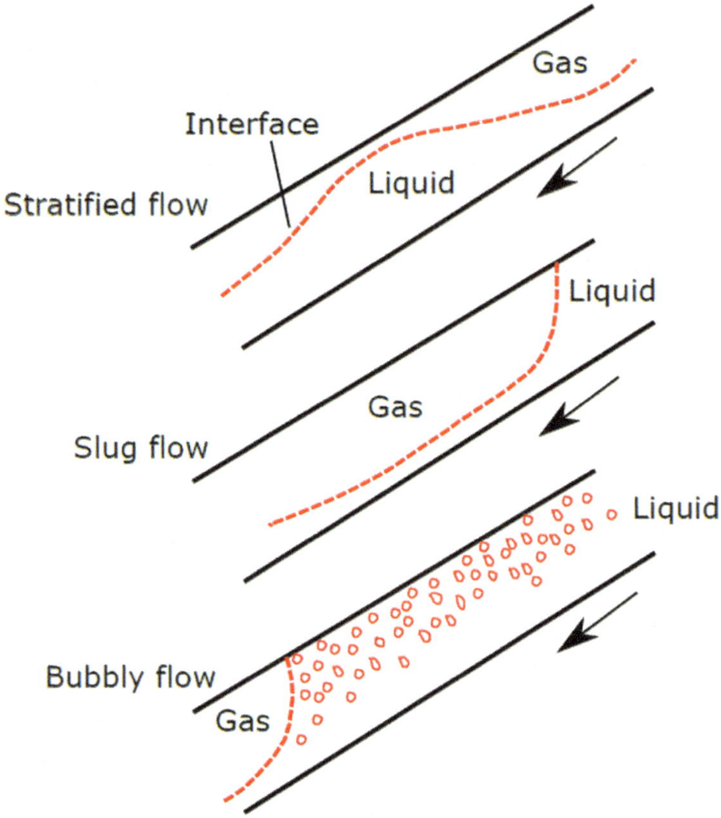

Fig. 14.12 Schematic of the general appearance of flow patterns during transient behavior of downward inclined gas-liquid two-phase flow. (Based on flow visualization done at the Two-Phase Flow Lab, OSU, Stillwater, OK)

stationary (hence also referred to as a pseudo-stationary slug) under the influence of buoyancy force acting on the gas phase in a direction opposite to that of the mean flow. The two-phase mixture coming from the upstream direction is accumulated over this pseudo-stationary slug, which further slips past the elongated gas bubble, and eventually forces the gas bubble to move in the downstream direction. During this process, the liquid phase churns the elongated gas bubble causing it to disintegrate and move further in the form of discrete bubbles. After the gas pocket is pushed downstream, the incoming two-phase flow is again in the form of stratified flow pattern, and this process repeats continuously showing stratified, slug, and bubbly flow patterns periodically. A change in flow pattern from one to another (with significantly different flow structure) causes acceleration and deceleration of the two-phase mixture that results in pressure loss and pressure recovery in the two-phase flow system and hence results in an abrupt change in the pressure drop signal. It is found that this transient nature of two-phase flow aggravates with

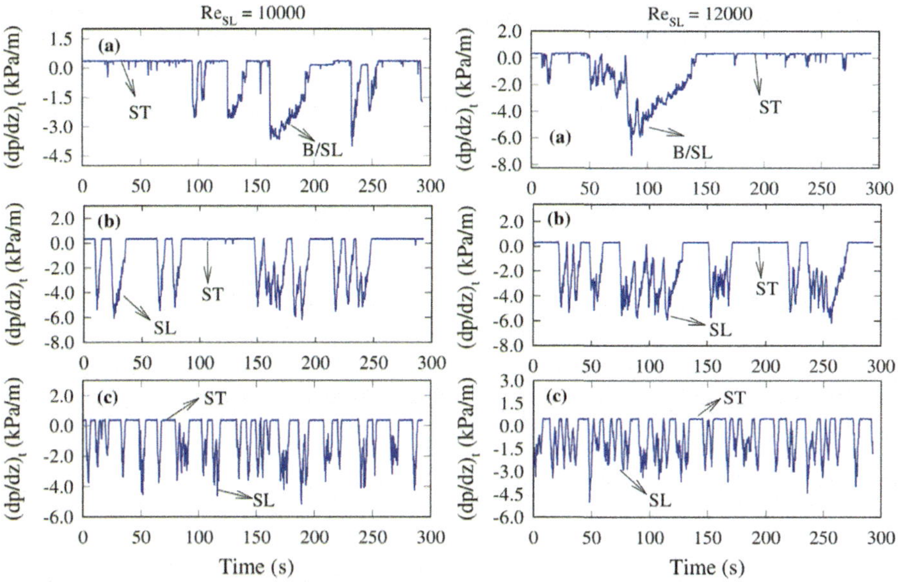

Fig. 14.13 Pressure drop signal to identify the transient behavior in downward inclined two-phase flow $\theta = -45^{\circ}$ (**a**) $Re_{SG} = 180$, (**b**) $Re_{SG} = 360$, and (**c**) $Re_{SG} = 450$ (B bubbly, SL slug, and ST stratified). (Based on the data collected at the Two-Phase Flow Lab, OSU, Stillwater, OK)

increase in downward pipe inclination and propagates toward slightly higher gas flow rates. At low gas flow rates, the time occupied by the gas slug in the test section between two consecutive stratified flow appearances is higher than that at higher gas flow rates. Thus, it is clear that this transient behavior is undesirable and must be avoided in practical applications involving downward inclined gas-liquid two-phase flow.

Flow Pattern Transitions

Two-phase flow literature reports both mechanistic and empirical models to predict transition between different flow patterns. However, these transition models are developed based on the visual observations and limited data bank and are based on several simplifying assumptions that render uncertainty in their practical applications specifically if the two-phase flow under consideration is on the verge of transition. The following section provides some of the existing flow pattern transition models available in the two-phase flow literature. Considering the vague definition of intermittent flow pattern, the transition models listed here are only for well-defined flow regimes such as bubbly, slug, stratified, and annular flows.

Transition Between Bubbly and Slug Flows The transition from bubbly to slug flow occurs when bubbles no longer move independently and tend to coalesce to form a larger bubble usually known as Taylor bubble. The transition from bubbly to slug flow can be achieved with decrease in liquid flow rate at a constant gas flow rate or with increase in gas flow rate at a constant liquid flow rate. In general, two-phase flow literature shows that this transition from bubbly to slug flow occurs for void fraction in a range of 0.25–0.3. The transition from bubbly to slug flow will occur if Eq. (14.1) suggested by Taitel et al. (1980) is satisfied. This relationship is obtained by assuming that the maximum void fraction during transition from bubbly to slug flow is approximately 0.25 and that the bubble velocity is equal to the slip velocity ($U_b \cong U_G - U_L$). The bubble velocity at different pipe inclinations represented by Eq. (14.2) is the modified form of Haramathy (1960) equation for bubble rise velocity suggested by Taitel et al. (1980) and Barnea (1987):

$$U_{SL} \leq 3U_{SG} - 1.32\left(\frac{\sigma g(\rho_L - \rho_G)}{\rho_L^2}\right)^{0.25} \sin\theta \quad \Big\} \quad (+60^\circ \leq \theta \leq +90^\circ) \tag{14.1}$$

$$U_b = 1.53\left[\frac{g\sigma(\rho_L - \rho_G)}{\rho_L^2}\right]^{0.25} \sqrt{1-\alpha_G}\sin\theta \tag{14.2}$$

Note that Eq. (14.1) is recommended for use for near vertical pipe orientations ($+60^\circ \leq \theta \leq +90^\circ$) since as per Taitel et al. (1980) and Barnea (1987), it is difficult for bubbly flow to exist in horizontal and near horizontal pipe orientations. Moreover, Taitel et al. (1980) also reported that the bubbly flow would exist if the pipe diameter of two-phase flow under consideration follows Eq. (14.3):

$$D > 19\left(\frac{(\rho_L - \rho_G)\sigma}{\rho_L^2 g}\right)^{0.5} \tag{14.3}$$

Following a similar concept, Mishima and Ishii (1984) presented a model to predict the transition between bubbly and slug flow (valid for vertical upward flow) of the form shown in Eq. (14.4). The transition from bubbly to slug flow would occur if Eq. (14.4) is satisfied. The parameter C_o is expressed as $C_o = 1.2 - 0.2\sqrt{\rho_G/\rho_L}$ such that for near atmospheric operating conditions, $C_o \approx 1.2$. The bubble velocity equation (Eq. 14.5) used by Mishima and Ishii (1984) is similar to Eq. (14.2), however with different multiplying factors and exponents:

$$U_{SL} \leq \left(\frac{3.33}{C_o} - 1\right)U_{SG} - \frac{0.76}{C_o}\left(\frac{\sigma g \Delta\rho}{\rho_L^2}\right)^{0.25} \quad \Big\} \quad (\theta = +90^\circ) \tag{14.4}$$

$$U_b = 1.41\left[\frac{g\sigma(\rho_L - \rho_G)}{\rho_L^2}\right]^{0.25}(1-\alpha_G)^{1.75} \tag{14.5}$$

For vertical upward flow, a comparison between Eqs. (14.1) and (14.4) shows that transition from bubbly to slug flow will occur at relatively lower U_{SL} using the criteria of Mishima and Ishii (1984). This discrepancy is essentially due to their assumption of $\alpha_G \approx 0.3$ during the transition and a slightly different form of bubble velocity (U_b) used by Mishima and Ishii (1984). As mentioned earlier, bubbly flow pattern can be further classified as bubbly-slug and dispersed bubbly flow. Thus, the bubbly flow predicted by Eqs. (14.1) and (14.4) may contain both bubbly-slug and dispersed bubbly two-phase flow. Unlike bubbly (elongated bubbly/slug bubbly) flow, the dispersed form of bubbly flow is known to exist in the entire range of pipe orientations, and its existence could be determined using the transition model of Barnea (1986). The model of Barnea (1986, 1987) requires calculation of three different diameters, namely, maximum bubble diameter (d_{max}), critical bubble diameter above which bubble is deformed (d_{def}), and critical bubble diameter below which migration of bubbles is prevented (d_{migr}). These different bubble diameters could be determined from Eqs. (14.6, 14.7, and 14.8), respectively. Thus, according to Barnea (1986), dispersed bubbly flow would exist if both $d_{max} < d_{def}$ and $d_{max} < d_{migr}$ conditions are satisfied. For these conditions typically at high liquid flow rates, dispersed bubbly flow exists since the turbulent fluctuations are capable to break down bubbles to spherical shape and also suppress the bubble coalesce that leads to slug formation:

$$d_{\max} = \left[0.725 + 4.15\left(\frac{U_{SG}}{U_M}\right)^{0.5}\right]\left(\frac{\sigma}{\rho_L}\right)^{0.6}\left(\frac{2f_M U_M^3}{D}\right)^{-0.4} \tag{14.6}$$

$$d_{\text{def}} = 2\left[\frac{0.4\sigma}{(\rho_L - \rho_G)g}\right]^{0.5} \tag{14.7}$$

$$d_{\text{migr}} = 0.375\left(\frac{f_M U_M^2 \rho_L}{g\cos\theta(\rho_L - \rho_G)}\right) \tag{14.8}$$

In Eq. (14.6), f_M is friction factor based on two-phase mixture velocity U_M and can be calculated using any single-phase friction factor correlation. Considering the packing density and packing configuration of bubbles, Barnea (1986) showed that this transition model for dispersed bubbly flow is valid for void fraction in a range of $0 < \alpha_G \leq 0.52$. For dispersed bubbly flow, the slippage at gas-liquid interface is quite small, and it can be assumed that $\alpha_G \approx \lambda$. Thus, the transition from dispersed bubbly to slug or intermittent flow would occur at $\lambda = U_{SG}/U_M > 0.52$.

Transition Between Stratified and Non-stratified Flows As established earlier, the transition boundary between stratified-slug and stratified-annular flow is very sensitive to the change in downward pipe inclination. Considering the dependency and practical difficulty in the modeling of two-phase flow variables such as void fraction, pressure drop, and non-boiling two-phase heat transfer in stratified two-phase flow, it is desirable to have a stratified flow pattern-specific model to predict these parameters. As such, it is first required to be able to correctly predict the existence of stratified flow as a function of phase flow rates, fluid properties, and pipe orientation. This section presents a brief review of the mechanistic flow pattern transition model of Taitel and Dukler (1976) and an empirical model of Bhagwat and Ghajar (2015a) to predict the transition to stratified flow in horizontal and downward pipe inclinations.

The most referred transition model in two-phase flow literature to separate out stratified flow from non-stratified flow is that of Taitel and Dukler (1976). Their transition model is based on the momentum balance between the two phases, assumes a smooth and flat gas-liquid interface such that $f_i/f_G \approx 1$, and gives the transition line equation that requires use of a graphical solution. As per the model of Taitel and Dukler (1976), stratified flow will exist if Eq. (14.9) is satisfied where F is defined by Eq. (14.10):

$$F^2\left[\frac{1}{\left(1-\widetilde{h}_L\right)^2}\frac{\widetilde{U}_G\sqrt{1-\left(2\widetilde{h}_L-1\right)^2}}{\widetilde{A}_G}\right]<1 \tag{14.9}$$

$$F=\sqrt{\frac{\rho_G}{\rho_L-\rho_G}}\frac{U_{SG}}{\sqrt{gD}\cos\theta} \tag{14.10}$$

The non-dimensional parameters $\widetilde{A}_G$ and $\widetilde{U}_G=(\pi/4)/\widetilde{A}_G$ are functions of non-dimensional liquid height ($\widetilde{h}_L=h_L/D$). The non-dimensional pipe cross-sectional area occupied by the gas phase is calculated using Eq. (14.11). As shown in Fig. 14.14, the non-dimensional liquid height $\widetilde{h}_L$ is obtained through the graphical solution of $\widetilde{h}_L$ vs X for different values of Y. The parameters X and Y are obtained from Eqs. (14.12) and (14.13), respectively:

$$\widetilde{A}_G=0.25\left[\cos^{-1}\left(2\widetilde{h}_L-1\right)-\left(2\widetilde{h}_L-1\right)\sqrt{1-\left(2\widetilde{h}_L-1\right)^2}\right] \tag{14.11}$$

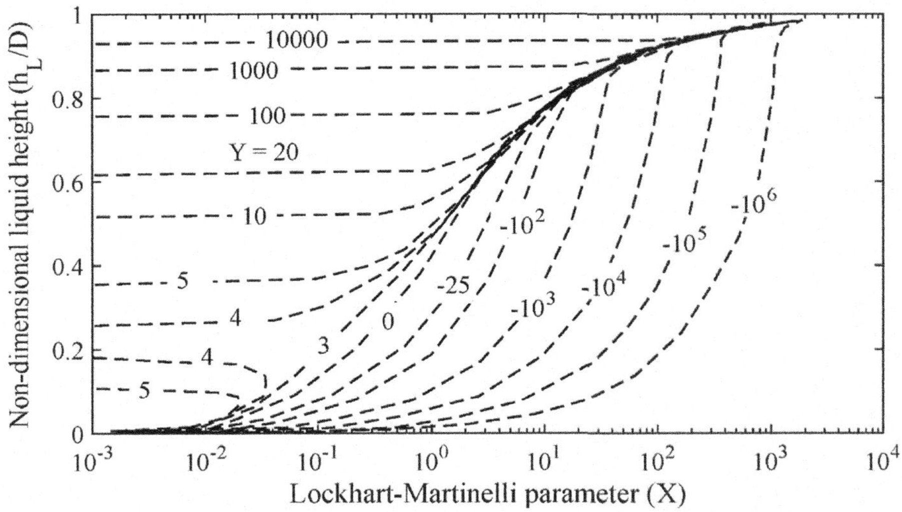

Fig. 14.14 Graphical solution to determine liquid-level height in stratified flow. (Adapted from Taitel and Dukler 1976)

$$X = \sqrt{\frac{(dp/dz)_L}{(dp/dz)_G}} \tag{14.12}$$

$$Y = \frac{(\rho_L - \rho_G)g\sin\theta}{(dp/dz)_G} \tag{14.13}$$

The application of Taitel and Dukler (1976) transition model to predict stratified flow is rather tedious due to the use of a graphical solution to determine non-dimensional liquid height $\widetilde{h}_L$ and consequently other parameters such as $\widetilde{U}_G$ and $\widetilde{A}_G$. The use of a graphical solution to determine h_L/D could be circumvented by adopting an iterative solution technique. For more details, refer to Taitel and Dukler (1976).

Taitel and Dukler (1976) further attempted to classify stratified flow as smooth stratified and wavy stratified flows. They suggested that the transition between smooth and wavy stratified flow is associated with phenomenon of wave generation at the gas-liquid interface and that the smooth stratified flow exists for superficial gas velocities (U_{SG}) that satisfy Eq. (14.14). In practice, smooth stratified flow exists for a very narrow range of two-phase flow conditions and is of less practical significance compared to other flow patterns.

$$U_{SG,t} \geq \left(\frac{4\nu_L g \Delta\rho \cos\theta}{0.01\rho_L U_L}\right)^{0.5} \tag{14.14}$$

A simplifying approximation to the Taitel and Dukler (1976) model has been presented by Cheng et al. (1988). They approximated the trend of Eq. (14.9) and presented a criterion given by Eq. (14.15) to identify the existence of stratified flow pattern in horizontal pipes in terms of Lockhart and Martinelli (1949) parameter applicable for turbulent-turbulent gas-liquid two-phase flow where Fr is the modified Froude number given by Eq. (14.16):

$$Fr_{SG} \leq \left(\frac{1}{0.65 + 1.11X_{tt}^{0.6}}\right)^2 \tag{14.15}$$

$$Fr_{SG} = \frac{Gx}{\sqrt{gD\rho_G(\rho_L - \rho_G)}} = \frac{U_{SG}}{\sqrt{gD}}\sqrt{\frac{\rho_G}{\rho_L - \rho_G}} \tag{14.16}$$

Combining Eqs. (14.15) and (14.16) gives the criteria for existence of stratified flow in terms of two-phase mixture mass flux. Equation (14.17) represents the maximum value of two-phase mixture mass flux (for a given quality and fluid properties) below which stratified flow will always exist and is recommended for use only in the case of horizontal two-phase flow:

$$\left.G_{max} = \frac{\sqrt{gD\rho_G(\rho_L - \rho_G)}}{x}\left(\frac{1}{0.65 + 1.11X_{tt}^{0.6}}\right)^2\right\} \quad (\theta = 0^\circ) \tag{14.17}$$

Using the concept of growth of disturbance waves on the gas-liquid interface of stratified flow, Mishima and Ishii (1980) presented a criterion to determine the limit of stratified flow in horizontal two-phase flow as given by Eq. (14.18). Note that use of this correlation also requires determination of liquid height (h_L) which could be obtained using graphical solution given by Fig. 14.14:

$$\left.U_G - U_L = 0.487\sqrt{\frac{g(D_h - h_L)(\rho_L - \rho_G)}{\rho_G}}\right\} \quad (\theta = 0^\circ) \tag{14.18}$$

The recent work of Bhagwat and Ghajar (2015a) expressed by Eq. (14.19) provides an empirical correlation to predict the existence of stratified flow in horizontal and downward pipe inclinations. Similar to the approach of Cheng et al. (1988) and Bhagwat and Ghajar (2015a), correlation attempts to approximate the trends of Taitel and Dukler (1976) in terms of non-dimensional parameters of Fr_{SG} and X and is developed based on the experimental data in horizontal and downward pipe inclinations. The Froude number in Eq. (14.19) is similar to that given by Cheng et al. (1988) in Eq. (14.16). The variables C_1, C_2, C_3, and C_4 required to solve Eq. (14.19) are expressed by Eqs. (14.20, 14.21, 14.22, and 14.23). The non-dimensional pipe diameter (D^+) is the normalized pipe diameter defined by $D^+ = D/0.0254$:

$$Fr_{SG} \le (0.6 + C_2)\exp\left(-C_1 C_2 X^{C_3}\right) X^{-C_4} \quad \left\} \begin{array}{l} \mathrm{D} = 8.9 - 300 \ \ \mathrm{mm} \\ \rho_L = 780 - 1420 \ \ \mathrm{kg/m^3} \\ \rho_G = 1.2 - 35 \ \ \mathrm{kg/m^3} \\ \theta = 0^\circ \text{ to } -90^\circ \\ \mu_L = 0.0002 - 0.08 \ \ \mathrm{Pa\ s} \end{array} \right. \tag{14.19}$$

$$C_1 = 1.3\ln\left(D^+\right) + 2.5 \tag{14.20}$$

$$C_2 = \frac{C_4^{0.65}}{\left[1 + 2\sin\left(2|\theta|\right) \times \left(1 + 10\tanh\left(1/|\theta|\right)\right)\right]} \tag{14.21}$$

$$C_3 = \begin{cases} 0.65\left(D^+\right)^{-0.15}\left(\left(\rho_L - \rho_G\right)/1000\right) & : \rho_L < 1000 \ \ \mathrm{kg/m^3} \\ 0.65\left(D^+\right)^{-0.15} & : \rho_L \ge 1000 \ \ \mathrm{kg/m^3} \end{cases} \tag{14.22}$$

$$C_4 = 0.2\sqrt{1/D^+} \tag{14.23}$$

Variables C_1 and C_2 account for the combined effect of pipe diameter and pipe orientation on the shift in transition line. The shift in the transition line by including these variables is in accordance with the observations of Nguyen (1975), Barnea et al. (1982), Crawford et al. (1985), and Ghajar and Bhagwat (2014a) that the increase in downward pipe inclination shifts the transition between stratified and non-stratified flow pattern toward higher values of liquid flow rates or alternatively toward higher values of X. Variable C_3 ensures that the transition line between stratified and non-stratified flow is shifted toward higher values of X with a decrease in liquid phase density. The variables C_1, C_2, C_3, and C_4 are adjusted such that C_1, C_2, and C_3 control the slope of the transition line in the buoyancy-driven region (large values of X) while C_4 controls the slope of the transition line in inertia-driven region (small values of X).

The accuracy of the model of Bhagwat and Ghajar (2015a) was compared against the model of Taitel and Dukler (1976) for both horizontal and downward inclinations, while the model of Cheng et al. (1988), as recommended by the authors, was

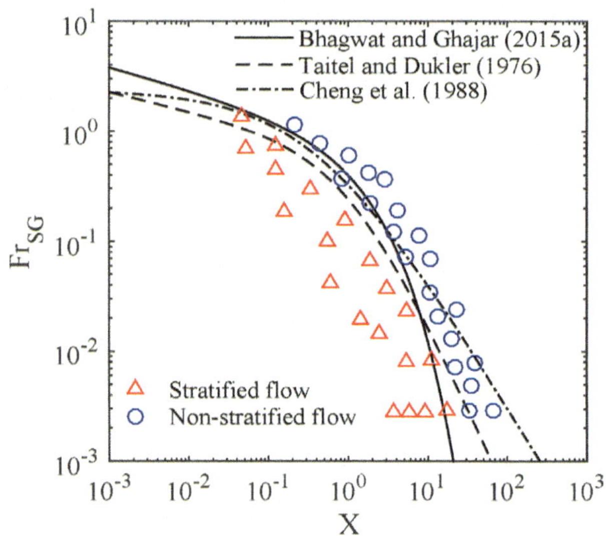

Fig. 14.15 Trends of different stratified flow transition models for horizontal two-phase flow of air-water in 12.7 mm I.D. pipe. (Based on data collected at the Two-Phase Flow Lab, OSU, Stillwater, OK)

only used for the horizontal orientation. The comparisons were made using flow visualization data collected from more than 16 data sources consisting of 8 fluid combinations (air-water, air-oil, R-134a, R-113, R22/R-410A, air-K_2CO_3, air-glycerin, air-kerosene), pipe diameter in a range of 8.9–300 mm, liquid density in a range of 780–1420 kg/m^3, gas density in a range of 1.2–35 kg/m^3, liquid dynamic viscosity in a range of 0.0002–0.08 Pa·s, and all downward pipe inclinations including horizontal. For the details of these comparisons, refer to Bhagwat and Ghajar (2015a).

A graphical comparison between Taitel and Dukler (1976), Cheng et al. (1988), and Bhagwat and Ghajar (2015a) for horizontal and downward inclined two-phase flow of air-water in 12.7 mm I.D. pipe is shown in Figs. 14.15 and 14.16, respectively. The experimental data for stratified flow indicated by symbol "Δ" consists of both smooth and wavy stratified flows, while the non-stratified flow indicated by symbol "○" may consist of any or all of the bubbly, slug/intermittent, and annular flow patterns.

Note that Bhagwat and Ghajar (2015a) model given by Eq. (14.19) is of empirical form and merely attempts to mimic the trends of Taitel and Dukler (1976) model using $Fr_{SG} - X$ coordinates. Nevertheless, considering the practical difficulty of using a graphical/iterative solution involved in Taitel and Dukler (1976) model, Eq. (14.19) is explicit in nature and may be regarded as a quick method for hand calculations and to get an estimate of the existence of stratified flow in horizontal and downward pipe inclinations.

Transition Between Annular and Non-annular Flows Typically, annular flow shares the transition boundary with intermittent (churn/wavy annular) flow patterns; see Figs. 14.8, 14.9, 14.10, and 14.11. Moreover, in the case of horizontal and downward pipe inclinations, annular flow also shares a transition boundary with

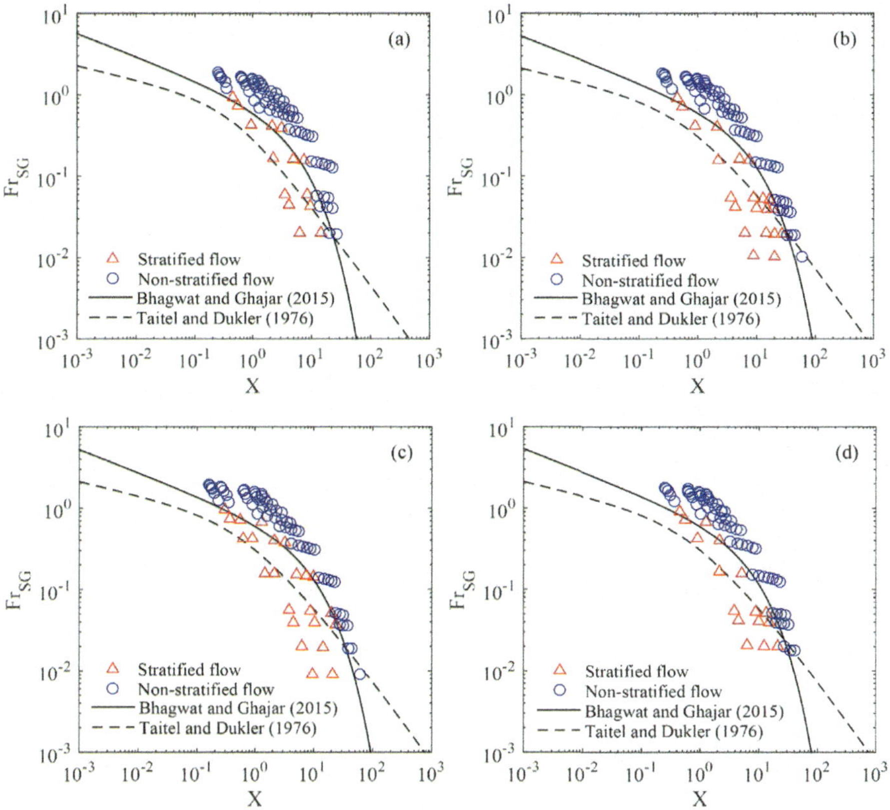

Fig. 14.16 Trends of different stratified flow transition models for downward inclined two-phase flow of air-water in 12.7 mm I.D. pipe (**a**) −10°, (**b**) −30°, (**c**) −45°, (**d**) −60° (Based on data collected at the Two-Phase Flow Lab, OSU, Stillwater, OK)

stratified flow pattern; see Figs. 14.10 and 14.11. In annular flow regime, the inertial forces greatly exceed the effect of gravitational forces, and hence it is usually assumed that the change in pipe orientation will have little effect on the transition between annular and other flow regimes. However, pipe diameter and fluid properties do affect the transition to annular flow regime, and hence, it is desired to have a transition equation for annular flow regime as a function of both pipe diameter and fluid properties. One such transition criterion proposed by Weisman and Kang (1981) is expressed by Eq. (14.24) and can be used for all pipe orientations. This criterion, however, at high liquid flow rates is found to incorrectly classify some region of the intermittent flow pattern as annular flow:

$$U_{SL} \geq U_{SG}\left[1.9\left(\frac{gD}{U_{SG}^{2}}\right)^{0.18}\left(\frac{g\Delta\rho\sigma}{U_{SG}\rho_{G}^{2}}\right)^{0.2}\right]^{8} \tag{14.24}$$

Another simple yet practical approach to predict the transition to annular flow regime is based on non-dimensional gas velocity (Froude number) corresponding to the point of pressure gradient minimum (concept discussed in Chap. 16). Two-phase flow literature for vertical upward flow reports that the pressure gradient minimum corresponds to the transition between churn/intermittent and annular flow regimes and the corresponding non-dimensional gas velocity satisfies Eq. (14.25). Based on the flow visualization and pressure drop measurements carried out at Two-Phase Flow Lab, OSU, it is found that the criteria given by Eq. (14.25) for transition to annular flow regime is a good approximation for the entire range of pipe orientations. The right-hand side of Eq. (14.25) gives a range between which the non-dimensional gas velocity can vary and for simplicity a value of 1.0 can be used. It also implies that the transition to annular flow regime would be independent of liquid velocity. Based on experience, it is recommended that in conjunction with Eq. (14.25), additional simple criteria of $\alpha_G > 0.75$ can be used to confirm the existence of annular flow regime:

$$\frac{U_{SG}}{\sqrt{gD}}\sqrt{\frac{\rho_G}{(\rho_L-\rho_G)}} \geq 0.9-1.1 \tag{14.25}$$

Taitel et al. (1980) present a criterion transition from dispersed bubbly to annular flow regime. Their transition criteria based on force balance between the drag on liquid droplet and its weight lead to Eq. (14.26):

$$\frac{U_{SG}\sqrt{\rho_G}}{(g\Delta\rho\sigma)^{0.25}} \geq 3.1 \tag{14.26}$$

A quick comparison between Eqs. (14.25) and (14.26) shows that the superficial gas velocity (U_{SG}) predicted by Eq. (14.26) will always be higher than that predicted by Eq. (14.25). This is possibly because Eq. (14.25) gives transition to annular flow regime from churn/intermittent flow regimes at lower liquid flow rates, while Eq. (14.26) predicts transition from dispersed bubbly flow at high liquid flow rates and hence would need higher gas flow rates for transition to annular flow regime. In addition to the transition theories, Table 14.1 presents a rule of thumb based on the range of void fraction and qualitative range of gas and liquid flow rates associated with different flow patterns. It must be mentioned that this qualitative and quantitative range of two-phase flow variables is an approximate range based on experience and may give slight deviations compared to the real two-phase flow scenario.

Table 14.1 Rule of thumb for qualitative classification of flow patterns

	Mass/Volumetric flow rate		
Void fraction (α_G)	Gas phase	Liquid phase	Flow regime
Low ($\alpha_G \leq 0.25$)	Low	Moderate/high	Bubbly
Low to moderate ($0 \leq \alpha_G \leq 0.5$)	Low	High	Dispersed bubbly
Moderate ($0.25 < \alpha_G \leq 0.75$)	Low	Low/moderate	Slug
Moderate ($0.25 < \alpha_G \leq 0.75$)	Moderate	Low/moderate	Intermittent
Moderate to high ($0.3 < \alpha_G \leq 0.9$)	Low/moderate	Low	Stratified
High ($0.75 < \alpha_G < 1$)	High	Low	Annular

Chapter 15
Void Fraction

Background

Void fraction (α_G) also referred to as liquid holdup ($\alpha_L = 1 - \alpha_G$) is defined based on its measurement technique such as local void fraction (using single-point probes), segmental void fraction (using gamma ray absorption method), cross-sectional void fraction (using capacitance probe), and volumetric void fraction (using quick closing valves). It must be noted that irrespective of the measurement technique, $\alpha_G + \alpha_L = 1$. Among these different methods, typically the cross-sectional ($\alpha_G = A_G/A$) and volumetric ($\alpha_G = V_G/V$) types of void fraction measurement are most preferred and of practical importance. Under the adiabatic two-phase flow condition over a short pipe length, it can be assumed that the two-phase flow structure remains unaltered throughout the pipe length and cross-sectional void fraction can be equated to volumetric void fraction. The void fraction as a stand-alone physical parameter is typically of no use unless embedded in other constitutive equations to calculate parameters such as actual phase velocity, two-phase mixture density, and hence hydrostatic two-phase pressure drop and heat transfer, liquid height in stratified flow, and liquid film thickness in annular flow regime.

The void fraction in gas-liquid two-phase flow is found to be influenced by several parameters such as pipe orientation, flow rates of individual phases, fluid properties, and pipe diameter. These influences on the void fraction will be discussed next.

A. J. Ghajar, *Single- and Two-Phase Flow Pressure Drop and Heat Transfer in Tubes*, Mechanical Engineering Series,
https://doi.org/10.1007/978-3-030-87281-6_15

Effect of Pipe Orientation on Void Fraction

The effect of pipe orientation on the void fraction is essentially due to the change in flow patterns and influence of buoyancy force acting on the gas phase. Most of the experimental void fraction data available in the literature is concentrated over a narrow range of pipe orientations and phase flow rates (or alternatively the flow patterns). In this study, the void fraction was measured at 17 different pipe orientations (0°, ± 5°, ± 10°, ± 20°, ± 30°, ± 45°, ±60°, ±75°, and ±90°) consisted of a wide range of void fraction ($0.09 \leq \alpha_G \leq 0.94$) and covered all important flow patterns observed in gas-liquid two-phase flow. The variation of void fraction with respect to change in pipe orientation is illustrated for different phase flow rates (Re_{SL} and Re_{SG}) in Figs. 15.1 and 15.2. The two liquid flow rates used in these figures are selected such that the representative data occupies the entire range of the void fraction and all flow patterns observed in this study.

Figure 15.1 shows the effect of pipe orientation on void fraction at low liquid flow rates ($Re_{SL} = 2000$) and a wide rage of gas flow rates ($Re_{SG} = 180 - 18,500$). As shown in the figure, the effect of pipe orientation on void fraction is significant (up to 100%) for low gas and liquid flow rates. These flow rates correspond to slug/

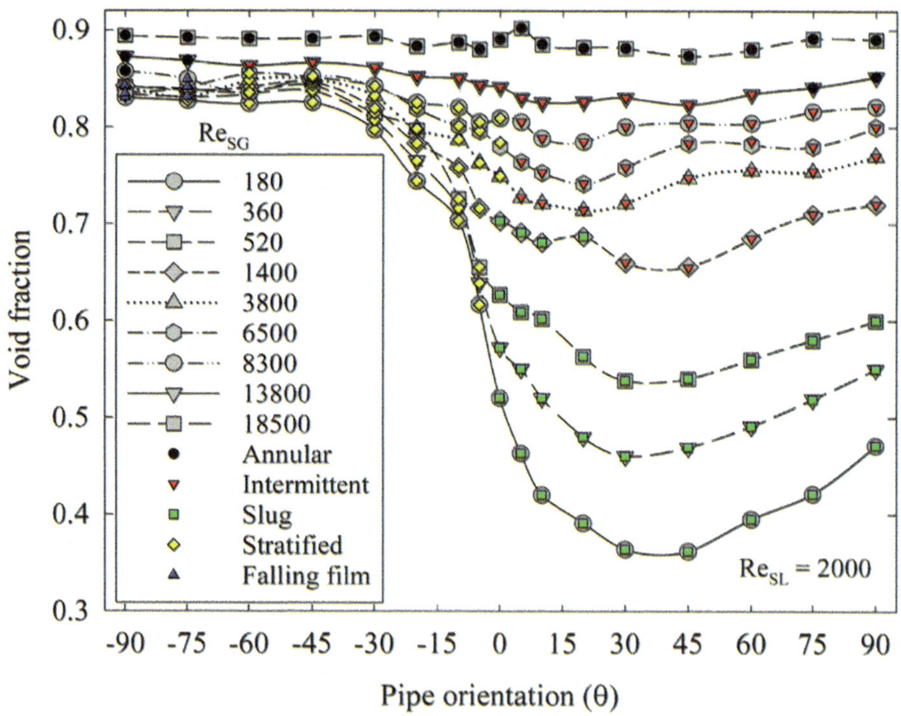

Fig. 15.1 Effect of pipe orientation on void fraction at low liquid flow rates. (Data measured at the Two-Phase Flow Lab, OSU, Stillwater, OK)

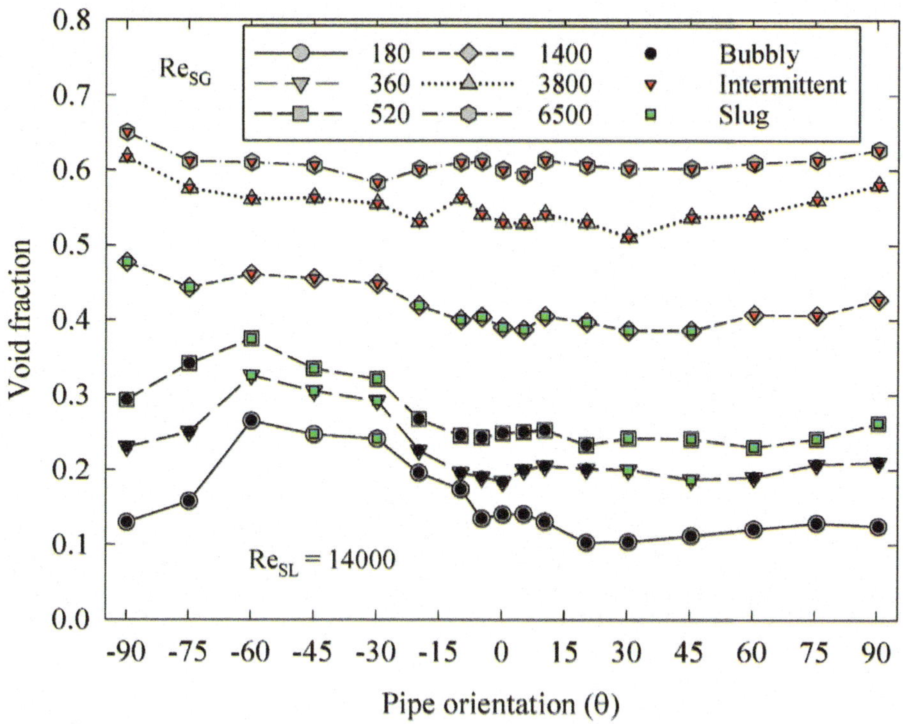

Fig. 15.2 Effect of pipe orientation on void fraction at high liquid flow rates. (Data measured at the Two-Phase Flow Lab, OSU, Stillwater, OK)

intermittent flow in upward pipe inclinations and stratified flow in downward pipe inclinations. At these flow rates, buoyancy acting on the gas phase aids the two-phase flow motion in upward pipe inclinations while resists its flow in downward pipe inclinations increasing its residence time in the pipe and hence the void fraction. For a fixed flow pattern (slug/intermittent) in upward inclined flow, it is seen that the void fraction initially decreases up to $+30^{\circ} < \theta < +45^{\circ}$ and then increases again with increase in the pipe orientation. This trend of change in void fraction with change in pipe orientation is due to the similar trend of the translational velocity of the gas slug (as a function of pipe orientation) that decides the residence time of the gas phase in the pipe and hence the void fraction. The effect of pipe orientation on void fraction is found to vanish for the inertia-driven region of intermittent (wavy annular) and annular two-phase flow. Moreover, the void fraction in stratified flow is apparently insensitive to the change in pipe orientation. At moderate gas and liquid flow rates, void fraction in pipe orientations steeper than -45° drops down due to the change in flow pattern from stratified to slug/intermittent at similar phase flow rates.

Figure 15.2 shows the effect of pipe orientation on void fraction at high liquid flow rates ($Re_{SL} = 14,000$) and a wide rage of gas flow rates ($Re_{SG} = 180 - 6500$). As shown in the figure, at high liquid flow rates, the effect of pipe orientation on void fraction gradually fades away; however, a similar type of flip in void fraction trend as mentioned above occurs after -60° of pipe orientations. This change in trend of void fraction is again due to dissimilar flow patterns that may exist for similar phase flow rates but different pipe orientations. It is also clear from these trends that the void fraction in downward pipe inclinations is always greater than that in upward pipe inclinations essentially due to the resistance offered by the buoyancy force to the gas phase.

Considering the overall effect of the pipe orientation on the void fraction, three different relationships between α_G and λ can be deduced such that for buoyancy-driven flows $\alpha_G > \lambda$ and for inertia-driven flows $\alpha_G < \lambda$ hold true. The third relationship would be $\alpha_G = \lambda$ when the change in buoyancy to inertia-driven flow takes place. The relationship $\alpha_G > \lambda$ implies that for buoyancy-driven two-phase flow in downward pipe inclinations, the gas phase travels at a lower velocity than the liquid phase and hence results into higher void fraction values. Note that for all upward pipe inclinations, gas phase moves faster than the liquid phase giving a positive slippage at the gas-liquid interface, and hence $\alpha_G < \lambda$ holds true. These three different cases of $\alpha_G - \lambda$ relationship that may exist in downward pipe inclinations are illustrated in Fig. 15.3. In general, the flow patterns/pipe orientations that may correspond to these different cases of $\alpha_G - \lambda$ are summarized in Table 15.1.

Effect of Phase Flow Rates on Void Fraction

The effect of phase flow rate on void fraction is indirectly through the change in flow pattern with change in phase flow rates. Figure 15.4 shows the effect of change in gas and liquid flow rate on void fraction. The void fraction data is plotted against a non-dimensional phase velocity ratio (U_{SG}/U_{SL}) to reduce the scatter of the data. Broadly, the void fraction as a function of phase flow rates can be divided into three regions shown in Fig. 15.4. In region I occupied by bubbly and slug flow patterns, for a fixed liquid flow rate, void fraction increases rapidly with a small increase in the gas flow rate, whereas in region III occupied by intermittent, stratified, and annular flow patterns, void fraction remains virtually independent of change in gas and liquid phase flow rates. Region II is a mixed bag of slug and intermittent and stratified flow patterns, and its slope merges smoothly with those in regions I and III. Compared to other flow patterns, the void fraction corresponding to stratified flow shows a different behavior in region II. Void fraction in stratified flow regime remains relatively insensitive to the change in phase flow rates for both regions II and III. Similar conclusions could be drawn by plotting the void fraction against two-phase flow quality, a well-adopted presentation style in high system pressure two-phase flows.

The variation of void fraction with change in two-phase flow quality is depicted in Fig. 15.5 for (a) constant system pressure of 1 MPa and varying slip ratios ($S = 1$ to 20) and (b) constant slip ratio ($S = 2$) and varying system pressure (0.5 MPa to 5 MPa). For fixed two-phase flow conditions such as system pressure and quality, an

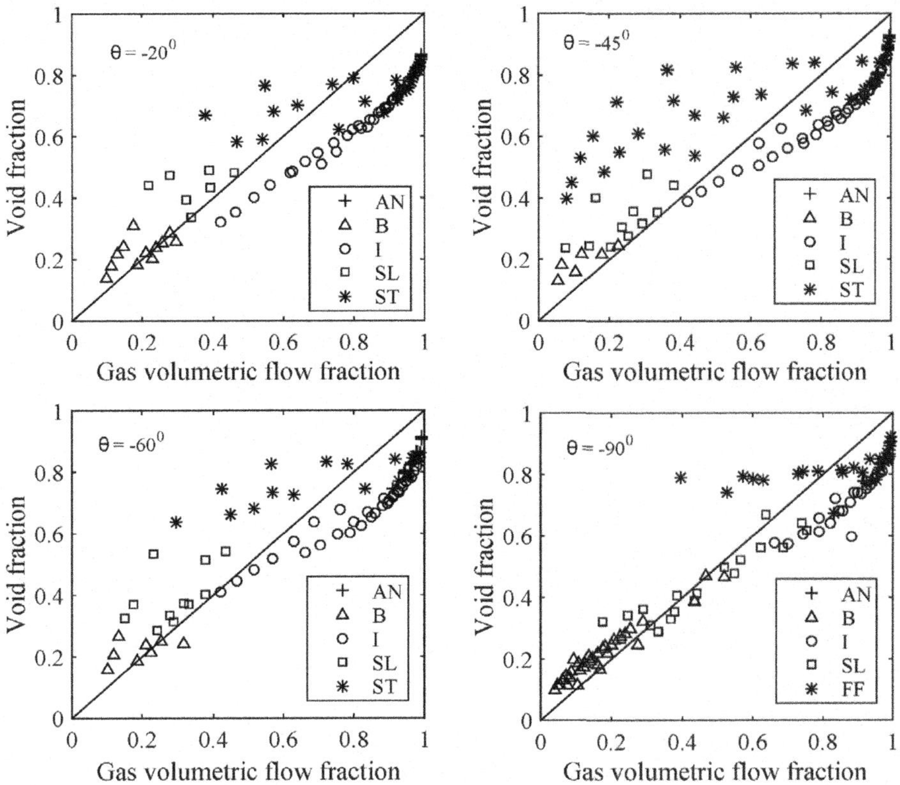

Fig. 15.3 Void fraction and gas volumetric flow fraction relationship in downward pipe inclinations (AN annular, B bubbly, FF falling film, I intermittent, SL slug, ST stratified). (Data from the Two-Phase Flow Lab, OSU, Stillwater, OK)

Table 15.1 Summary of two-phase flow situations corresponding to $\alpha_G - \lambda$ relationships

Condition	Flow patterns/Pipe orientation
$\alpha_G < \lambda$ ($U_G > U_L$, $S > 1$)	All flow patterns in upward pipe inclination
$\alpha_G = \lambda$ ($U_G \approx U_L$, $S \approx 1$)	Homogeneous flow (dispersed bubbly, annular mist)
$\alpha_G > \lambda$ ($U_G < U_L$, $S < 1$)	Stratified, slug flow $Fr_{SG} < 0.1$ and $\theta < 0^o$

increase in slip ratio reduces the void fraction; see Fig. 15.5a. For these fixed conditions, the maximum void fraction is essentially the homogeneous void fraction obtained for no-slip condition ($S = 1$). In the case of fixed slip ratio, an increase in system pressure decreases the specific volume occupied by the gas phase (due to increase in gas phase density) and hence reduces the void fraction; see Fig. 15.5b.

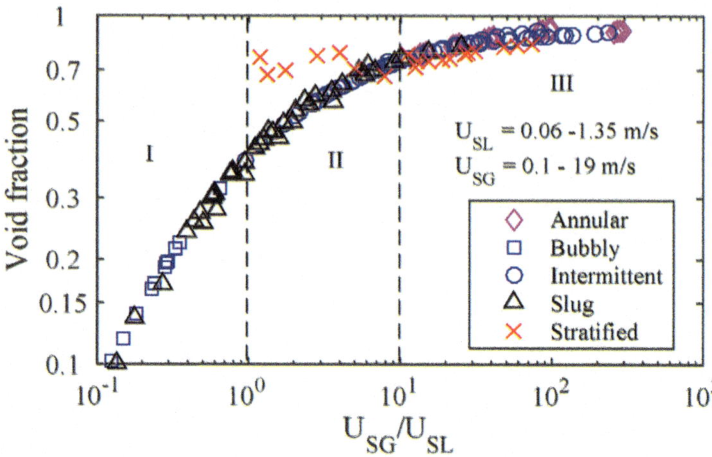

Fig. 15.4 Variation of void fraction with change in gas and liquid flow rates. (Data measured at the Two-Phase Flow Lab, OSU, Stillwater, OK)

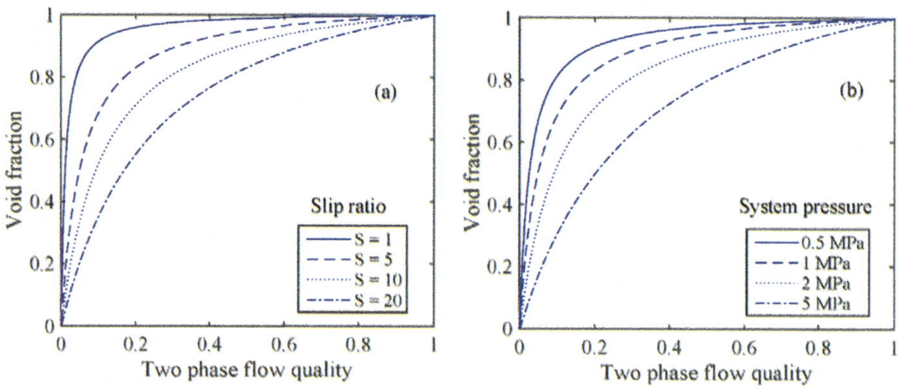

Fig. 15.5 Variation of void fraction with change in two-phase flow quality. (**a**) System pressure = 1 MPa. (**b**) Slip ratio = 2. (Data measured at the Two-Phase Flow Lab, OSU, Stillwater, OK)

Effect of Fluid Properties on Void Fraction

The two important fluid properties that are known to affect the void fraction are the gas phase density and the liquid phase dynamic viscosity. The effect of gas phase density on the void fraction is essentially in a case of high system pressure flow such as that observed in nuclear cooling applications and the refrigeration industry, whereas the effect of liquid phase dynamic viscosity on the void fraction is of significant interest in chemical engineering processes and the oil and gas industry applications that involve viscous oils having dynamic viscosity up to an order of 500 compared to that of liquid water at room temperature and pressure. The increase in system pressure or alternatively increase in the gas phase density reduces the

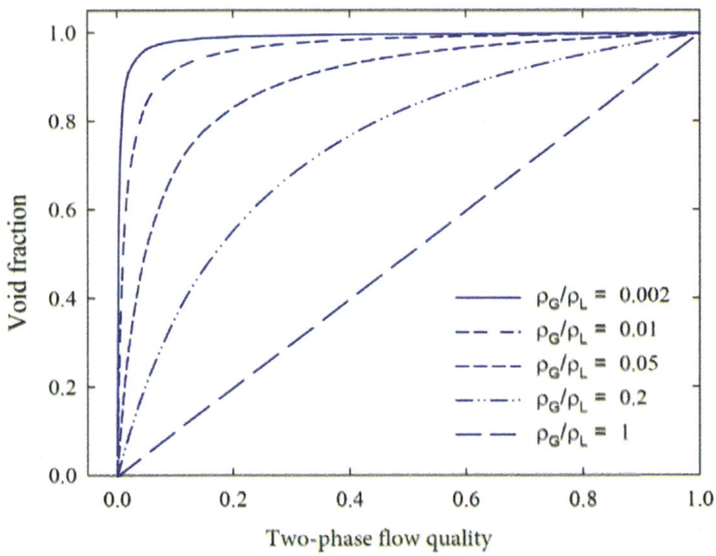

Fig. 15.6 Variation of void fraction with respect to change in quality as a function of phase density ratio. (Data measured at the Two-Phase Flow Lab, OSU, Stillwater, OK)

slippage between the two phases and also reduces the void fraction measured at similar mass flow rate as was shown in Fig. 15.5b. A denser gas phase (lower specific volume) will occupy less volume in the test section and hence would result in reduced void fraction. As shown in Fig. 15.6, the slope of void fraction vs quality is a function of the phase density ratio, and in the limiting case of $\rho_G \approx \rho_L$, there is a negligible slippage between the two phases, and void fraction varies linearly with change in two-phase flow quality ($\alpha_G = x$). In this case, the two-phase flow void fraction becomes equal to the homogeneous flow void fraction. A similar trend of variation of void fraction as a function of gas phase density for two-phase flow of refrigerants has been reported by Shedd (2010) and Keinath (2012). They found the void fraction to decrease with increase in refrigerant saturation temperature (increase in saturation or system pressure).

In addition to the gas phase density, void fraction is also known to be affected by the change in liquid dynamic viscosity. Gokcal (2008) and Jeyachandra (2011) measured the void fraction in a 50.2 mm I.D. pipe using air and oil of dynamic viscosity in a range of 0.18 Pa-s to 0.58 Pa-s. Overall they found the void fraction to slightly decrease with increase in the liquid dynamic viscosity. Similar results are reported by Oshinowo (1971) who measured void fraction in vertical upward and downward pipes using air-water and air and four different concentrations of ethylene glycol. Mukherjee (1979) measured void fraction using air-kerosene and air-oil fluid combinations in upward and downward pipe inclinations. The results of these experiments revealed that the void fraction is inversely proportional to the liquid dynamic viscosity as shown in Fig. 15.7. From the results of these studies, it was

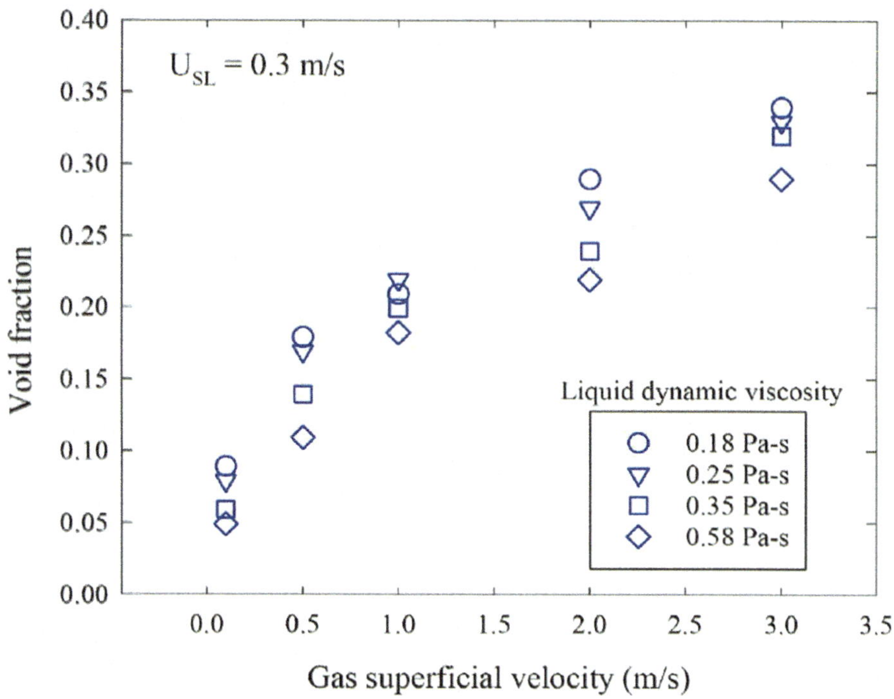

Fig. 15.7 Effect of liquid dynamic viscosity on void fraction. (Adapted from Gokcal 2008)

concluded that the increase in liquid viscosity increases the viscous drag on liquid phase and reduces its velocity causing it to accumulate in the pipe and hence results in decrease in void fraction due to increase in liquid holdup.

Effect of Pipe Diameter on Void Fraction

Effect of pipe diameter on void fraction is reflected in the form of pipe wall-induced drag exerted on the gas phase. In comparison to large pipes, small diameter pipes offer more resistance to the motion of gas phase increasing its residence time and hence the local and averaged void fraction in pipes. Experimental data of Bowers and Hrnjak (2010) in Fig. 15.8 show about 20–80% increase in measured void fraction for a decrease in pipe diameter from 15.3 to 7.2 mm. The void fraction as a function of pipe diameter also depends upon phase flow rates (flow patterns/mass flux/quality) such that the effect of change in pipe diameter on void fraction is most prominent for bubbly/slug/intermittent flows, whereas, for the high values of void fraction corresponding to that in annular flow regime (shear-driven flow) typically $\alpha_G > 0.8$, the void fraction is relatively insensitive to the pipe diameter. This observation is in agreement with the work of Kaji and Azzopardi (2010) who

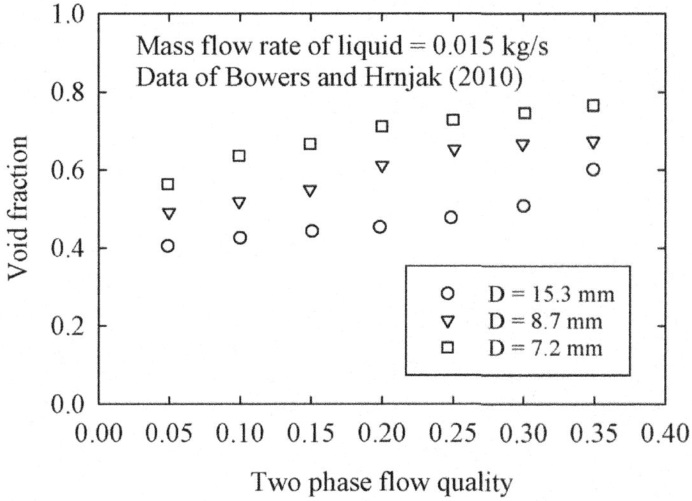

Fig. 15.8 Effect of pipe diameter on void fraction. (Adapted from Bowers and Hrnjak 2010)

investigated the effect of pipe diameter in a range of 5–70 mm on void fraction in vertical upward annular flow regime for non-boiling two-phase flow for similar gas and liquid superficial velocities. They found that the effect of pipe diameter on void fraction in the annular flow regime is negligible. The higher values of void fraction in small diameter pipes can be justified considering the flow structure and wall effects on the gas phase. In the case of two-phase flow through small diameter pipes, typically at low gas flow rates, the two-phase flow is in the form of bubble train (quasi-spherical or equal length elongated bubbles that occupy the entire pipe cross section), whereas the flow structure in large diameter pipes may be in the form of dispersed bubbles. This relationship between void fraction and pipe diameter may not be applicable to micro-scale two-phase flow.

The parametric analysis of void fraction reveals that the void fraction must be modeled to consider the effect of flow patterns, pipe orientation, pipe diameters, and fluid properties. What follows is a brief review of some of the void fraction modeling methods available in the two-phase flow literature and recommendation of well-scrutinized void fraction correlations applicable for a wide range of two-phase flow conditions.

Modeling of Void Fraction

Two-phase flow literature reports plethora of void fraction models/correlations; however, most of these models are confined to certain flow patterns, pipe orientation, and fluid properties. Based on the general physical form of these correlations, they can be broadly classified as those based on separated flow models, drift flux models, and empirical models. A comprehensive review of these different types of models is presented by Woldesemayat and Ghajar (2007), Godbole et al. (2011), and Ghajar and Bhagwat (2014b). Among all these models, the drift flux model was identified to be more versatile, accurate, and flexible. It was also one of the most preferred models for several industrial applications. For these reasons, the drift flux model was used in our studies. Following is a brief summary of the physical structure of the drift flux model-based void fraction correlation.

Drift Flux Model (DFM) The general concept of the DFM was first envisaged by Zuber and Findlay (1965) and later developed and contributed by Wallis (1969) and Ishii (1977). The DFM assumes one phase dispersed in other continuous phase and requires the determination of distribution parameter (C_o) and drift velocity (U_{GM}) as variables to calculate the void fraction. The DFM is best applicable for one-dimensional flows and is usually not recommended for flow patterns involving significant interfacial slippage such as stratified and annular flow patterns. The flow patterns such as bubbly flow, slug flow, and mist flow are the preferred flow patterns to be modeled using the concept of drift flux. However, with appropriate formulations that consider the two-phase flow physics, the application of DFM can also be extended to other flow patterns such as annular flow. The general form of DFM to calculate void fraction is presented by Eq. (15.1) where $U_M = U_{SL} + U_{SG}$:

$$\langle \alpha_G \rangle = \frac{\langle U_{SG} \rangle}{C_o \langle U_M \rangle + \langle\langle U_{GM} \rangle\rangle} \quad (15.1)$$

The terms involving $\langle\rangle$ represent cross-sectional averaged properties, while $\langle\langle\rangle\rangle$ indicate cross-sectional and void fraction weighted averaged properties. The distribution parameter C_o is a representation to account for the distribution of the gas phase across the pipe cross section (concentration profile). It also serves as a correction factor to the homogeneous flow theory (which assumes no local slip between the two phases) to acknowledge the fact that the concentration profile and the two-phase flow velocity profile can vary independent of each other across the pipe cross section, whereas the physical interpretation of drift velocity (U_{GM}) is the cross-sectional void fraction weighted average of the local relative velocity of the gas phase with respect to the two-phase mixture velocity at the pipe volume center. The local relative motion between the gas and the two-phase mixtures is considerable and uniform across the pipe cross section when there is a strong coupling between the two phases as in the case of the flow of dispersed bubbles in continuous liquid medium, that is, dispersed bubbly flow. As the two-phase flow transits to annular flow regime, the local relative velocity of the gas phase with respect to the two-phase mixture at the pipe volume center becomes negligible and so does the drift velocity. Although the one-dimensional DFM is used in context of cross-sectional averaged void fraction, there are several DFM-based correlations available in the literature that can be practically implemented for void fraction calculations based on volumetric void fraction experimental data instead of the

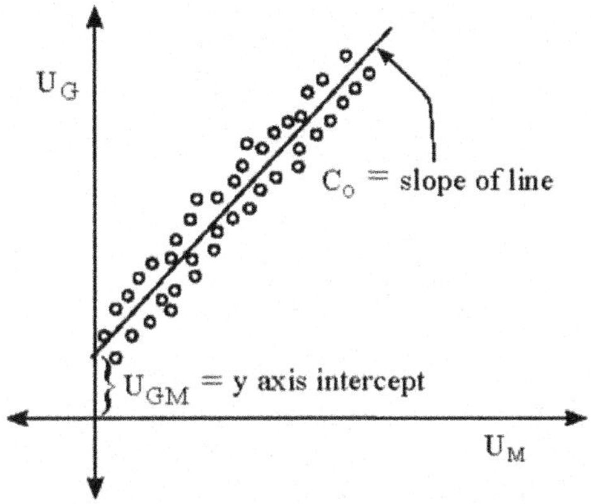

Fig. 15.9 Graphical representation of drift flux model

cross-sectional void fraction. The parity between the cross-sectional and volumetric void fraction holds true in the case of non-boiling two-phase flow (two-component two-phase flow) where the cross-sectional distribution of the gas phase with respect to the liquid phase remains virtually unaltered over a short length of pipe. Henceforth, without making any distinction between the cross-sectional and volumetric void fraction, the void fraction is simply expressed as $\langle \alpha_G \rangle = \alpha_G$. Similar justification is applicable for all the cross-sectional averaged quantities involved in Eq. (15.1). The physical form of Eq. (15.1) resembles the equation of a straight line ($y = mx + c$) shown in Fig. 15.9 such that the distribution parameter C_o is analogous to the slope "m" and the drift velocity is analogous to the y-intercept "c." Thus, the distribution parameter and the drift velocity could be obtained by plotting the data with $U_G = U_{SG}/\alpha_G$ and U_M coordinates and then finding the slope and intercept of the best-fitting line. However, this method is highly impractical since the slope and intercept of line would depend on the scatter of data and cannot be used in the case of real-time applications where void fraction information is not known beforehand. Obviously, another approach to determine C_o and U_{GM} is to use empirical correlations (as a function of two-phase flow variables) available in the two-phase flow literature. What follows next is a brief presentation of some of the well-scrutinized and validated flow pattern-specific and flow pattern-independent correlations to determine the distribution parameter C_o and the drift velocity U_{GM} used in the general form of DFM, Eq. (15.1), to calculate void fraction.

Hibiki and Ishii (2003) Correlation for Vertical Upward Bubbly Flow The distribution parameter and drift velocity equations given by Hibiki and Ishii (2003) for vertical upward bubbly flow are given in Eq. (15.2). To use this equation, the existence of bubbly flow could be determined by first using flow pattern transition equation given by Eq. (14.4).

$$\left.\begin{aligned} C_o &= \left(1.2 - 0.2\sqrt{\frac{\rho_G}{\rho_L}}\right)(1 - \exp(-18\alpha_G)) \\ U_{GM} &= \sqrt{2}\left(\frac{\sigma g(\rho_L - \rho_G)}{\rho_L^2}\right)^{0.25}(1 - \alpha_G)^{1.75} \end{aligned}\right\} \begin{array}{l} \text{(bubbly flow)} \\ \theta = +90^{\circ} \end{array} \tag{15.2}$$

Gomez et al. (2000) Correlation for Bubbly Flow Another most referred void fraction model for bubbly flow is that of Gomez et al. (2000) given by Eq. (15.3). This model is essentially developed for upward pipe inclinations, and it may not reproduce correct results if used for horizontal two-phase flow since Eq. (15.3) would give $U_{GM} = 0$. Moreover, Gomez et al. (2000) have recommended use of a constant to represent distribution parameter that does not account for the effect of fluid properties, flow properties, and pipe orientation on the distribution parameter. Although their void fraction model is developed for bubbly flow, our experience shows that this correlation also works well for the slug flow regime. Note that to use Eq. (15.3), the existence of bubbly flow can be confirmed by using bubbly-slug flow transition model given by Eq. (14.1).

$$\left.\begin{aligned} C_o &= 1.15 \\ U_{GM} &= 1.53\left(\frac{g\sigma(\rho_L - \rho_G)}{\rho_L^2}\right)^{0.25}\sqrt{1 - \alpha_G}\sin\theta \end{aligned}\right\} \begin{array}{l} \text{(bubbly, slug flow)} \\ 0^{\circ} < \theta \leq +90^{\circ} \end{array} \tag{15.3}$$

Hibiki and Ishii (2003) Correlation for Slug Flow For slug flow regime, Hibiki and Ishii (2003) proposed expressions for distribution parameter and drift velocity given by Eq. (15.4). The drift velocity used in their equations is essentially the modified form of proportionality (between pipe diameter and slug translational velocity) proposed by Nicklin et al. (1962). These correlations given by Eqs. (15.2 and 15.4) are valid for small diameter pipes ($10 < D_h < 50$ mm). For large diameter pipes, Kataoka and Ishii (1987) have proposed sets of equations given by Eq. (15.5) to account for the effect of pipe diameter and fluid properties on the drift velocity. The non-dimensional hydraulic pipe diameter (D_h^+), non-dimensional drift velocity (U_{GM}^+), and viscosity number (N_μ) are defined by Eqs. (15.6, 15.7 and 15.8), respectively. Their equation is developed based on data of air-water and steam-water vertical upward two-phase flow consisting of data with $20 < D_h < 240$ mm and $0.1 \leq p_{SYS} \leq 18$ MPa. The distribution parameter of Kataoka and Ishii (1987) correlation is the same as that of Eq. (15.4). In the case of two-phase flow through rectangular ducts, the constants "1.2" and "0.2" in Eq. (15.4) are replaced by "1.35" and "0.35," respectively:

$$\left.\begin{aligned} C_o &= \left(1.2 - 0.2\sqrt{\frac{\rho_G}{\rho_L}}\right) \\ U_{GM} &= 0.35\sqrt{\frac{gD(\rho_L - \rho_G)}{\rho_L}} \end{aligned}\right\} \begin{array}{l} \text{(slug flow)} \\ \theta = +90^\text{o} \end{array} \tag{15.4}$$

$$U_{GM}^{+} = \begin{pmatrix} 0.00019\left(D_h^{+}\right)^{0.809}\left(\frac{\rho_G}{\rho_L}\right)^{-0.157} N_\mu^{-0.562} & : N_\mu \leq 2.25 \times 10^{-3}, D_h^{+} > 30 \\ 0.030\left(\frac{\rho_G}{\rho_L}\right)^{-0.157} N_\mu^{-0.562} & : N_\mu \leq 2.25 \times 10^{-3}, D_h^{+} \leq 30 \\ 0.92\left(\frac{\rho_G}{\rho_L}\right)^{-0.157} & : N_\mu > 2.25 \times 10^{-3} \end{pmatrix} \tag{15.5}$$

$$D_h^{+} = \frac{D_h}{\sqrt{\sigma/(g\Delta\rho)}} \tag{15.6}$$

$$U_{GM}^{+} = \frac{U_{GM}}{\left((\sigma g \Delta\rho)/\rho_L^2\right)^{0.25}} \tag{15.7}$$

$$N_\mu = \frac{\mu_L}{\left(\rho_L \sigma \sqrt{\sigma/(g\Delta\rho)}\right)^{0.5}} \tag{15.8}$$

Rouhani and Axelsson (1970) Correlation The DFM-based void fraction correlation of Rouhani and Axelsson (1970) given by Eqs. (15.9 and 15.10) is one of the most preferred correlations used in refrigeration industry to estimate void fraction in evaporators and condensers. The term $(1 - x)$ ensures decrease of distribution parameter with increase in two-phase flow quality or alternatively shift of flow pattern from bubbly/slug to annular flow:

$$C_o = \begin{pmatrix} 1 + 0.2(1 - x)\left(\frac{gD\rho_L^2}{G^2}\right)^{0.25} & : \theta = +90^\text{o}, 0.1 \leq p_\text{sys} \leq 14\ \text{MPa} \\ 1 + 0.12(1 - x) & : \theta = 0^\text{o}, 0.1 \leq p_\text{sys} \leq 14\ \text{MPa} \end{pmatrix} \tag{15.9}$$

$$U_{GM} = 1.18\left[\frac{g\sigma(\rho_L - \rho_G)}{\rho_L^2}\right]^{0.25} \tag{15.10}$$

Woldesemayat and Ghajar (2007) Correlation To get rid of the flow pattern dependency, Woldesemayat and Ghajar (2007) proposed equations for distribution parameter and drift velocity as a function of fluid properties, pipe diameter, and pipe

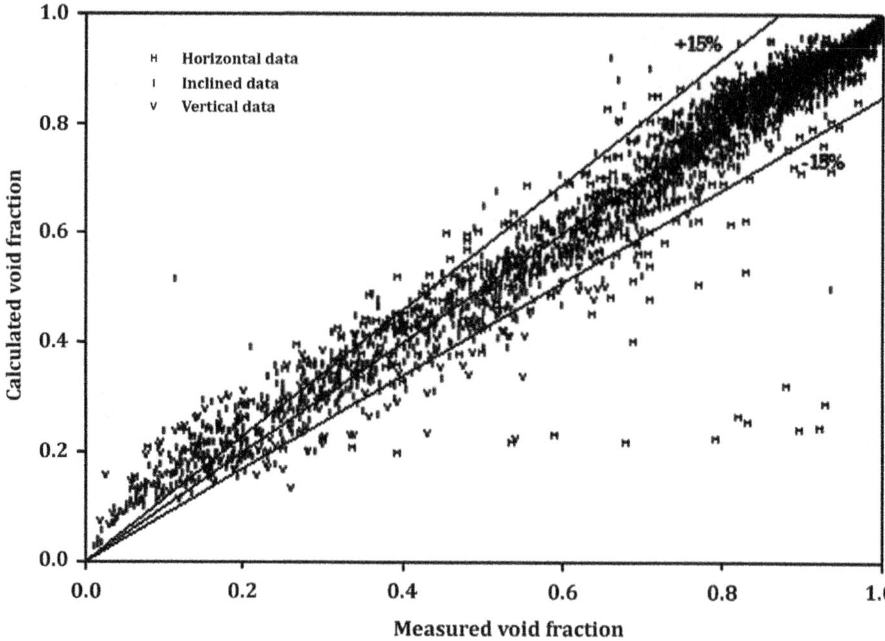

Fig. 15.10 Comparison of Woldesemayat and Ghajar (2007) void fraction correaltion with 2845 experimental data points for horizontal, upward inclined, and upward verical pipe inclinations

orientation shown in Eqs. (15.11 and 15.12). Their correlation is a modification of Dix (1971) correlation for two-phase flow over rod bundles and is independent of flow patterns. Correlation of Woldesemayat and Ghajar (2007) is verified against 2845 data points from 8 different sources for horizontal (900 data points), upward inclined (1542 data points), and upward vertical (403 data points) pipe inclinations. The data covered pipe diameters in a range of 12–80 mm and different fluid combinations (natural gas-water, air-water, air-kerosene). Figure 15.10 shows the comparison of their correlation with the entire set of experimental data for horizontal, upward inclined, and upward vertical pipe inclinations. Correlation of Woldesemayat and Ghajar (2007) predicted 79% and 86% of the entire data set within an accuracy of ± 10 and $\pm$ 15, respectively. In the case of unknown flow pattern in horizontal and upward pipe inclinations, Woldesemayat and Ghajar (2007) correlation can be used with a reasonable accuracy (except for stratified flow). Note that in Eq. (15.12), the multiplying factor of 2.9 has units of m $^{-0.25}$:

$$C_o = \lambda\left(1 + \left[\frac{U_{SL}}{U_{SG}}\right]^{(\rho_G/\rho_L)^{0.1}}\right) \tag{15.11}$$

$$U_{GM} = 2.9\left[\frac{gD\sigma(1+\cos\theta)(\rho_L-\rho_G)}{\rho_L^2}\right]^{0.25}(1.22+1.22\sin\theta)^{p_{atm}/p_{sys}} \quad (15.12)$$

Bhagwat and Ghajar (2014) Correlation A fairly recent void fraction correlation developed by Bhagwat and Ghajar (2014) considers the effect of flow patterns (in the form of phase flow rates), pipe orientation, pipe diameter, and fluid properties on the distribution parameter (C_o) and drift velocity (U_{GM}) as shown in Eqs. (15.13 and 15.16), respectively. The experimental database used for performance verification of their void fraction correlation consists of 8255 data points, with 4277 data points for air-water, natural gas-water, and argon-water fluid combinations, 1110 data points for liquid refrigerants and their vapors (R11, R12, R22, R134a, R114, R410A, R290, and R1234yf), 570 data points for steam-water, 567 data points for air-oil with liquid phase dynamic viscosity ranging from 0.005 Pa-s to 0.6 Pa-s, and 1731 data points for other fluid combinations such as air-kerosene, air-glycerin, argon-ethanol, argon-acetone, and argon-alcohol. The experimental data listed above belongs to more than 60 sources in the two-phase flow literature and consists of pipe orientations in a range of -90° to $+90^o$ and hydraulic pipe diameters from 0.5 mm to 305 mm covering the micro- to macro-scale two-phase flow phenomenon. The experimental void fraction data for refrigerants and steam-water consists of both adiabatic and diabatic (boiling and condensing) flow conditions. In fact, about 36% of the refrigerant data and 62% of the steam-water data are for diabatic conditions. Moreover, the diabatic two-phase flow data for steam-water consists of noncircular (rectangular) pipes with dimensions 11.1 mm × 93.6 mm, 6.35 mm × 50.8 mm, and 12.7 mm × 50.8 mm. Their void fraction correlation is based on the most comprehensive data bank and is the most robust correlation applicable for a wide range of two-phase flow conditions. Table 15.2 provides the application range of their correlation, and Fig. 15.11 depicts the distribution of the experimental data over different two-phase flow parameters. Since the experimental data for very large pipe diameter ($D_h = 305$ mm) and very high system pressure ($p_{sys} \approx 18$ MPa) is limited compared to other experimental data, they are not included in the histograms shown in Fig. 15.11.

Table 15.2 Application range of two-phase flow parameters for Bhagwat and Ghajar (2014) void fraction correlation

Parameter	Range
Pipe diameter (mm)	0.5–305
Pipe orientation (θ)	$-90^o \leq \theta \leq +90^o$
Density ratio (ρ_L/ρ_G)	6–875
Liquid dynamic viscosity (Pa·s)	0.0005–0.6
Mixture mass flux (kg/m^2·s)	10–8450
Two-phase quality (x)	0–1
Two-phase Reynolds number (Re_M)	$10 - 5 \times 10^6$
Pipe geometry	Circular, rectangular, annular
Flow patterns	All (except stratified flow with $Fr_{SG} \leq 0.1$ in $\theta < 0^o$)

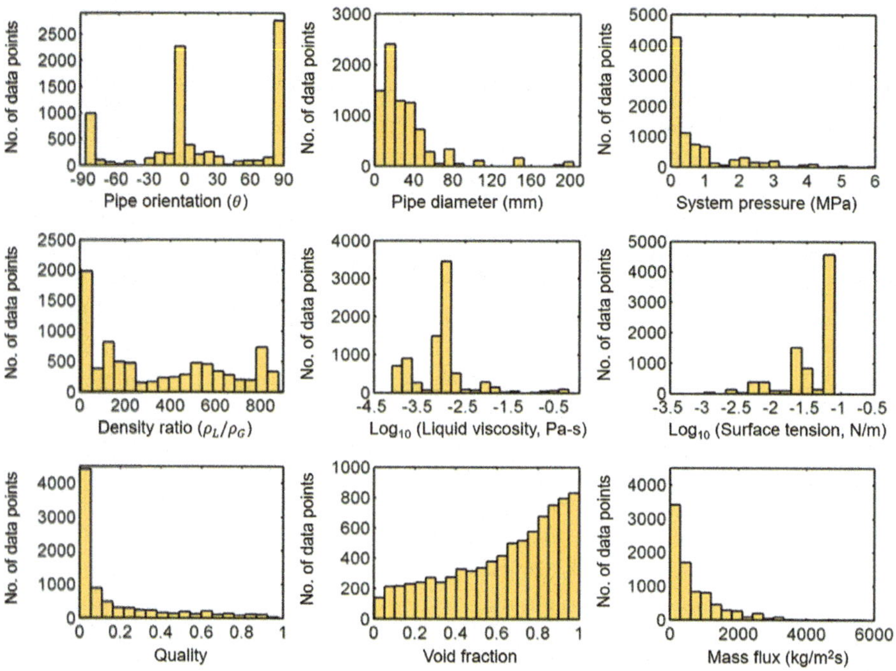

Fig. 15.11 Distribution of the experimental data used in the development of Bhagwat and Ghajar (2014) void fraction correlation

The variable $C_{o,\ 1}$ used in Eq. (15.13) for the distribution parameter (C_o) is defined by Eq. (15.14) and is modeled as a variable such that it may vary between 0.2 and 0 to ensure the variation of distribution parameter for bubbly flow $C_o \approx 1.2$ to annular flow where $C_o \rightarrow 1.0$. More details on the development of Eqs. (15.13 and 15.14) may be obtained from Bhagwat and Ghajar (2014):

$$C_o = \frac{2 - (\rho_G/\rho_L)^2}{1 + (Re_M/1000)^2} + \frac{\left[\sqrt{\left(1 + (\rho_G/\rho_L)^2 \cos\theta\right)/(1 + \cos\theta)}^{(1-\alpha_G)}\right]^{2/5} + C_{o,1}}{1 + (1000/Re_M)^2} \tag{15.13}$$

where two-phase mixture Reynolds number (Re_M) is given by Eq. (15.15).

$$C_{o,1} = \left(c_1 - c_1\sqrt{\rho_G/\rho_L}\right)\left[(2.6 - \lambda)^{0.15} - \sqrt{f_M}\right](1 - x)^{1.5} \tag{15.14}$$

The constant c_1 is 0.2 for circular pipes and 0.4 for rectangular and annular ducts. Note that the two-phase friction factor (f_M) in Eq. (15.14) is based on two-phase mixture Reynolds number (Re_M) given by Eq. (15.15) and could be calculated using appropriate single-phase friction factor equation for laminar and turbulent flows:

$$Re_M = \frac{(U_{SL} + U_{SG})\rho_L D_h}{\mu_L} \tag{15.15}$$

The expression for drift velocity (U_{GM}) used by Bhagwat and Ghajar (2014) as given in Eq. (15.16) is essentially based on the translational velocity of the gas slug in different pipe inclinations suggested by Bendiksen (1984). However, instead of using the constant of 0.54 as originally proposed by Bendiksen (1984), a proportionality constant of 0.45 is used since it gave a better accuracy in overall prediction of the void fraction. The correction factors c_2 and c_3 used in Eq. (15.16) account for the effect of increase in liquid viscosity and pipe diameter on the reduction in drift velocity. These correction factors are expressed by Eqs. (15.17 and 15.18). More details and justification about the effect of these parameters on drift velocity and hence the use of these correction factors could be obtained from Gokcal et al. (2009), Kataoka and Ishii (1987), and Bhagwat and Ghajar (2014). The pipe diameter beyond which the correction to drift velocity in the form of c_3 is deemed necessary is identified using Laplace number (non-dimensional pipe diameter) defined by Eq. (15.19):

$$U_{GM} = (0.35\sin\theta + 0.45\cos\theta)\sqrt{\frac{gD_h(\rho_L - \rho_G)}{\rho_L}}(1 - \alpha_G)^{0.5} c_2 c_3 \tag{15.16}$$

$$c_2 = \left(\begin{array}{ll} \left(\frac{0.434}{\log_{10}(\mu_L/0.001)}\right)^{0.15} & : (\mu_L/0.001) > 10 \\ 1 & : (\mu_L/0.001) \le 10) \end{array} \right. \tag{15.17}$$

$$c_3 = \left(\begin{array}{ll} (La/0.025)^{0.9} & : La < 0.025 \\ 1 & : La \ge 0.025 \end{array} \right. \tag{15.18}$$

$$La = \sqrt{\frac{\sigma}{g(\rho_L - \rho_G)}}\frac{1}{D_h} \tag{15.19}$$

Graphical performance of void fraction correlation of Bhagwat and Ghajar (2014) for OSU's air-water data in horizontal, vertical upward, and vertical downward pipe orientation is shown in Figs. 15.12 and 15.13. As shown in Fig. 15.12, for horizontal flow, their correlation predicts more than 90% of data points within ±20% error

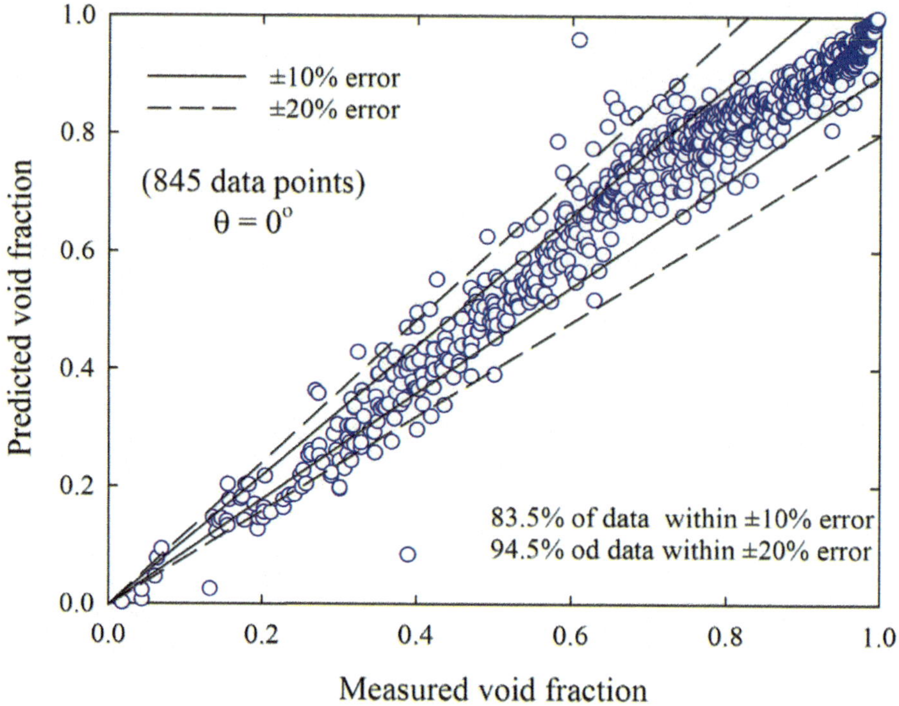

Fig. 15.12 Performance of the Bhagwat and Ghajar (2014) void fraction correlation against air–water data for horizontal pipe inclination. (Data measured at the Two-Phase Flow Lab, OSU, Stillwater, OK)

bands, while for vertical upward and downward flow (see Fig. 15.13), their correlation predicts 88.3% and 94.6% of data within ±20% error bands, respectively.

The graphical performance of Bhagwat and Ghajar (2014) void fraction correlation for air-water void fraction data in upward and downward inclined flow is depicted in Figs. 15.14 and 15.15. It is seen that the performance of their correlation deteriorates for downward pipe inclinations. The inaccuracies involved in the prediction of void fraction in downward inclinations are essentially due to the flow stratification that causes the slip ratio to decrease ($\alpha_G > \lambda$). It is found that all drift flux and separated flow model-based correlations exhibit an underprediction tendency for all downward pipe inclinations when $\alpha_G > \lambda$.

The prediction of Bhagwat and Ghajar (2014) void fraction correlation against experimental void fraction data for different refrigerants is graphically presented in Fig. 15.16. For details about the sources of the data sets used, refer to Bhagwat and Ghajar (2014). With the exception of R22 and R134a, for all other refrigerants, their void fraction correlation predicts majority of the data within ±20% error bands. It should be noted that the data points for refrigerant R22 that deviate significantly from predicted values of Bhagwat and Ghajar (2014) correlation belong to the data of Sacks (1975) whose uncertainty analysis is not available. Other void fraction

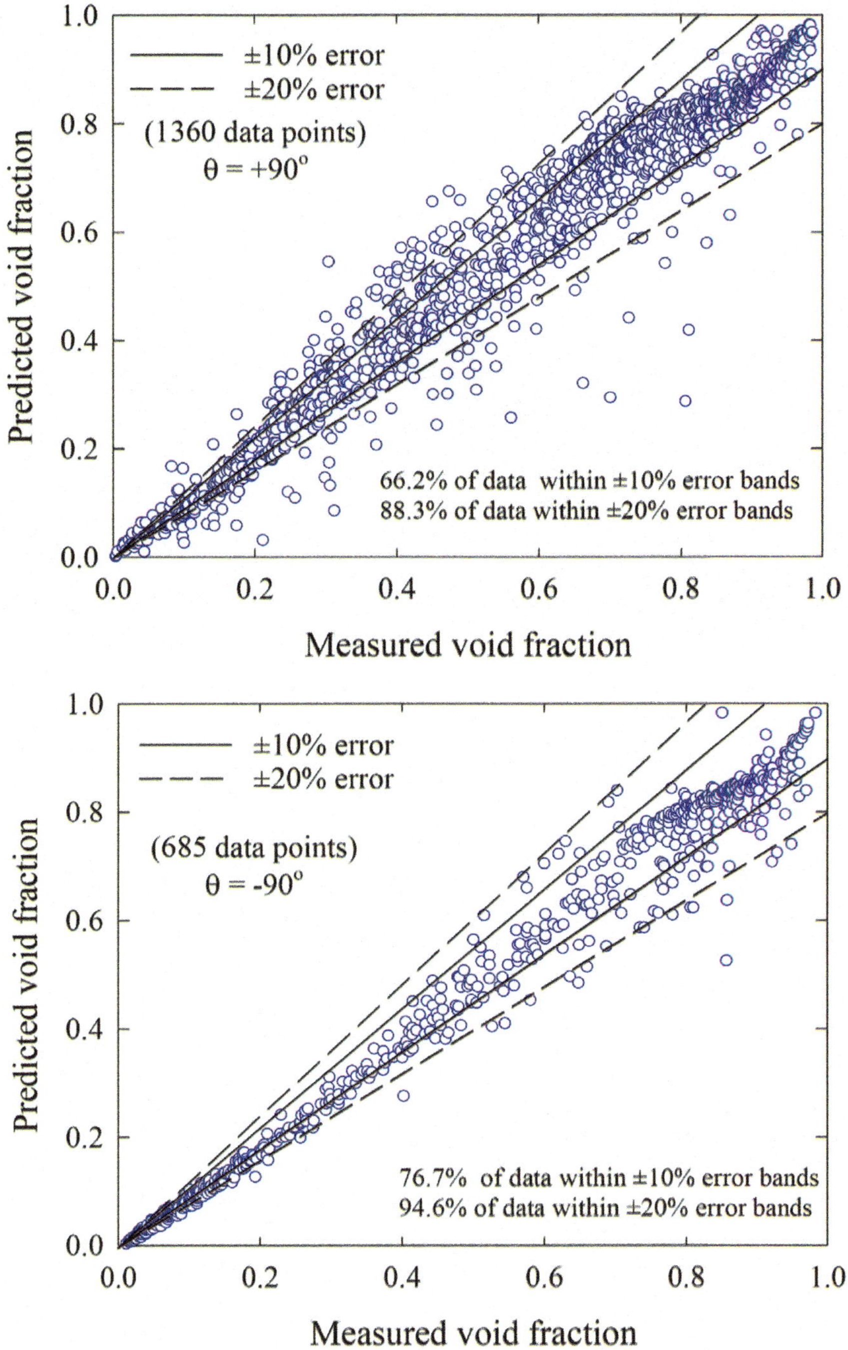

Fig. 15.13 Performance of the Bhagwat and Ghajar (2014) void fraction correlation against air–water data for vertical upward and downward pipe inclinations. (Data measured at the Two-Phase Flow Lab, OSU, Stillwater, OK)

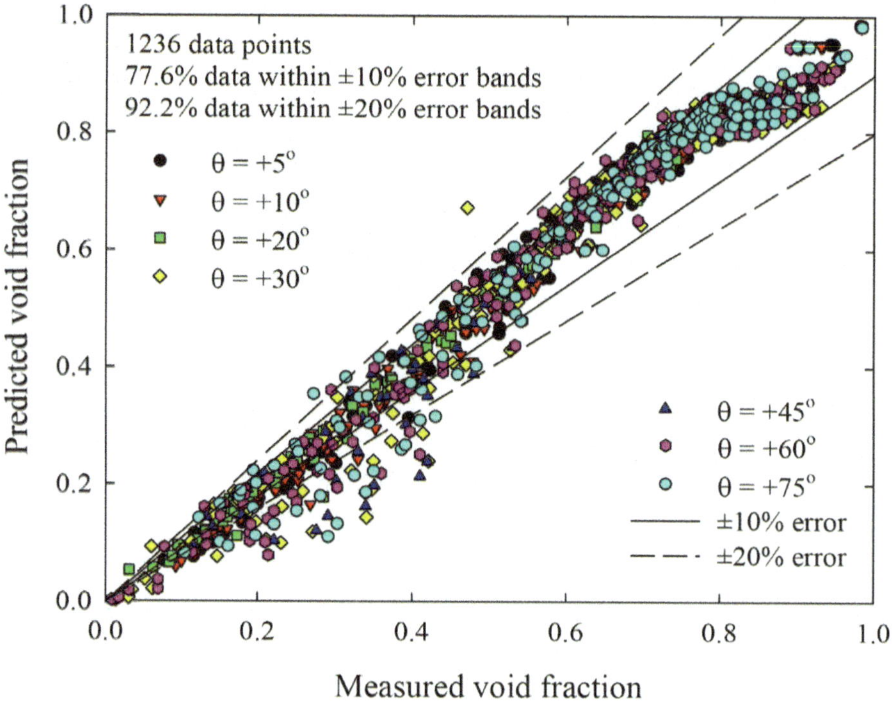

Fig. 15.14 Performance of the Bhagwat and Ghajar (2014) void fraction correlation against air–water data for upward inclined pipe inclinations. (Data measured at the Two-Phase Flow Lab, OSU, Stillwater, OK)

correlations available in the literature predicted these data points of R22 with more or less same error. In the case of R134a, the overpredicted data belong to the experimental work of Bowers and Hrnjak (2010) who measured the void fraction in pipe diameters in a range of 7–15 mm through the pixel analysis of the images obtained by high-speed visualization. The reliability of these measurements needs further examination since they did not report uncertainty of the measured void fraction data and did not make any comparison regarding its agreement with other measurement techniques or with the predictions of available void fraction correlations. Bhagwat and Ghajar (2014) considered these data points of R22 and R134a as outliers; however, these data points were included in the figures and performance analysis to avoid any bias in the experimental data used for the development of their void fraction correlation.

Figure 15.17 shows the excellent predictions of void fraction by Bhagwat and Ghajar (2014) void fraction correlation in mini- and microchannels. Their correlation predicts the R410A experimental data of Shedd (2010) for pipe diameters in a range of 0.5–2.96 mm well within ±20% error bands. A close observation shows that for $0.75 < \alpha_G < 1$, their correlation predicts 79% of data within ±5% error bands.

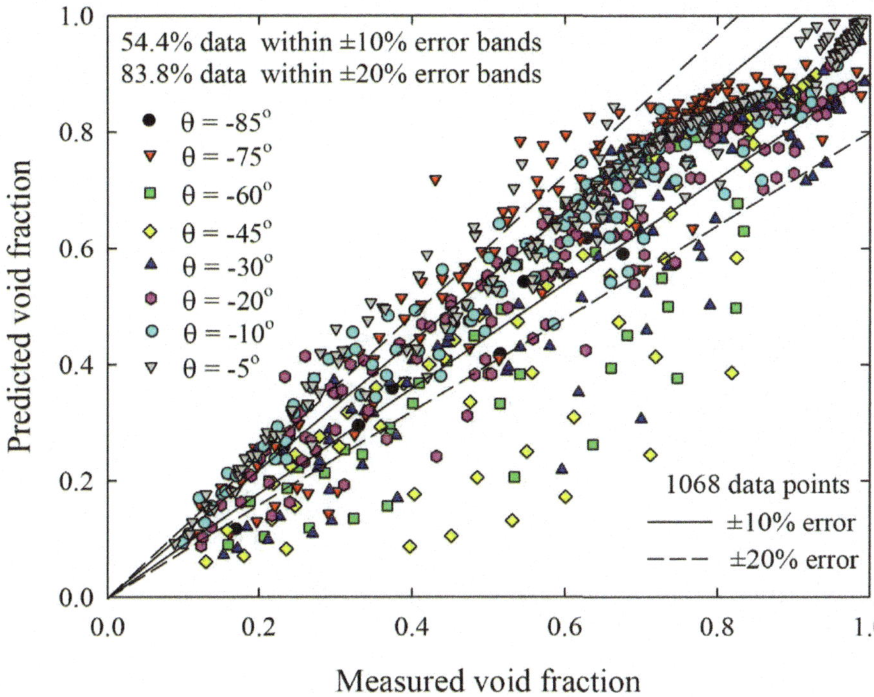

Fig. 15.15 Performance of the Bhagwat and Ghajar (2014) void fraction correlation against air–water data for downward inclined pipe inclinations. (Data measured at the Two-Phase Flow Lab, OSU, Stillwater, OK)

Bhagwat and Ghajar (2014), in order to verify the ability of their void fraction correlation to handle high system pressures, compared its outcome with the measured void fraction data for high system pressures as shown in Fig. 15.18. Their correlation predicts the void fraction within ±7.5% of the measured values for different mass flux at high system pressure of 7.2 MPa and 18.1 MPa, respectively. For 7.2 MPa system pressure, their correlation predicts 76% of data points of Inoue et al. (1995) within ±10% error bands. In the case of 18.1 MPa system pressure, their correlation is found to predict 92% of data points of Liu et al. (2013) within ±10% error bands.

The performance of the Bhagwat and Ghajar (2014) void fraction correlation against air-oil void fraction data for different liquid dynamic viscosities is shown in Fig. 15.19. It is seen that for a wide range of liquid phase dynamic viscosities, their correlation predicts majority of the data within ±20% error bands. For details about the sources of the data sets used, refer to Bhagwat and Ghajar (2014).

Since the expression for drift velocity (U_{GM}) used by Bhagwat and Ghajar (2014) as given in Eq. (15.16) incorporates a correction factor (c_3) to account for the reduced wall drag for large diameter pipes, it is also of interest to see how well their void fraction correlation predicted the void fraction data for large diameter

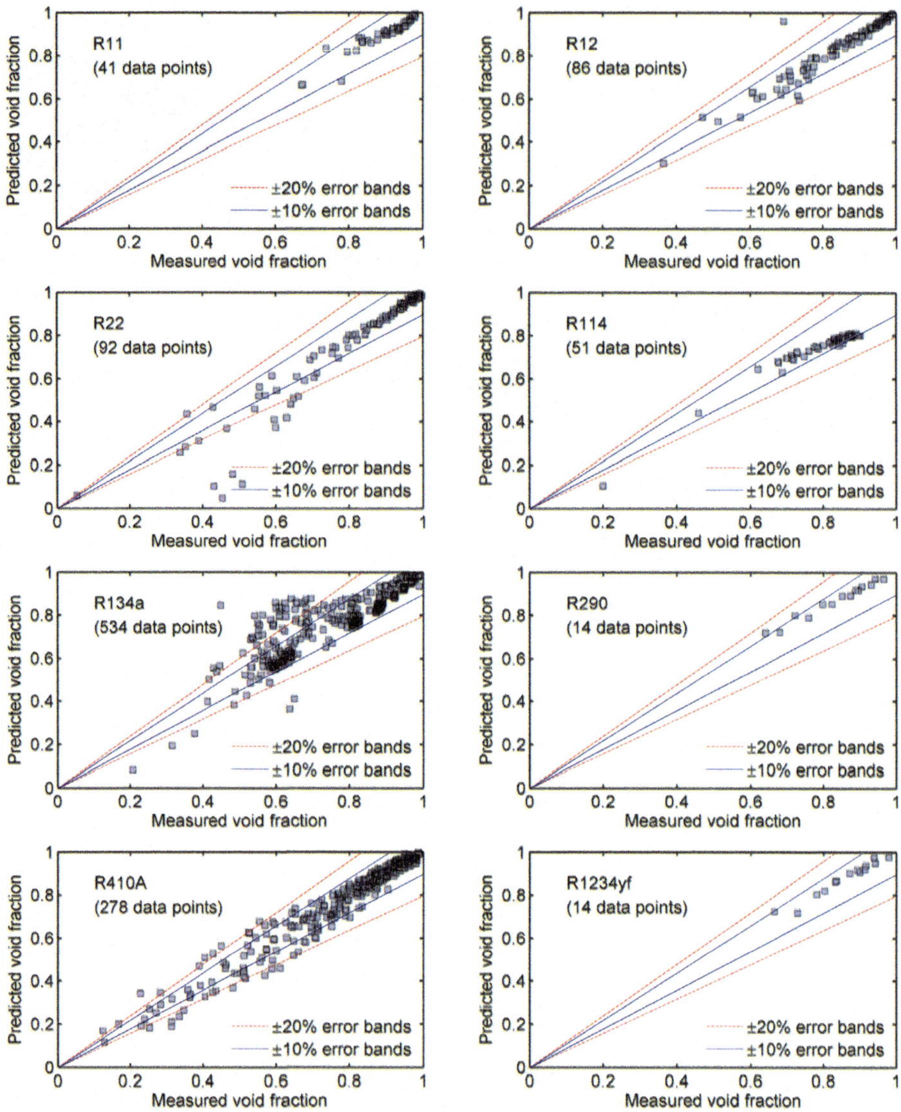

Fig. 15.16 Performance of the Bhagwat and Ghajar (2014) void fraction correlation against refrigerant void fraction data (0.5 mm $\leq D_h \leq$ 650 mm)

pipes corresponding to $La < 0.025$. Based on the Laplace variable, the pipe diameters that classify as large diameter pipes are in a range of 0.067–0.305 m. The experimental data for large diameter pipes is obtained from Abdulkadir et al. (2010) for air-oil fluid combination and from Schlegel et al. (2010), Hills (1976), Inoue (2001), Shoukri et al. (2003), and Hashemi et al. (1986) for air-water fluid combination in vertical upward two-phase flow. With the exception of very few outliers for the data of Hills (1976) and Shoukri et al. (2003), the void fraction correlation of Bhagwat

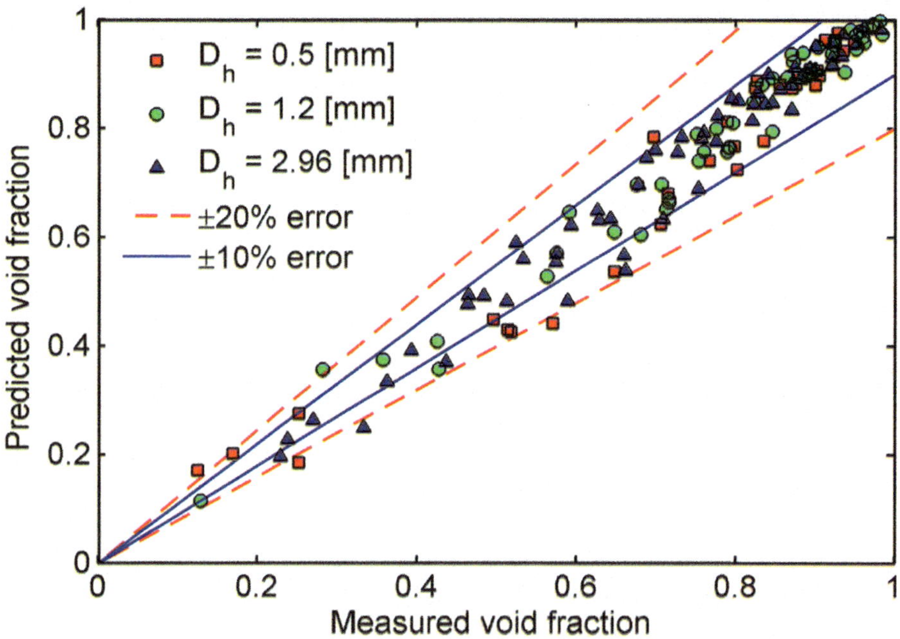

Fig. 15.17 Performance of the Bhagwat and Ghajar (2014) void fraction correlation against void fraction data of Shedd (2010) in mini and micro size channels using R410A

and Ghajar (2014) predicts the void fraction data for $La < 0.025$ well within ±20% error bands as shown in Fig. 15.20. The pipe diameter of 0.067 m for $La = 0.022$ corresponds to air-oil data, while all other data sets belong to air-water fluid combination in vertical upward pipe orientation.

In addition to the large diameter pipes, the void fraction correlation of Bhagwat and Ghajar (2014) is also valid for noncircular pipe geometries. The performance of their correlation against the void fraction data in annular and rectangular geometries is illustrated in Fig. 15.21. Their correlation for the distribution parameter (C_o) given by Eq. (15.13) recommends use of Eq. (15.14) with constant $c_1 = 0.4$ for rectangular and annular ducts. The void fraction data for annular pipes belongs to air-water data (123 data points from Wongwises and Pipathttakul (2006) for $D_o = 12.5$ mm and $D_i = 8$ mm and 103 data points from Das et al. (2002) for $D_o = 50.8$ mm and $D_i = 25.4$ mm where D_o and D_i are the outer and inner pipe diameter of annulus, respectively). The void fraction data in rectangular channels belong to steam-water data (157 data points from Marchaterre (1956) for 11.1 mm × 93.6 mm, 101 data points from Marchaterre et al. (1960) for 6.35 mm × 50.8 mm, and 98 data points from Marchaterre et al. (1960) for 12.7 mm × 50.8 mm rectangular channels).

The performance of the void fraction correlation of Bhagwat and Ghajar (2014) was also compared with some of the existing top performing void fraction correlations based on the drift flux and separated flow models. As detailed in Bhagwat and Ghajar (2014), their correlation consistently gave better performance over the entire range of parameters shown in Table 15.2 and the entire range of void fraction

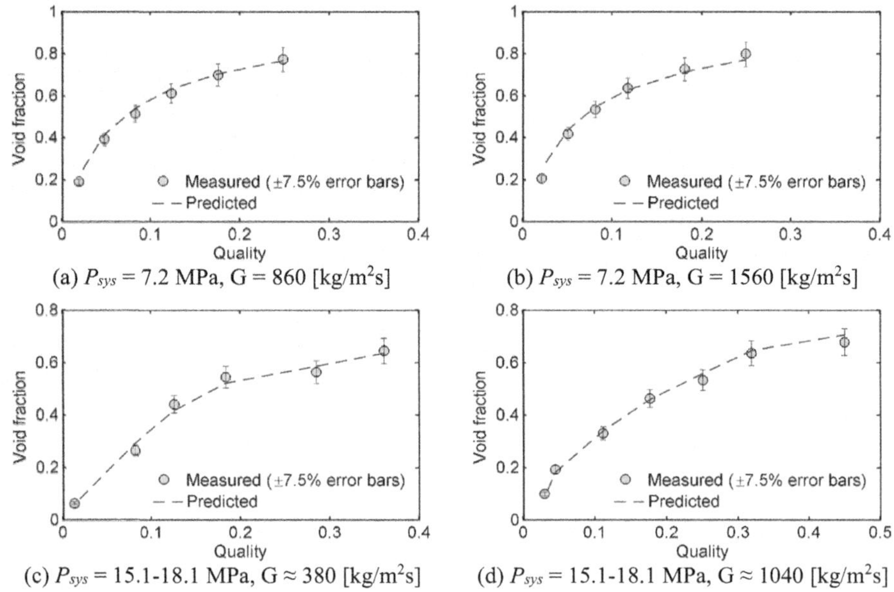

(a) P_{sys} = 7.2 MPa, G = 860 [kg/m^{2}s]

(b) P_{sys} = 7.2 MPa, G = 1560 [kg/m^{2}s]

(c) P_{sys} = 15.1-18.1 MPa, G ≈ 380 [kg/m^{2}s]

(d) P_{sys} = 15.1-18.1 MPa, G ≈ 1040 [kg/m^{2}s]

Fig. 15.18 Performance of the Bhagwat and Ghajar (2014) void fraction correlation against high pressure steam–water void fraction data. (Experimental data for (**a**) and (**b**) from Inoue et al. (1995) and (**c**) and (**d**) from Liu et al. (2013))

($0 < \alpha_G < 1$) without any reference to flow regime map. It is evident from Table 15.2 that their correlation for void fraction covers pretty much the entire range of two-phase flow parameters that would exist in practical applications.

Note that the correlation of Bhagwat and Ghajar (2014) is not validated against data for microchannels with pipe diameters less than 0.5 mm. For gas-liquid two-phase flow in small size pipes less than 0.5 mm, correlation of Xiong and Chung (2006) could be used. Their correlation is based on the experimental data for pipe diameters in a range of 0.1–0.6 mm and consists of bubbly, slug, annular flow, and entire range of gas volumetric flow fraction (λ). Their void fraction correlation, Eqs. (15.20 and 15.21), accounts for the nonlinear relationship between void fraction and gas volumetric flow fraction as a function of hydraulic pipe diameter (D_h) measured in mm:

$$\alpha_G = \frac{a_1 \lambda^{0.5}}{1 - (1 - a_1)\lambda^{0.5}} \left\} \begin{array}{l} 0.1 \leq D_h \leq 0.6 \text{ mm} \\ 4 \leq Re_{SL} \leq 1670, 4 \leq Re_{SG} \leq 1725 \end{array} \right. \tag{15.20}$$

$$a_1 = \frac{0.266}{1 + 13.8 \exp(-6.88 D_h)} \tag{15.21}$$

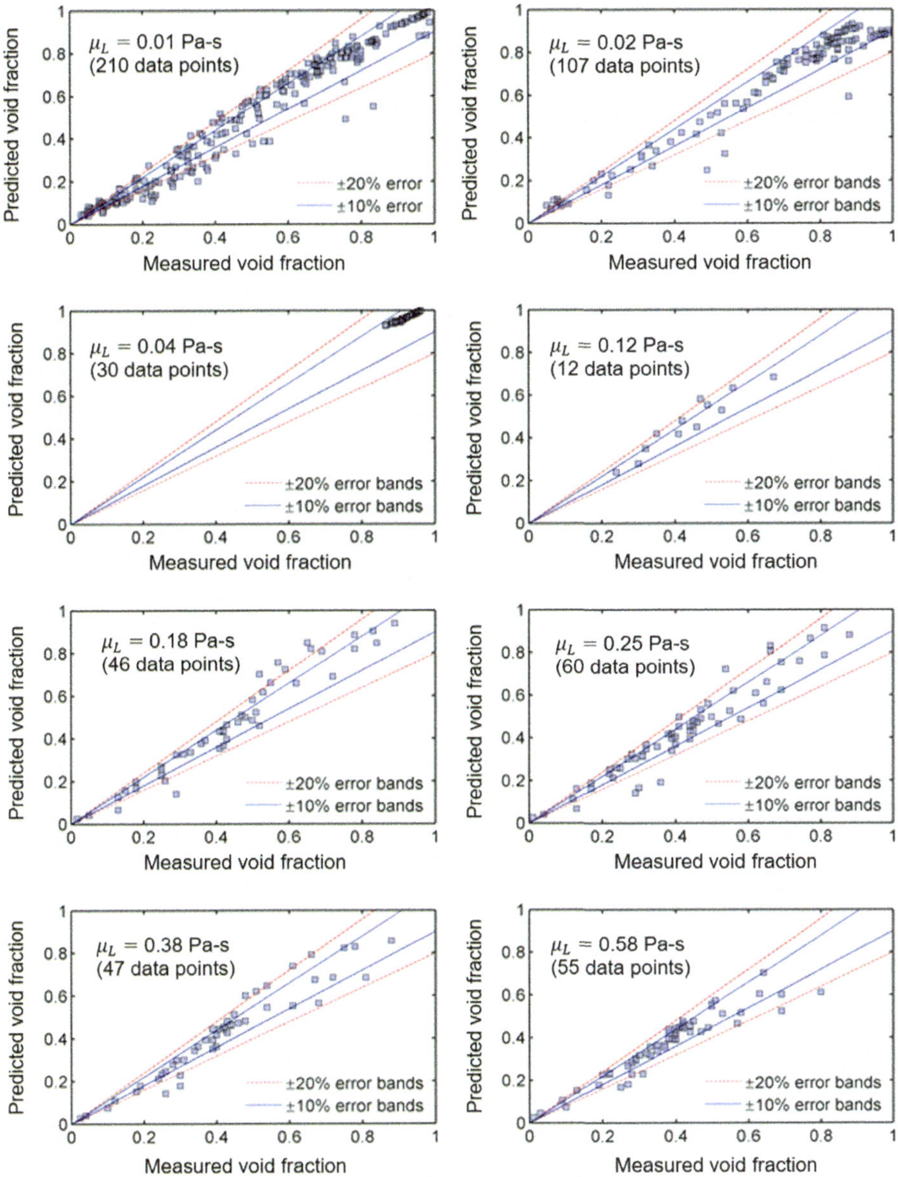

Fig. 15.19 Performance of the Bhagwat and Ghajar (2014) void fraction correlation against air–oil void fraction data (0.025 m $\leq D_h \leq$ 0.19 m)

It must be noted that drift flux model-based void fraction correlations are not suitable to model stratified type of flow. Compared to drift flux models, separated flow models typically work better for stratified flow. However, these models too cannot provide accurate prediction of the void fraction with desired accuracy.

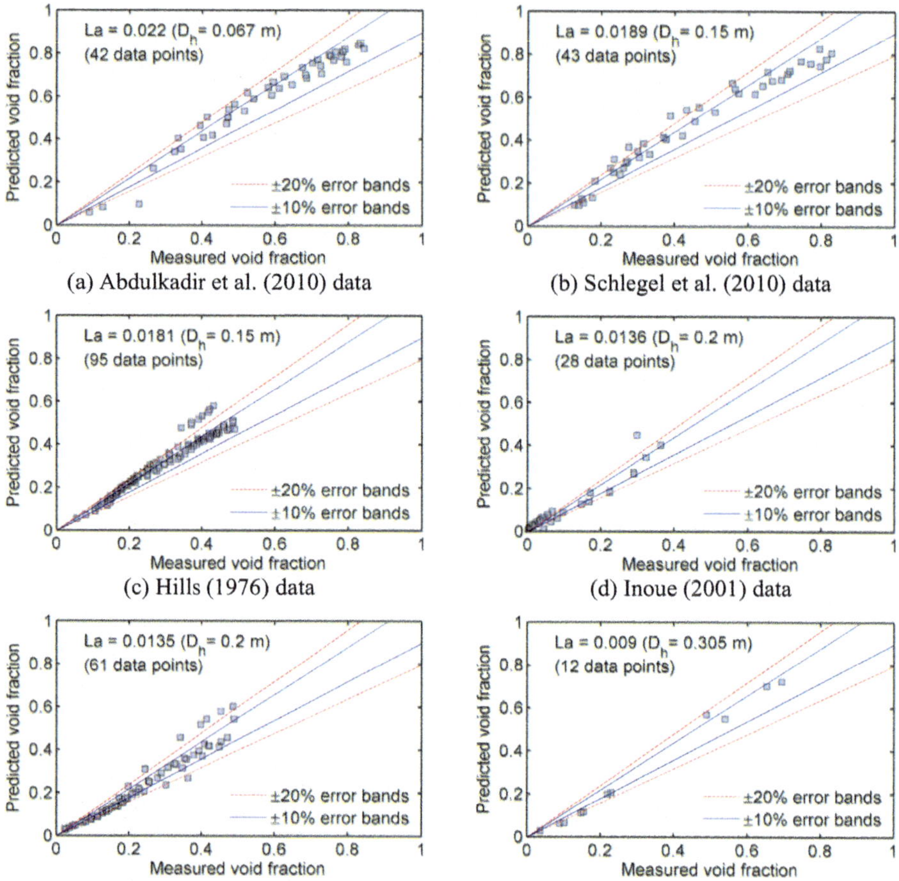

(a) Abdulkadir et al. (2010) data

(b) Schlegel et al. (2010) data

(c) Hills (1976) data

(d) Inoue (2001) data

(e) Shoukri et al. (2003) data

(f) Hashemi et al. (1986) data

Fig. 15.20 Performance of the Bhagwat and Ghajar (2014) void fraction correlation against void fraction data in large diameter pipes ((**a**) air–oil data and (**b–f**) air–water data)

Stratified flow pattern in horizontal and downward pipe inclinations has a peculiar flow structure, and it is practically difficult to accurately model it with existing flow pattern-independent two-phase flow models. Apparently, the stratified flow pattern needs to be modeled using mechanistic flow model approach introduced by Taitel and Dukler (1976) and its derivatives. In addition to Taitel and Dukler (1976) model, two-phase literature also reports semi-mechanistic models called "apparent rough surface model" first introduced by Hamersma and Hart (1987) and Hart et al. (1989) and "double circle model" proposed by Chen et al. (1997), to calculate void fraction and two-phase pressure drop in stratified two-phase flow. It must be mentioned that the apparent rough surface and double circle models are developed only for stratified flow in horizontal pipe orientation and need to be scrutinized against stratified flow data in downward pipe inclinations. For a detailed discussion on these three models for stratified flow, refer to Ghajar and Bhagwat (2017).

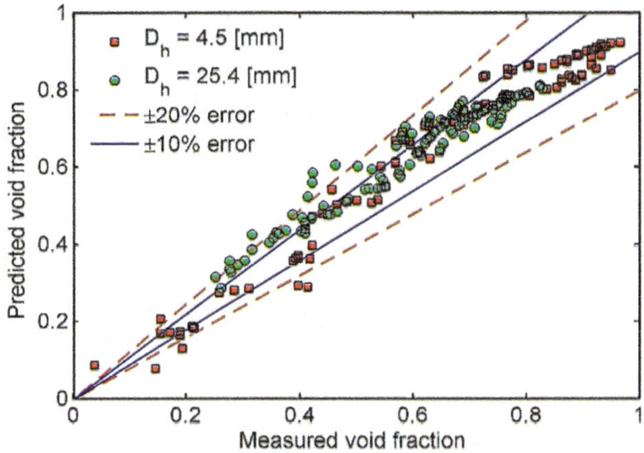

(a) □: D_o = 12.5 mm, D_i = 8 mm, ○: D_o = 50.8 mm, D_i = 25.4 mm

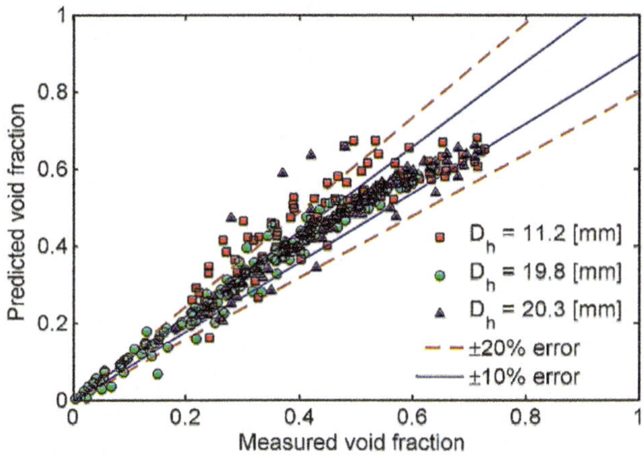

(b) □: 6.35 mm × 50.8 mm, ○: 11.1 mm × 93.6 mm and Δ: 12.7 mm × 50.8 mm

Fig. 15.21 Performance of the Bhagwat and Ghajar (2014) void fraction correlation against void fraction data in noncircular geometries ((**a**) annular channels and (**b**) rectangular channels)

An Illustrative Example

Use of Woldesemayat and Ghajar (2007) and Bhagwat and Ghajar (2014) Void Fraction Correlations Consider the vertical upward two-phase flow of air and silicone oil in a 12 mm I.D. stainless-steel pipe having a surface roughness of 0.002 mm. The mass flow rate of gas and liquid phase is 0.0015 kg/s and 0.9 kg/s, respectively. The fluid thermophysical properties may be taken as follows: $\rho_G = 1.2$ kg/m^3, $\rho_L = 920$ kg/m^3, $\mu_L = 0.005$ Pa-s, and $\sigma = 0.02$ N/m. The measured value of void fraction (α_G) for this flow is 0.5. Compare the measured value of α_G with the predicted values of α_G using (1) the explicit void fraction correlation of

Woldesemayat and Ghajar (2007) and (2) the implicit void fraction correlation of Bhagwat and Ghajar (2014).

Solution

1. Calculation of α_G using Woldesemayat and Ghajar (2007) explicit void fraction correlation, given by Eqs. (15.11 and 15.12):

Referring to Table 12.1, the superficial gas (U_{SG}) and liquid (U_{SL}) velocities and the two-phase gas volumetric flow fraction (λ) are first calculated as follows:

$$U_{SG} = \frac{4\dot{m}_G}{\pi D^2 \rho_G} = \frac{4 \times 0.0015}{\pi \times 0.012^2 \times 1.2} = 11.05 \text{ m/s}$$

$$U_{SL} = \frac{4\dot{m}_L}{\pi D^2 \rho_L} = \frac{4 \times 0.9}{\pi \times 0.012^2 \times 920} = 8.65 \text{ m/s}$$

$$\lambda = \frac{U_{SG}}{U_{SG} + U_{SL}} = \frac{11.05}{11.05 + 8.65} = 0.561$$

Next, calculate the distribution parameter (C_o) using Eq. (15.11) and the drift velocity (U_{GM}) using Eq. (15.12). In using Eq. (15.12), assume $p_{atm} = p_{sys}$:

$$C_o = \lambda\left(1 + \left[\frac{U_{SL}}{U_{SG}}\right]^{(\rho_G/\rho_L)^{0.1}}\right) = (0.561)\left(1 + \left[\frac{8.65}{11.05}\right]^{(1.2/920)^{0.1}}\right) = 1.0556$$

$$U_{GM} = 2.9\left[\frac{gD\sigma(1+\cos\theta)(\rho_L - \rho_G)}{\rho_L^2}\right]^{0.25}(1.22 + 1.22\sin\theta)^{p_{atm}/p_{sys}}$$

$$= 2.9\left[\frac{(9.81)(0.012)(0.02)(1+\cos 90)(920 - 1.2)}{920^2}\right]^{0.25}(1.22 + 1.22\sin 90) = 0.283$$

Finally, the void fraction (α_G) is calculated from Eq. (15.1) where

$$\alpha_G = \frac{U_{SG}}{C_o U_M + U_{GM}} = \frac{11.05}{1.0556\,(19.7) + (0.283)} = 0.52$$

where $U_M = U_{SL} + U_{SG} = 8.65 + 11.05 = 19.7$ m/s

2. Calculation of α_G using Bhagwat and Ghajar (2014) implicit void fraction correlation, given by Eqs. (15.13) to (15.19):

To calculate the distribution parameter (C_o) using Eq. (15.13), the following calculations should first take place:

Referring to Table 12.1, the two-phase quality (x) is

$$x = \frac{\dot{m}_G}{\dot{m}_G + \dot{m}_L} = \frac{0.0015}{0.0015 + 0.9} = 0.00166$$

From Eq. (15.15), $Re_M = 4.35 \times 10^4$, and from the following equation for the Colebrook (1939) fanning friction factor correlation with $\varepsilon/D = 1.67 \times 10^{-4}$, the two-phase fanning friction factor is $f_M = 0.0055$:

$$\frac{1}{\sqrt{f_M}} = -4.0 \log\left(\frac{\varepsilon/D}{3.7} + \frac{1.256}{Re_M\sqrt{f_M}}\right)$$

Now, with $c_1 = 0.2$ for circular pipes and $\lambda = 0.561$ (see part 1 of this problem), the value of $C_{o,1}$ from Eq. (15.14) can be determined.

To calculate U_{GM} from Eq. (15.16), the following calculations should first take place:

From Eq. (15.17), since $\mu_L/0.001 \leq 10 \Rightarrow c_2 = 1$
From Eqs. (15.18 and 15.19), since $La = 0.124 \geq 0.025 \Rightarrow c_3 = 1$

Now, from Eq. (15.13) for the distribution parameter (C_o) and Eq. (15.16) for the drift velocity (U_{GM}), the iterative calculations yield, $C_o = 1.2$ and $U_{GM} = 0.088$ m/s. Finally, the void fraction (α_G) is calculated from Eq. (15.1) where

$$\alpha_G = \frac{U_{SG}}{C_o U_M + U_{GM}} = \frac{11.05}{1.2\,(19.7) + (0.088)} = 0.47$$

where $U_M = U_{SL} + U_{SG} = 8.65 + 11.05 = 19.7$ m/s

Discussion The explicit void fraction correlation of Woldesemayat and Ghajar (2007), Eqs. (15.11 and 15.12), overpredicted the measured value of α_G by 4% (0.52 vs 0.5). The implicit void fraction correlation of Bhagwat and Ghajar (2014), Eqs. (15.13, 15.14, 15.15, 15.16, 15.17, 15.18, and 15.19), underpredicted the measured value of α_G by 6% (0.47 vs 0.5). Both correlations did an excellent job of predicting the experimental value of void fraction for this flow.

Chapter 16
Pressure Drop

Background

Similar to single-phase flow, total pressure drop in gas-liquid two-phase flow is comprised of hydrostatic, frictional, and accelerational components as was shown in Eq. (13.1). However, unlike the single-phase flow, these components of two-phase pressure drop in gas-liquid flows may depend upon several parameters such as void fraction, pipe orientation, flow patterns, and liquid entertainment fraction. The hydrostatic or gravitational component of two-phase pressure drop due to the pipe inclination depends on the two-phase mixture density which in turn is a function of flow patterns and void fraction. Accelerational pressure drop is due to the change in momentum of the two-phase mixture and can be neglected for non-boiling (adiabatic) two-phase flows within a short pipe length. However, for two-phase flow undergoing phase change process or even for adiabatic two-phase flow in long and inclined pipes, depending upon the flow pattern, the accelerational pressure drop can contribute up to 20% of the total two-phase pressure drop. Finally, the frictional component of two-phase pressure drop is a result of the interfacial friction between the two phases and wall friction with both or either phase and depends upon the flow pattern structure, pipe diameter, and fluid properties.

The two-phase pressure drop is influenced by several parameters such as pipe orientation, flow rates of individual phases, pipe diameter, fluid properties, and surface roughness. These influences on the two-phase pressure drop will be discussed next.

A. J. Ghajar, *Single- and Two-Phase Flow Pressure Drop and Heat Transfer in Tubes*, Mechanical Engineering Series,
https://doi.org/10.1007/978-3-030-87281-6_16

Effect of Pipe Orientation on Pressure Drop

The effect of pipe orientation on total two-phase pressure drop is primarily due to the effect of pipe orientation on the hydrostatic component of the two-phase pressure drop. Two-phase literature reports very few studies focused on measurement of two-phase pressure drop over the entire range of pipe orientations. Some of these studies include work of Beggs (1972), Mukherjee (1979), Spedding et al. (1982), and Lips and Meyer (2012). The total two-phase pressure drop measurements for different pipe orientations ranging from +90° to −90° carried out at the Two-Phase Flow Lab, OSU, and reported by Lips and Meyer (2012) are presented in Figs. 16.1 and 16.2, respectively.

A comparison between Figs. 16.1 and 16.2 shows that the variation of the total two-phase pressure drop as a function of pipe orientation is similar for both non-boiling (Fig. 16.1) and condensing (Fig. 16.2) two-phase flows. It is evident that at low gas and liquid flow rates, the total two-phase pressure drop is relatively insensitive to the change in downward pipe inclination. This is due to the fact that at these flow rates, stratified flow exists and its physical structure and associated void fraction are relatively insensitive to the change in pipe orientation. In upward pipe inclinations, the total two-phase pressure drop increases rapidly with increase in the pipe orientation, and it is fascinating to see that the total two-phase pressure drop at lower gas flow rates is greater than that at higher gas flow rates. This is because of the reversal of liquid film that induces decreasing trend of pressure gradient with increase in gas flow rates. For a detailed discussion on the pressure gradient minimum and flow reversal in upward inclined flow, refer to Ghajar and Bhagwat (2017). At higher liquid flow rates, negative values of total two-phase pressure drop in downward pipe inclinations represent partial pressure recovery due to gain in hydrostatic component of two-phase pressure drop.

Effect of Phase Flow Rates on Pressure Drop

The effect of change in phase flow rates on the two-phase pressure drop is essentially due to the change in physical structure of the flow patterns and is of different nature for the hydrostatic and frictional components of two-phase pressure drop. Since the hydrostatic component of two-phase pressure drop depends only on the void fraction (see Eq. 13.2), for a fixed liquid flow rate, increase in gas flow rate decreases the contribution of hydrostatic pressure drop (due to decrease in mixture density) where it increases the frictional pressure drop component. The relation between change in phase flow rates and the frictional pressure drop could be construed in a better way by presenting their relationship in different coordinate systems. As shown in Fig. 16.3, for low liquid and low gas flow rates, the two-phase frictional pressure drop remains virtually constant and increases drastically only for moderate to high gas flow rates (intermittent and annular flow regimes). For intermittent type of flow,

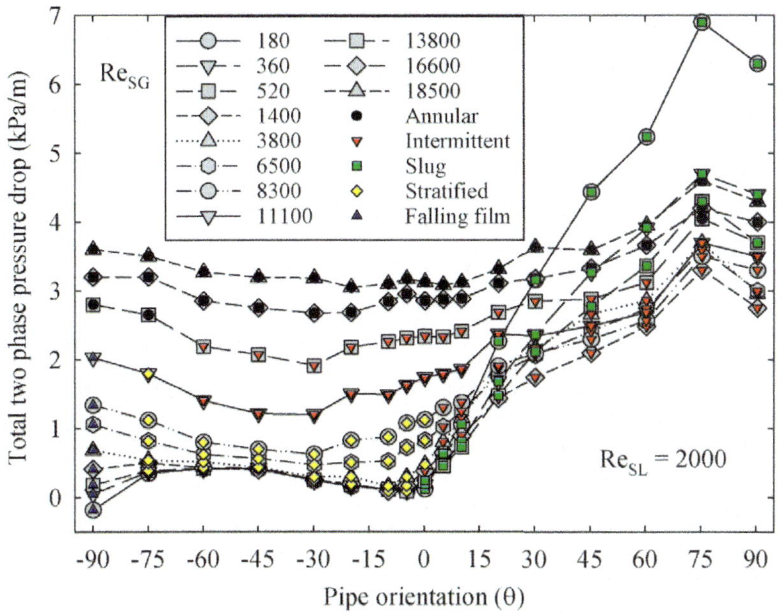

(a)

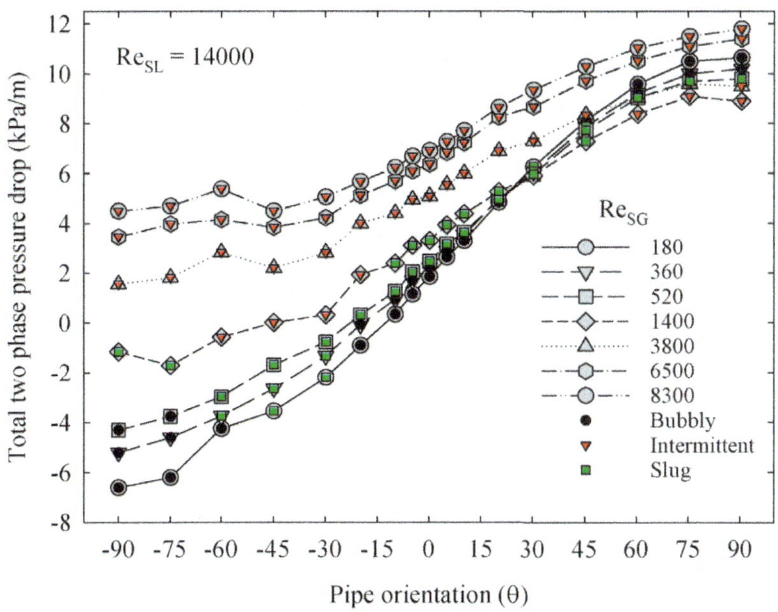

(b)

Fig. 16.1 Effect of pipe orientation on total two-phase pressure drop. (**a**) Re_{SL}= 2000 and Re_{SG} = 180–18,500. (**b**) Re_{SL} = 14,000 and Re_{SG} = 180–8300. (Data measured at the Two-Phase Flow Lab, OSU, Stillwater, OK)

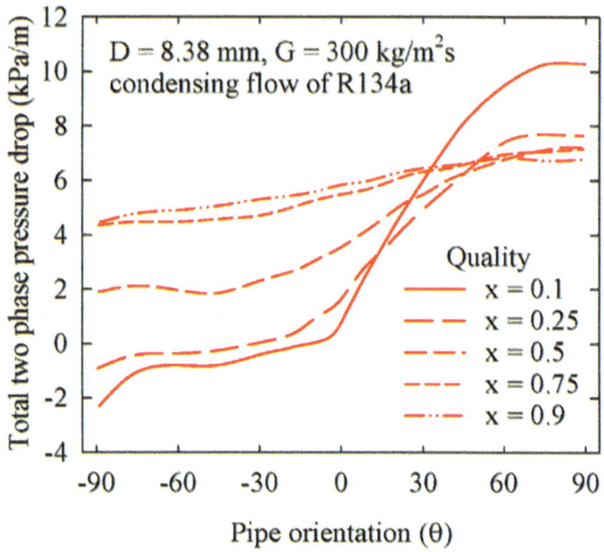

Fig. 16.2 Effect of pipe orientation on two-phase pressure drop. (Data of Lips and Meyer 2012)

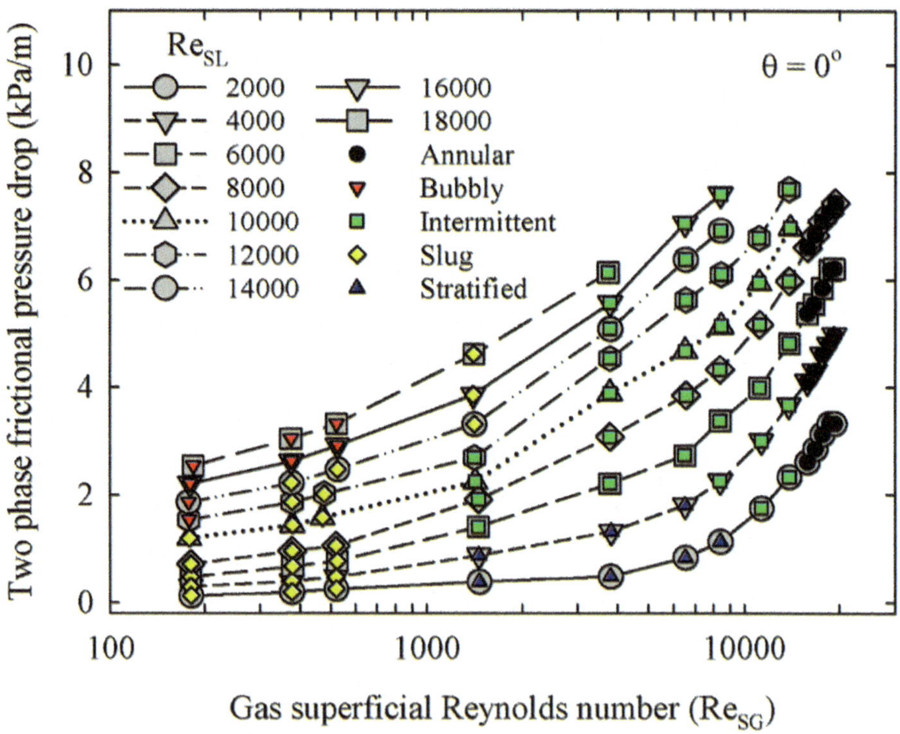

Fig. 16.3 Two-phase frictional pressure drop for varying gas and liquid flow rates. (Data measured at the Two-Phase Flow Lab, OSU, Stillwater, OK)

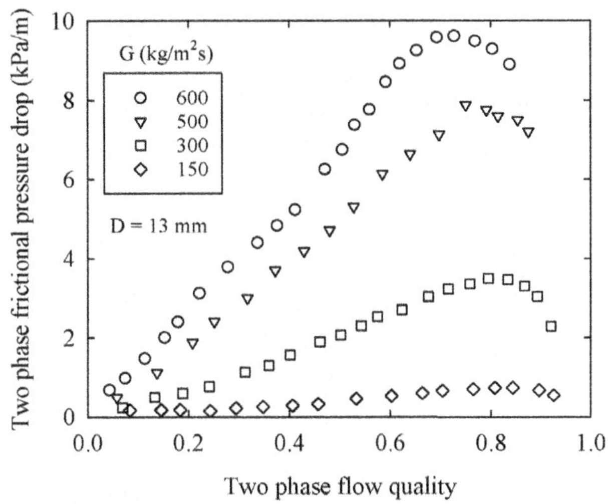

Fig. 16.4 Two-phase pressure drop as a function of phase flow rates. (Data of Quiben and Thome 2007)

sharp increase in the frictional pressure drop is essentially due to the turbulent and chaotic nature of the two-phase mixture. For annular flow, the pressure gradient increases sharply with increase in the gas and liquid flow rates. At a fixed gas flow rate, increase in the liquid flow rate increases the liquid film thickness and thus reduces the cross-sectional area available for the gas phase. This increases the actual gas velocity and exerts higher shear on the gas-liquid interface to increase the two-phase frictional pressure drop. Additionally, the gas-liquid interface offers a rough surface to the gas flow, and the interface gets roughened progressively with the increase in the liquid film thickness or alternatively the liquid flow rate that augments the frictional pressure drop.

Variation of two-phase pressure drop with change in two-phase flow rates could also be presented in another perspective using two-phase flow quality data of Quiben and Thome (2007). This type of presentation shown in Fig. 16.4 is usually adopted for high-pressure systems (often encountered in refrigeration and nuclear applications) which tend to occupy the entire range of two-phase flow quality ($0 < x < 1$). It is evident that the two-phase pressure drop increases systematically with increase in the mass flux (G) and quality (x). An inflection point that flips the trend of increase in two-phase pressure drop with increase in quality is quite noticeable. This change in the trend of two-phase pressure drop could be explained based on the change in structure of the two-phase flow pattern at the inflection point. By the time the two-phase flow system reaches the point of maximum pressure drop (inflection point), the flow pattern attains annular flow structure; however, beyond that point due to severe entrainment process, the liquid phase in contact with the pipe wall is gradually moved to the central gas core reducing the liquid film thickness and consequently reducing the effective viscosity in the near-wall region. This ultimately results into a reduced frictional pressure drop. This type of trend is not evident in Fig. 16.3 since the data presented there does not contain two-phase flow with

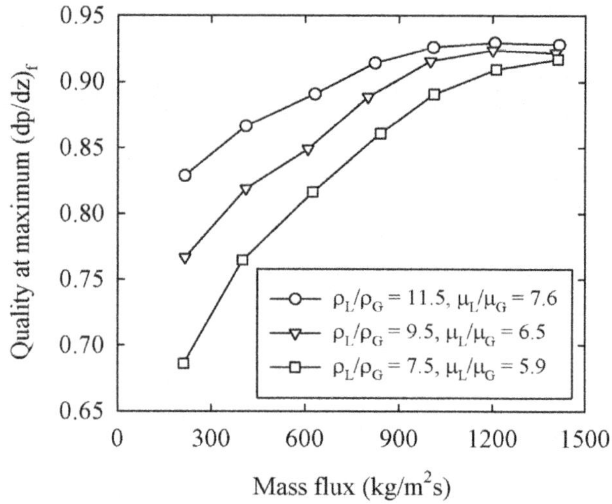

Fig. 16.5 Influence of mass flux and fluid properties on quality corresponding to maximum pressure drop. (Adapted from Ducoulombier et al. 2011)

significant entrainment. It is also evident from Fig. 16.4 that the quality at the point of maximum pressure drop shifts toward lower qualities with increase in the two-phase mixture mass flux. Additionally, Ducoulombier et al. (2011) reported that quality associated with maximum frictional pressure drop also depends upon the fluid properties. The combined effect of mass flux and fluid thermophysical properties on the quality at maximum pressure drop is reported in Fig. 16.5. It is clear that the two-phase flow quality corresponding to the point of maximum pressure drop may vary approximately between 0.65 and 0.95.

Effect of Pipe Diameter on Pressure Drop

Similar to the single-phase pressure drop, the effect of decrease in pipe diameter is to increase the two-phase frictional pressure drop. However, this effect of pipe diameter is usually in conjunction with two-phase flow patterns. The two-phase frictional pressure drop in annular flow regime depends on the pipe diameter to a great extent where it is relatively less sensitive to the pipe diameter in bubbly flow regime. It is evident from Fig. 16.6a that for a fixed mass flux of R134a, the two-phase frictional pressure drop for different pipe diameters deviates significantly for higher values of two-phase flow quality (annular flow pattern), whereas for low values of two-phase flow quality (bubbly flow), the effect of pipe diameter on two-phase frictional pressure drop gradually diminishes. Similar conclusions could be drawn for air-water two-phase flow system from the work of Kaji and Azzopardi (2010) who studied the effect of pipe diameters in a range of 10 to 50 mm on two-phase pressure drop in annular flow regime. As shown in Fig. 16.6b, for a fixed liquid flow rate, the effect of change in pipe diameter is most significant at higher gas flow rates (i.e.,

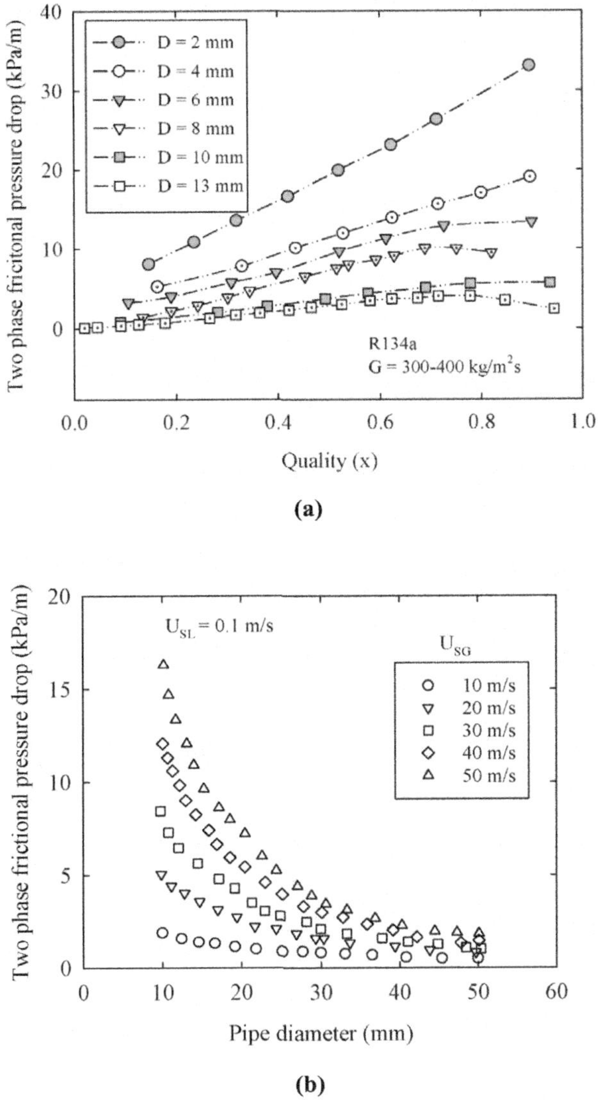

Fig. 16.6 Effect of pipe diameter on two-phase frictional pressure drop. (**a**) Two-phase flow of R-134a refrigerants. (**b**) Two-phase flow of air-water. (Adapted from Kaji and Azzopardi 2010)

higher two-phase flow qualities). The two-phase frictional pressure drop is due to the friction at the pipe wall as well as the gas-liquid interface, and the gas-liquid interfacial area increases with increase in the pipe diameter. Thus, it is evident that any two-phase frictional pressure drop correlation must definitely account for the pipe diameter effect in annular flow regime.

Effect of Fluid Properties on Pressure Drop

The important fluid properties that show a noticeable effect on two-phase frictional pressure drop are the gas phase density and liquid phase dynamic viscosity. The increase in gas phase density decreases the slippage at the gas-liquid interface and hence decreases the frictional component of two-phase pressure drop, whereas the increase in liquid dynamic viscosity increases the shear in liquid phase in contact with the pipe wall and also the shear at gas-liquid interface resulting in increase in frictional pressure drop. Intuitively, it can be said that the effect of fluid properties on two-phase frictional pressure drop is more prominent for shear-driven flows compared to buoyancy-driven two-phase flow. For more details on the effect of fluid properties on frictional pressure drop, refer to Oshinowo (1971), Fukano and Furukawa (1998), Abduvayat et al. (2003), Hlaing et al. (2007), and Gokcal (2008).

Effect of Surface Roughness on Pressure Drop

The effect of pipe surface roughness on two-phase pressure drop is crucial in applications involving two-phase flow through steel or micro-finned tubes. The internally ribbed or micro-finned tubes are used in air conditioning and refrigeration applications to improve the tube side heat transfer, however, at the expense of enhanced pressure drop. Most of the two-phase flow research reported in the literature is carried out in a transparent (smooth) pipe, while the effect of pipe surface roughness on the frictional two-phase pressure drop is a relatively less investigated issue. Work of Wongs-ngam et al. (2004) and Shannak (2008) reported a considerable effect (20–60%) of wall roughness on pressure drop especially for the two-phase flow at high mass flux and quality. Figure 16.7 shows that the effect of pipe wall roughness on two-phase pressure drop is considerable at high gas velocities and increases with increase in liquid velocity. Obviously, similar to single-phase flow, the effect of wall roughness on two-phase pressure drop increases with decrease in the pipe diameter. Experiments performed at the Two-Phase Flow Lab,

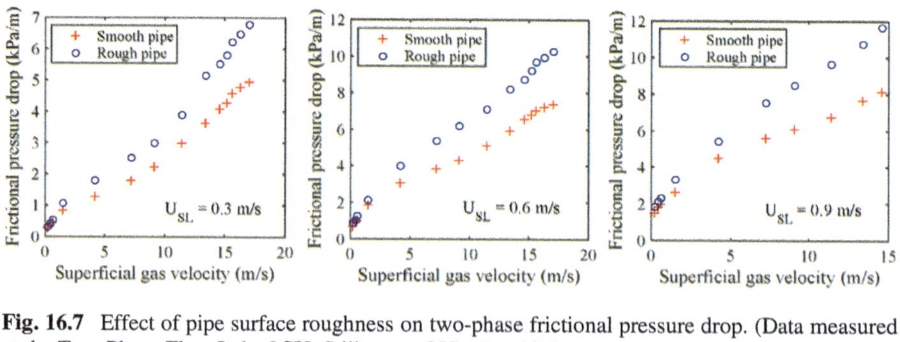

Fig. 16.7 Effect of pipe surface roughness on two-phase frictional pressure drop. (Data measured at the Two-Phase Flow Lab, OSU, Stillwater, OK) (D= 12.5 mm, ε = 20 μm for rough pipe)

OSU, confirm these observations. Moreover, their work also shows that the effect of pipe roughness on two-phase pressure drop is independent of the pipe orientation. The effect of pipe surface roughness on two-phase frictional pressure drop is maximum for annular flow (at high gas flow rates) since, in addition to the effect of pipe wall roughness acting on the liquid phase, the structure of this flow pattern also offers a continuous and rough gas-liquid interface responsible for enhanced pressure drop.

Pressure Gradient Minimum and Flow Reversal in Upward Inclined Flow

In cocurrent gas-liquid upward inclined two-phase flow, at low liquid and moderate gas flow rates, although the net two-phase flow is observed in the upward direction, reversal of liquid film in contact with the pipe wall is observed under the influence of gravity forces. This phenomenon is a consequence of the interaction between the interfacial shear stress (exerted by gas phase on liquid film) and the gravitational forces acting on the liquid film. The liquid film in contact with the pipe wall may undergo a partial reversal (fluctuating wall shear stress) or complete reversal (negative wall shear stress) depending upon the gas and liquid flow rates and exhibits a decreasing trend of two-phase pressure gradient.

As long as there is a balance between interfacial shear stress and hydrostatic pressure drop, liquid film in contact with the pipe wall will either remain stationary or oscillate giving a net zero wall shear stress. Thus, when the interfacial shear stress is large enough to supersede the gravity forces (in the form of hydrostatic pressure drop), the net flow of two-phase mixture will be in upward direction without any reversal of liquid film.

Experimentally, this condition could be determined by analyzing the pressure drop signal as a function of gas and liquid flow rates. At a fixed liquid flow rate and with an increase in gas flow rate, a point is reached where the gas phase does not have enough potential to carry liquid phase along with it in the downstream direction. At this point, the liquid phase in contact with the pipe wall appears to oscillate or move in the downward direction. The point corresponding to the inception of decreasing pressure gradient trend (in spite of increase in gas flow rate) is regarded as the *flow reversal point*, while the condition at which the pressure gradient commence to increase again with increase in gas flow rate is identified as *pressure gradient minimum point*. The two-phase literature provides several instances of decreasing pressure gradient trends in context to churn-annular transition in vertical upward two-phase flow. As a matter of fact, the decreasing trend of pressure gradient minimum also exists at lower gas flow rates during slug-churn transition in upward inclined pipes. Nevertheless, the flow physics governing these two separate regions of pressure gradient minimum is significantly different. A typical example of decreasing trends of pressure gradient minimum in vertical

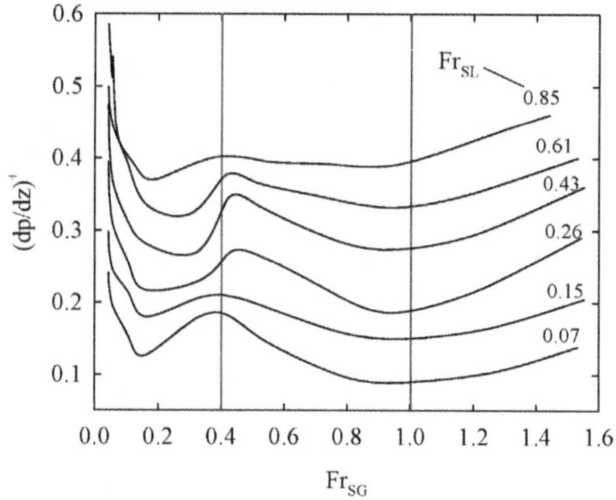

Fig. 16.8 Non-dimensional pressure drop for varying non-dimensional gas and liquid flow rates. (Data measured at the Two-Phase Flow Lab, OSU, Stillwater, OK)

upward pipe inclination is illustrated in Fig. 16.8. The measured two-phase pressure drop and corresponding gas and liquid flow rates are made non-dimensional as shown in Eqs. (16.1, 16.2, and 16.3), respectively:

$$\left(\frac{dp}{dz}\right)_t^+ = \frac{(dp/dz)_t}{(\rho_L - \rho_G)g} \tag{16.1}$$

$$Fr_{SG} = \frac{U_{SG}}{\sqrt{gD}}\sqrt{\frac{\rho_G}{(\rho_L - \rho_G)}} \tag{16.2}$$

$$Fr_{SL} = \frac{U_{SL}}{\sqrt{gD}}\sqrt{\frac{\rho_L}{(\rho_L - \rho_G)}} \tag{16.3}$$

It is seen from Fig. 16.8 that for a fixed liquid flow rate, when the gas flow rate is increased, the two-phase pressure gradient exhibits first minimum at $Fr_{SG} \approx 0.2 - 0.3$ and a maximum at $Fr_{SG} \approx 0.4$. The first instance of decreasing pressure drop is due to falling film surrounding the gas slug. As the gas slug rises in downstream direction, it sheds liquid phase surrounding it to maintain the continuity. The sudden increase in pressure drop between $Fr_{SG} \approx 0.2 - 0.4$ is due to the high level of turbulence caused by disintegration of gas slug during slug to churn/intermittent flow transition. Further this point of maximum pressure gradient at $Fr_{SG} \approx 0.4$, churn/intermittent flow is known to commence, and again a decreasing trend of pressure gradient followed by a pressure gradient minimum is observed at $Fr_{SG} \approx 0.9 - 1$. Beyond this point, annular flow is known to exist. Note that this value of the non-dimensional gas flow rate is essentially the criteria given by Eq. (14.25).

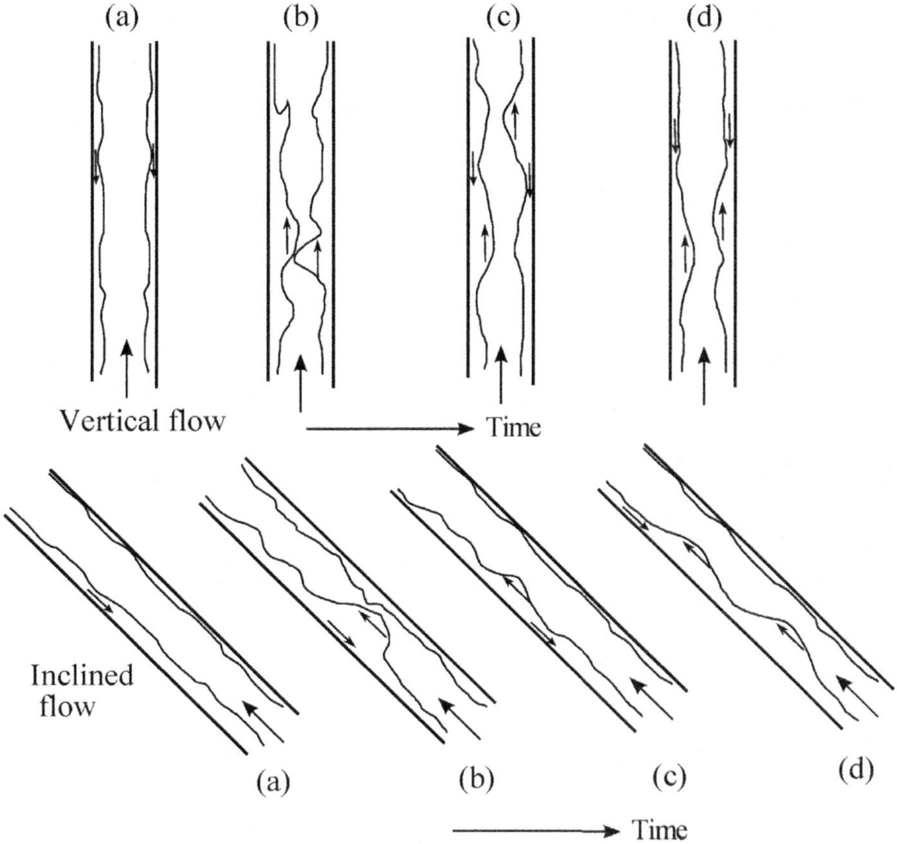

Fig. 16.9 Schematics of mechanism governing pressure gradient minimum phenomenon

The second trend of decreasing pressure drop could be explained using Fig. 16.9. During churn-annular flow transition, large interfacial disturbance waves are generated that travel in downstream direction (Fig. 16.9-sketch b). During the swiping action of disturbance wave, liquid film travels in downstream direction under the influence of interfacial drag. However, once the disturbance wave passes by a certain pipe cross section, there is no driving potential for the liquid film, and it tends to fall back under the influence of gravity (Fig. 16.9-sketch c and d). The direction of travel of liquid film for different cases is depicted by small arrows inside the film. During and just after the encounter of liquid film with the disturbance wave, the portion of liquid film near interface may travel in upward direction, while that away from interface (or near pipe wall) may travel in downward direction leading to severe fluctuations in shear stress and velocity distributions in liquid film. The falling liquid film is again lifted up by upcoming disturbance wave, and the process continues until the frequency of disturbance waves is high enough to prevent reversal of liquid film. More details about this phenomenon and its dependency on disturbance waves could be found in Hewitt et al. (1965,1985), Hewitt and Lacey (1965), and Owen (1986).

When it is desirable to operate a two-phase flow system in the region *not affected* by reversal of liquid film, it is crucial to identify the gas and liquid flow rates that belong to the region *affected* by flow reversal. Compared to the first trend of decreasing pressure gradient, its second instance is more important since this region is marked by severe turbulence and instability of the liquid film. The two-phase flow literature reports that the point of second pressure gradient minimum occurs at $Fr_{SG} \approx 1$. However, this criterion does not provide any idea on the range of liquid flow rates for which this trend would exist. Experiments carried out at the Two-Phase Flow Lab, OSU, reveal that the liquid flow rates at which this trend of second pressure gradient minimum would exist depend upon the pipe orientation such that the relationship between non-dimensional liquid flow rates and pipe orientation corresponding to second pressure gradient minimum is given by Eq. (16.4). Note that this relationship need not be used for pipe orientations less than +10° from horizontal since the effect of pipe orientation on pressure gradient is negligibly small. For a fixed pipe orientation, Eq. (16.4) gives a threshold value of non-dimensional liquid flow rate (Fr_{SL}) above which decreasing pressure gradient trends or alternatively reversal of liquid film will no longer exist. Independent of pipe orientation and liquid flow rates, the non-dimensional gas flow rates (Fr_{SG}) corresponding to this point are in the range of 0.9–1:

$$Fr_{SL} = 0.6575(\sin\theta)^{1.1175} \quad \} \quad +10^{\circ} \leq \theta \leq +90^{\circ} \tag{16.4}$$

Two-Phase Pressure Drop Modeling

As was shown in Eq. (13.1), the total pressure drop in gas-liquid two-phase flow consists of three main components: hydrostatic pressure drop (due to pipe elevation), pressure drop due to friction at the pipe wall and gas-liquid interface, and pressure drop due to acceleration (due to change in specific volume of the two-phase mixture). The contribution of each of these components to the total pressure drop depends upon the flow patterns, fluid properties, pipe orientation, and pipe diameter. For example, in the case of bubbly flow in vertical upward orientation, the hydrostatic pressure drop component can be dominant in comparison to the frictional pressure drop depending upon the pipe diameter, while, for annular flow regime, the frictional pressure drop has a prominent share in the total pressure drop, and the hydrostatic pressure drop component is negligible. Moreover, the contribution of the frictional pressure drop to the total pressure drop is observed to increase with decreasing pipe diameter. The total two-phase pressure drop per unit pipe length is the summation of its three components as expressed by Eq. (13.1). The right-hand side of Eq. (13.1) is typically modeled using homogeneous and separated flow models. A brief description of these models will be presented next.

Homogeneous Flow Model The correlations based on homogenous flow model assume that the two phases are well mixed with each other and move with identical velocities (there is no slippage at the gas-liquid interface). The homogenous flow model thus considers the two-phase mixture as a pseudo single-phase mixture having two-phase physical properties. Under the consideration of homogeneous flow model, the hydrostatic two-phase pressure drop is calculated using homogeneous two-phase mixture density (ρ_M) as shown in Eq. (16.5). The two-phase mixture density can be expressed in terms of either two-phase flow quality (x) or the gas volumetric flow fraction (λ) as shown in Eq. (16.6). See Table 12.1 for interchangeability between λ and x:

$$\left(\frac{dp}{dz}\right)_h = \rho_M g \sin\theta \tag{16.5}$$

$$\rho_M = \left(\frac{x}{\rho_G} + \frac{1-x}{\rho_L}\right)^{-1} = \lambda\rho_G + (1-\lambda)\rho_L \tag{16.6}$$

The pressure drop due to acceleration of the two-phase mixture is calculated from Eq. (16.7). For the case of two-component two-phase flow (non-boiling, non-condensing), the two-phase flow quality remains constant for a relatively short length of pipe, and the magnitude of pressure drop due to change in specific volume of two-phase mixture is negligible and hence is ignored:

$$\left(\frac{dp}{dz}\right)_a = G^2\frac{dv_M}{dz} = G^2\left(v_{LG}\frac{dx}{dz} + x\frac{dv_G}{dp}\frac{dp}{dz}\right) \tag{16.7}$$

Finally, the frictional component of two-phase pressure drop is expressed in terms of the two-phase friction factor as shown in Eq. (16.8). The two-phase friction factor (f_M) is calculated based on two-phase Reynolds number (Re_M) which in turn is the function of two-phase mixture dynamic viscosity (μ_M) as shown in Eq. (16.9). Two-phase flow literature provides several models to calculate two-phase dynamic viscosity as a function of two-phase flow quality. For more details on the different two-phase dynamic viscosity models, refer to Ghajar and Bhagwat (2017):

$$\left(\frac{dp}{dz}\right)_f = \frac{2f_M G^2}{D\rho_M} \tag{16.8}$$

$$Re_M = \frac{GD}{\mu_M} \tag{16.9}$$

The two-phase friction factor (f_M) is calculated using single-phase fanning friction factor correlations such as Blasius (1913) and Churchill (1977b).

Thus, the total two-phase pressure drop is the summation of the hydrostatic, accelerational, and frictional two-phase pressure drops as given by Eqs. (16.5), (16.7), and (16.8), respectively. For a more detailed discussion on the homogenous flow model, refer to Ghajar and Bhagwat (2017).

Separated Flow Model The correlations based on separated flow model assume the two phases to flow separately and share a definite and continuous interface between them, and unlike homogeneous flow model, it considers the slippage at the gas-liquid interface. Thus, using two-phase separated flow model, the two-phase mixture density (ρ_M) and hence the two-phase hydrostatic pressure drop calculation are based on the void fraction (α_G) as given by Eq. (16.10) where the two-phase mixture density is given by Eq. (16.11):

$$\left(\frac{dp}{dz}\right)_h = \rho_M g \sin\theta = (\alpha_G\rho_G + (1-\alpha_G)\rho_L)g\sin\theta \tag{16.10}$$

$$\rho_M = \alpha_G\rho_G + (1-\alpha_G)\rho_L \tag{16.11}$$

The accelerational component of two-phase pressure drop is expressed by Eq. (16.12). This expression is valid for all but annular flow pattern with considerable liquid entrainment. In particular, for annular flow, when the entrainment of liquid droplets into the central gas core is significant, the error associated with the assumption of negligible accelerational pressure drop can be severe. In the case of annular flow with liquid entrainment, the process of entrainment also contributes to the two-phase pressure drop and hence cannot be ignored. For such a specific case (annular flow with liquid entrainment), Eq. (16.12) can be expressed in the form of Eq. (16.13). See Chap. 18 for a detailed discussion on the entrainment mechanisms and the different correlations used to predict liquid entrainment fraction (E) in gas-liquid annular flow:

$$\left(\frac{dp}{dz}\right)_a = G^2\frac{d}{dz}\left[\frac{x^2}{\alpha_G\rho_G} + \frac{(1-x)^2}{(1-\alpha_G)\rho_L}\right] \tag{16.12}$$

$$\left(\frac{dp}{dz}\right)_a = G^2\frac{d}{dz}\left[\frac{x^2}{\alpha_G\rho_G} + \frac{(1-E)^2(1-x)^2}{\beta\rho_L} + \frac{E^2(1-x)^2}{(1-\alpha_G-\beta)\rho_L}\right] \tag{16.13}$$

The volume fraction occupied by the liquid film (β) defined by Eq. (16.14) does not consider the fraction of liquid droplets entrained in the gas core and hence may not be confused with liquid holdup (α_L). However, it is evident that under the condition of $E \approx 0$, the volume fraction of liquid film is equal to the liquid holdup, i.e., $\beta = \alpha_L$. Equation (16.14) is based on a valid assumption that entrained liquid droplets travel at a velocity same as that of the gas:

$$\beta = 1 - \alpha_G - \frac{\alpha_G E(1-x)}{x(\rho_L/\rho_G)} \tag{16.14}$$

It must be reiterated that this is required only in the case of annular flow with considerable liquid entrainment. For the case of annular flow at low system pressures ($E \approx 0$) and at flow patterns other than annular flow ($E = 0$), hence, it is simplified back to Eq. (16.12) such that the accelerational pressure drop at a given location can be determined based on void fraction and mass flux of each phase. For the case of adiabatic two-phase flow over a short pipe length, two-phase flow quality and void fraction can be assumed to remain constant and hence $(dp/dz)_a \approx 0$.

Finally, the frictional component of two-phase pressure drop is calculated as

$$\left(\frac{dp}{dz}\right)_f = \Phi_j^2 \left(\frac{dp}{dz}\right)_j \tag{16.15}$$

The two-phase frictional multiplier Φ^2 can be expressed in several different forms depending upon the assumption of flow of either single phase through the pipe. Broadly, four different cases of two-phase frictional multiplier can be defined as shown in Eqs. (16.16, 16.17, 16.18, and 16.19). In these equations, single-phase pressure drop is the single-phase frictional pressure drop unless otherwise specified, and $(dp/dz)_f$ would always represent frictional component of the two-phase pressure drop:

$$\left(\frac{dp}{dz}\right)_f = \Phi_{LO}^2 \left(\frac{dp}{dz}\right)_{LO} \text{ where } \left(\frac{dp}{dz}\right)_{LO} = \frac{2f_{Lo}G^2}{D\rho_L} \tag{16.16}$$

$$\left(\frac{dp}{dz}\right)_f = \Phi_L^2 \left(\frac{dp}{dz}\right)_L \text{ where } \left(\frac{dp}{dz}\right)_L = \frac{2f_L G^2 (1-x)^2}{D\rho_L} \tag{16.17}$$

$$\left(\frac{dp}{dz}\right)_f = \Phi_{GO}^2 \left(\frac{dp}{dz}\right)_{GO} \text{ where } \left(\frac{dp}{dz}\right)_{GO} = \frac{2f_{Go}G^2}{D\rho_G} \tag{16.18}$$

$$\left(\frac{dp}{dz}\right)_f = \Phi_G^2 \left(\frac{dp}{dz}\right)_G \text{ where } \left(\frac{dp}{dz}\right)_G = \frac{2f_G G^2 x^2}{D\rho_G} \tag{16.19}$$

Two-phase flow literature provides several frictional pressure drop correlations based on these different two-phase frictional multipliers (Φ^2); see Eqs. (16.16, 16.17, 16.18, and 16.19). Next, some of the top-performing pressure drop correlations used to determine frictional component of the two-phase pressure drop will be briefly discussed. For a more detailed discussion on the separated flow model, refer to Ghajar and Bhagwat (2017).

Table 16.1 Values of constant C to be used in Lockhart and Martinelli (1949) correlation, Eq. (16.20), for different single-phase flow regimes

Liquid phase	Gas phase	C
Turbulent	Turbulent	20
Laminar	Turbulent	12
Turbulent	Laminar	10
Laminar	Laminar	5

Lockhart and Martinelli (1949) Correlation They proposed one of the first such correlations. Their correlation expresses two-phase frictional multipliers Φ_L^2 and Φ_G^2 as a function of X parameter and is expressed by Eq. (16.20). The Lockhart and Martinelli (1949) parameter (X) is given by Eq. (16.21) where the single-phase friction factors (f_L and f_G) are found using appropriate single-phase friction factors such as those of Blasius (1913) and Churchill (1977b):

$$\Phi_L^2 = 1 + C/X + 1/X^2 \quad \text{or} \quad \Phi_G^2 = 1 + CX + X^2 \tag{16.20}$$

$$X = \left[\frac{(dp/dz)_{f,L}}{(dp/dz)_{f,G}}\right]^{0.5} = \frac{1-x}{x}\left[\frac{f_L}{f_G}\frac{\rho_G}{\rho_L}\right]^{0.5} = \frac{U_{SL}}{U_{SG}}\left[\frac{f_L}{f_G}\frac{\rho_L}{\rho_G}\right]^{0.5} \tag{16.21}$$

The value of C used in Eq. (16.20) given by Chisholm (1967) is given in Table 16.1, and it depends on the laminar or turbulent nature of the gas and liquid phase, respectively. A worked-out example problem illustrating the use of Lockhart and Martinelli (1949) correlation is provided at the end of Chap. 20.

Muller-Steinhagen and Heck (1986) Correlation They developed a correlation to extrapolate between single-phase liquid and single-phase gas flow. Their correlation can be used for both adiabatic and evaporating two-phase flow conditions (provided the two-phase flow quality distribution across the pipe length is known). The physical form of their correlation is as shown in Eq. (16.22). The parameter Y defined in Eq. (16.23) is the same as that given by Chisholm (1973). The correlation is verified against 9300 measurements including data of air-oil, air-water, and steam-water fluid combinations and pipe diameters ranging from 4 to 39 mm I.D. The application of Eq. (16.22) is restricted to $Re_{LO} > 100$ and $Y^2 > 1$. Note that for determination of single-phase friction factor f_j, their correlation separates single-phase laminar and turbulent regions using a threshold value of $Re_j = 1187$ where "j" can be gas or liquid phase. Muller-Steinhagen and Heck (1986) claimed their correlation to predict the two-phase frictional pressure drop data with an average error of 41.7% with only 49.5% of the data points predicted within ±30% error bands:

$$\Phi_{LO}^2 = (1-x)^{0.33}\left(1+2x\left(Y^2-1\right)\right)+Y^2x^3 \tag{16.22}$$

$$Y = \sqrt{\frac{(dp/dz)_{GO}}{(dp/dz)_{LO}}} \tag{16.23}$$

Xu and Fang (2012) Correlation They proposed a two-phase frictional pressure drop correlation for evaporating two-phase flow of refrigerants in horizontal pipe. Their correlation is based on 2622 data points of 14 refrigerants (R11, R12, R22, R134a, R32, R407C, R507, R507A, R410A, CO_2, R404A, R32/R125, R123, and ammonia) for $1 < D_h < 19$ mm (circular and rectangular pipe geometries) and $25 < G < 1150$ kg/m ^{2}s. They found that the pipe hydraulic diameter and gas-liquid interface surface tension affect the two-phase frictional pressure drop significantly than any other two-phase flow parameter, and hence their correlation accounts for these two variables through the inclusion of non-dimensional Laplace number. The physical form of their correlation is as expressed in Eq. (16.24):

$$\Phi_{LO}^2 = \left[Y^2x^3+(1-x)^{0.33}\left(1+2x\left(Y^2-1\right)\right)\right]\left(1+1.54(1-x)^{0.5}La\right) \tag{16.24}$$

In this equation, Y is calculated from Eq. (16.23), while La is the Laplace number defined by Eq. (15.19). Their correlation is claimed to predict the two-phase frictional pressure drop data with a mean absolute relative deviation of 25.2%. It should be noted that Xu and Fang (2012) recommend use of Fang et al. (2011) correlation to calculate single-phase fanning friction factors f_{LO} and f_{GO}. The correlations of Fang et al. (2011) for single-phase liquid and gas fanning friction factors are reported in Eqs. (16.25a) and (16.25b):

$$f_{LO} = 0.0625\left[\log\left(\frac{150.39}{Re_{LO}^{0.98865}}-\frac{152.66}{Re_{LO}}\right)\right]^{-2} \tag{16.25a}$$

$$f_{GO} = 0.0625\left[\log\left(\frac{150.39}{Re_{GO}^{0.98865}}-\frac{152.66}{Re_{GO}}\right)\right]^{-2} \tag{16.25b}$$

Zhang et al. (2010) Correlation For mini–/microchannels having pipe diameters in a range of 0.1 to 6 mm, Zhang et al. (2010) proposed a two-phase frictional pressure drop correlation (for adiabatic two-phase flow) based on the experimental data of two-phase flow of refrigerants, air-water, and air-ethanol. Their correlation for two-phase frictional multiplier is the same as Eq. (16.20) proposed by Lockhart and Martinelli (1949), however with the parameter C as a variable given by Eq. (16.26):

$$C = \begin{cases} 21[1 - \exp(-0.674/La)] & : (\text{two} - \text{component two} - \text{phase flow}) \\ 21[1 - \exp(-0.142/La)] & : (\text{one} - \text{component two} - \text{phase flow}) \end{cases} \tag{16.26}$$

Bhagwat (2015) Correlation Ghajar and Bhagwat (2014b) and later on Bhagwat (2015) did a comprehensive study of 16 top-performing two-phase frictional pressure drop correlations based on separated flow model available in the two-phase flow literature. The performance of these correlations was assessed using a consolidated database of 9623 data points (horizontal and upward inclined two-phase flow) amassed from 60 data sources in the two-phase flow literature and consisted of air-water, air-oil, refrigerants (R12, R22, R123, R134a, R236fa, R245fa, R404A, R410A, R507, R32 + R125, R290 (propane), R1234yf, R422D, R717 (ammonia), R744 (CO_2)), and other miscellaneous fluid combinations such as air-glycerin and air-kerosene. The analysis of the two-phase frictional pressure drop correlations against the experimental data bank showed that the frictional pressure drop correlation of Muller-Steinhagen and Heck (1986) given by Eq. (16.22) performed relatively better than the other correlations; see Bhagwat (2015) for additional details. As mentioned earlier, the correlation of Muller-Steinhagen and Heck (1986) is restricted to $Re_{LO} > 100$ and $Y^2 > 1$ and does not have the flexibility of accommodating a wide range of fluid properties, flow rates, pipe diameters, and pipe orientations. To expand on the performance range of correlation of Muller-Steinhagen and Heck (1986), Bhagwat (2015), using the consolidated database of 9623 data points explained earlier, modified the Muller-Steinhagen and Heck (1986) correlation in the form of Eq. (16.27) by introducing three new variables B_1, B_2, and B_3 in Eq. (16.22) given by Eqs. (16.28, 16.29, and 16.30). Note that in Eq. (16.27) by setting $B_1 = 2$, $B_2 = 1$, and $B_3 = 0$, the original Muller-Steinhagen and Heck (1986) correlation given by Eq. (16.22) is obtained. Recall that the parameter Y used in Eq. (16.27) was defined in Eq. (16.23):

$$\Phi_{LO}^2 = \left(\left[(1-x)^{0.33}\left(1 + B_1 x\left(Y^2 - 1\right)\right)\right] + B_2 Y^2 x^3\right)\left(1 + B_3(1-x)^2\right) \tag{16.27}$$

$$B_1 = [0.85 + 1.703(1 - \exp(-6.25\ \zeta \times Bo))] \times \Pi_1 \Pi_2 \Pi_3 \tag{16.28}$$

$$B_2 = 1 - \sqrt{\rho_G/\rho_L} \tag{16.29}$$

$$B_3 = \begin{cases} -0.3(1 + \sin\theta)^{-16.25} + 0.3 & : 0^\circ \le \theta \le +20^\circ \\ -0.012(1 + \sin\theta)^{4.1} + 0.34 & : +20^\circ < \theta \le +90^\circ \end{cases} \tag{16.30}$$

$$\Pi_1 = \left(1 + 2.65\left[1 - \exp\left(-1.677N_{\mu L}\right)\right]\right) \tag{16.31}$$

$$\Pi_2 = \begin{pmatrix} 0.55 & : \zeta \leq 1.0, & Bo \geq 1 \\ 1 & : \zeta > 1.0, & Bo < 1 \end{pmatrix} \tag{16.32}$$

$$\Pi_3 = \left[1 + 0.005\left(\frac{1-x}{x}\right)\right]^{0.5} \tag{16.33}$$

$$\zeta = 2.5\sqrt{\frac{\rho_L}{\rho_{ref}}}\left(\frac{\mu_G}{\mu_L}\right)^{0.25} \tag{16.34}$$

$$Bo = \frac{g(\rho_L - \rho_G)(D_h/2)^2}{\sigma} \tag{16.35}$$

$$N_{\mu L} = \frac{\mu_L}{\left[\rho_L \sigma \sqrt{\frac{\sigma}{g(\rho_L - \rho_G)}}\right]^{0.5}} \tag{16.36}$$

Referring to Eq. (16.27), the parameters B_1 (multiplying factor to the term $Y^2 - 1$) given by Eq. (16.28) and B_2 (multiplying factor to the term Y^2) given by Eq. (16.29) control the slope of Φ_{LO}^2 vs x for lower and higher values of quality, respectively, whereas parameter B_3 given by Eq. (16.30) simply acts as a multiplier to Φ_{LO}^2 that modifies the overall value of two-phase pressure drop as a function of pipe orientation. As expressed by Eq. (16.28), the parameter B_1 is modeled as a function of pipe hydraulic diameter (D_h), fluid properties (ρ_L, ρ_G, μ_L and σ), and the two-phase flow quality (x). The effect of these two-phase flow variables on B_1 and hence Φ_{LO}^2 is accounted in the form of non-dimensional numbers such as Bond number (Bo) and viscosity number ($N_{\mu L}$) given by Eqs. (16.35) and (16.36), respectively. The variables Π_1, Π_2, and Π_3 used in parameter B_1 are expressed by Eqs. (16.31), (16.32), and (16.33), respectively. Parameter B_2 given by Eq. (16.29) is modeled as a function of the density ratio (ρ_G/ρ_L) and is found to vary in a range of 0.5 to 1.0. As shown in Eq. (16.30), parameter B_3 is modeled as a function of pipe orientation such that its value varies in a range of 0 to 0.3. Based on the experimental data used, the two different equations (of the form $y = ax^b + c$) are chosen for near horizontal and steeper pipe inclinations in the upward direction. Note that there is a good agreement (within 1%) between the extrapolated values of B_3 for two cases of $0^o \leq \theta \leq +20^o$ ($B_3 = 0.299$) and $+20^o < \theta \leq +90^o$ ($B_3 = 0.297$). The term

$(1-x)^2$ used in conjunction of B_3 ensures that the predicted Φ^2_{LO} is weighted such that it gradually decreases with increase in the two-phase flow quality or alternatively as the flow patterns become inertia driven in nature and hence relatively insensitive to the change in the pipe orientation.

Getting back to parameter B_1, let us look at its physical significance and need for including parameters Π_1, Π_2, and Π_3. The parameter B_1 is modeled as a variable such that it is sensitive to the increase in Bond number in the vicinity of $Bo \approx 1$ ($0.7 < Bo < 1.3$) and thereafter follows a saturation trend. Note that this threshold value of Bo after which saturated trend ensues is subject to a small change with variation in property group ζ given by Eq. (16.34). The inclusion of Bond number accounts for the balance between body and surface tension forces on the two-phase frictional multiplier. For small diameter pipes, effect of surface tension on two-phase flow is more significant compared to the effect of gravity. Accordingly, for $Bo < 1$, the surface tension effects are prominent in comparison to the body forces, whereas for $Bo > 1$, the effect of surface tension on the two-phase flow is inconsequential.

The data bank used by Bhagwat (2015) showed that the condition of $Bo < 1$ corresponds to pipe diameters in a range of 0.069–4.5 mm and consists of refrigerants (1822 data points) as well as air-water (645 data points) fluid combinations. According to the observations of Kandlikar (2002), approximately this range of pipe diameter ($0 < D \leq 3$ mm) represents the micro- to mini-scale two-phase flow phenomenon. The variable B_1 as a function of Bo increases the value of Φ^2_{LO} with increase in the pipe diameter (due to increase in the gas-liquid interfacial area).

Variable B_1 also accounts for the effect of liquid dynamic viscosity through parameter Π_1 which is modeled as a function of viscosity number ($N_{\mu L}$); see Eq. (16.36). Increase in liquid phase viscosity increases the shear in the liquid phase and hence increases the two-phase frictional multiplier and consequently the two-phase pressure drop.

The parameter Π_2 switches between a constant value of 0.55 and 1.0 for two different cases based on Bo and ζ as shown in Eq. (16.32). According to Bhagwat (2015), analysis of the experimental data used showed that in comparison to liquid refrigerant and its vapor, data for air-water, air-oil, air-kerosene, and air-glycerin fluid combinations is overpredicted (parallel shift) while using Eq. (16.28). To reduce the magnitude of this overprediction, use of multiplying factor less than unity (0.55 in this case) was found to give the best fit. Using the property group of ζ defined by Eq. (16.34), these fluid combinations could be separated from the refrigerant data. Note that ρ_{ref} used in definition of ζ is reference value of the density equal to the density of liquid water taken as 1000 kg/m^3.

Finally, the variable Π_3 as a function of two-phase flow quality (x) is modeled such that for small values of x, $\Pi_3 > 1$, whereas for $x \to 1$, $\Pi_3 \approx 1$; see Eq. (16.33). The small values of x typically correspond to bubbly and slug flow patterns with liquid and gas phase in laminar-laminar or turbulent-laminar flow regimes. Thus, the predicted values of two-phase pressure drop are sensitive to Π_3 only in these flow regimes without affecting the predicted data in laminar-turbulent and turbulent-turbulent flow regimes.

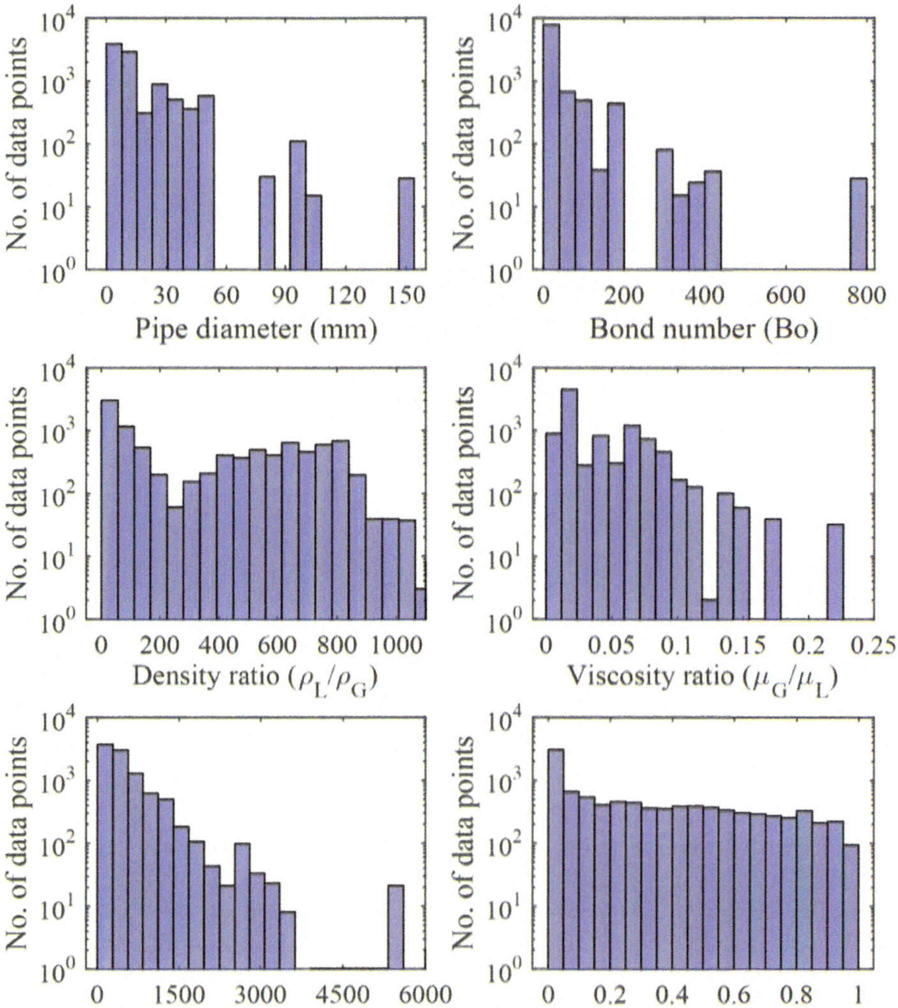

Fig. 16.10 Distribution of experimental data bank (9623 data points) used by Bhagwat (2015)

The distribution of experimental data at varying flow conditions used for the development of Bhagwat (2015) correlation is shown in Fig. 16.10. The distribution of hydraulic pipe diameter is such that it occupies micro- to macro-scale two-phase flow ($0.069 \leq D_h \leq 152$ mm) that corresponds to a wide range of Bond numbers (0.0015 to 800). It is seen that the two-phase flow quality is fairly distributed over the entire range ($0 < x < 1$) and hence is expected to occupy the entire range of flow patterns typically observed in gas-liquid two-phase flow. The experimental data is also observed to be distributed over a wide range of two-phase mixture mass flux. The experimental database also consists of a wide range of density ratio (ρ_L/ρ_G). The

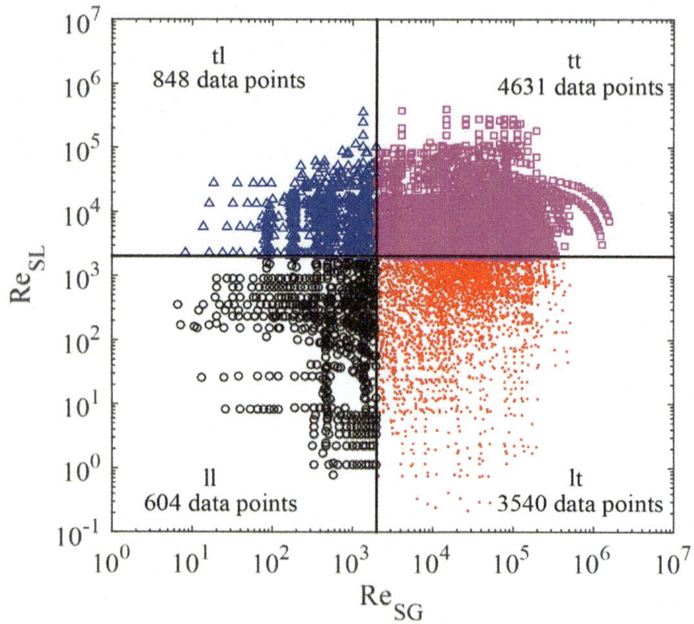

Fig. 16.11 Distribution of experimental data in different flow regimes (ll = laminar-laminar, lt = laminar-turbulent, tl = turbulent-laminar, tt = turbulent-turbulent)

low values of density ratio correspond to high system pressure data such as that for refrigerants (typically for CO_2, propane, ammonia, and refrigerants at higher saturation temperature), while high-density ratio data consist of air-water and air-glycerin fluid combinations. In addition to the density ratio, a wide range of viscosity ratio (μ_G/μ_L) is also observed. The lower and higher values of (μ_G/μ_L) typically correspond to the air-oil and refrigerants' two-phase flow data, respectively. Additionally, the distribution of experimental data for four different cases of laminar-laminar (ll), laminar-turbulent (lt), turbulent-laminar (tl), and turbulent-turbulent (tt) two-phase flow is shown in Fig. 16.11. From experience, it can be said that the laminar-laminar and turbulent-laminar flow regions typically consist slug/plug and bubbly/bubbly-slug flow patterns, respectively, whereas the laminar-turbulent and turbulent-turbulent flow regions may consist of wavy slug (intermittent) and annular flow patterns. A brief summary of the range of different two-phase flow parameters used in the development of Bhagwat (2015) correlation is reported in Table 16.2.

The performance of the Bhagwat (2015) correlation is evaluated based on the statistical parameters such as mean relative deviation (MRD) and mean absolute relative deviation (MARD). Note that achieving a low value of MRD or MARD is not the absolute measure to gauge the accuracy of any correlation since these parameters (MRD and MARD) are strongly influenced by the number of data points and distribution of the error. It is also important for the correlation to be able to

Table 16.2 Range of experimental data used for development of Bhagwat (2015) two-phase frictional pressure drop correlation

Parameter	Range
Hydraulic pipe diameter (mm)	0.1–152
Mixture mass flux (kg/m ^{2}s)	5–7000
Two-phase flow quality	0.0001–0.99
Liquid to gas density ratio (ρ_L/ρ_G)	5–920
Liquid to gas viscosity ratio (μ_L/μ_G)	10–5000
Surface tension (N/m)	0.002–0.075
Pipe orientation (θ)	$0^\circ \leq \theta \leq +90^\circ$
Bond number (Bo)	0.0015–800
Liquid only Reynolds number (Re_{LO})	1–2.85 × 10^5
Gas only Reynolds number (Re_{GO})	1800–1.5 × 10^7
Viscosity number ($N_{\mu L}$)	0.0008–4
Lockhart-Martinelli parameter (X)	0.002–800
Gas volumetric flow fraction (λ)	0.005–0.99
Flow patterns	All except stratified

predict maximum number of data points within certain error bands. Accordingly, the accuracy of the correlation is also compared based on the percentage of data points predicted within ±30% (N_{30}) and ±50% (N_{50}) error bands. Considering the practical difficulty in modeling of two-phase pressure drop (as a function of several two-phase flow variables), use of higher error bands up to ±50% in the prediction of two-phase pressure drop is considered as acceptable and has become a norm in the two-phase flow literature.

The goodness of the correlation of Bhagwat (2015) for different cases of single-phase flow regimes (laminar/turbulent) is shown in Fig. 16.12. The figure shows the prediction of correlation of Bhagwat (2015) against the measured data for different combinations of single-phase flow regimes. For laminar-laminar ($Re_{SL} \leq 2000$, $Re_{SG} \leq 2000$), laminar-turbulent ($Re_{SL} \leq 2000$, $Re_{SG} > 2000$), and turbulent-turbulent ($Re_{SL} > 2000$, $Re_{SG} > 2000$) flow regimes, the correlation of Bhagwat (2015) successfully predicts more than 70% and 90% of data points within ±30% and ±50% error bands, respectively. For the laminar-turbulent and turbulent-turbulent cases, MRD is less than 4%, and MARD is 25% or less. Accuracy of the correlation of Bhagwat (2015) slightly drops down for the case of turbulent-laminar flow ($Re_{SL} > 2000$, $Re_{SG} \leq 2000$). A careful observation of the experimental data shows that a majority of data in this flow regime belongs to the slug/churn (intermittent flow) transition in upward inclined two-phase flow where the pressure drop increases suddenly due to slug breakup resulting into vigorous mixing and chaotic two-phase flow behavior. Consequently, all two-phase flow models underpredict the frictional pressure drop for this case.

Figure 16.13 shows the performance of Muller-Steinhagen and Heck (1986) correlation for different combinations of single-phase flow regimes. Similar to the Bhagwat (2015) correlation, accuracy of Muller-Steinhagen and Heck (1986) is found to deteriorate for the case of turbulent-laminar flow. Considering the overall accuracy in terms of N_{30}, N_{50}, MRD, and MARD, the Bhagwat (2015) correlation

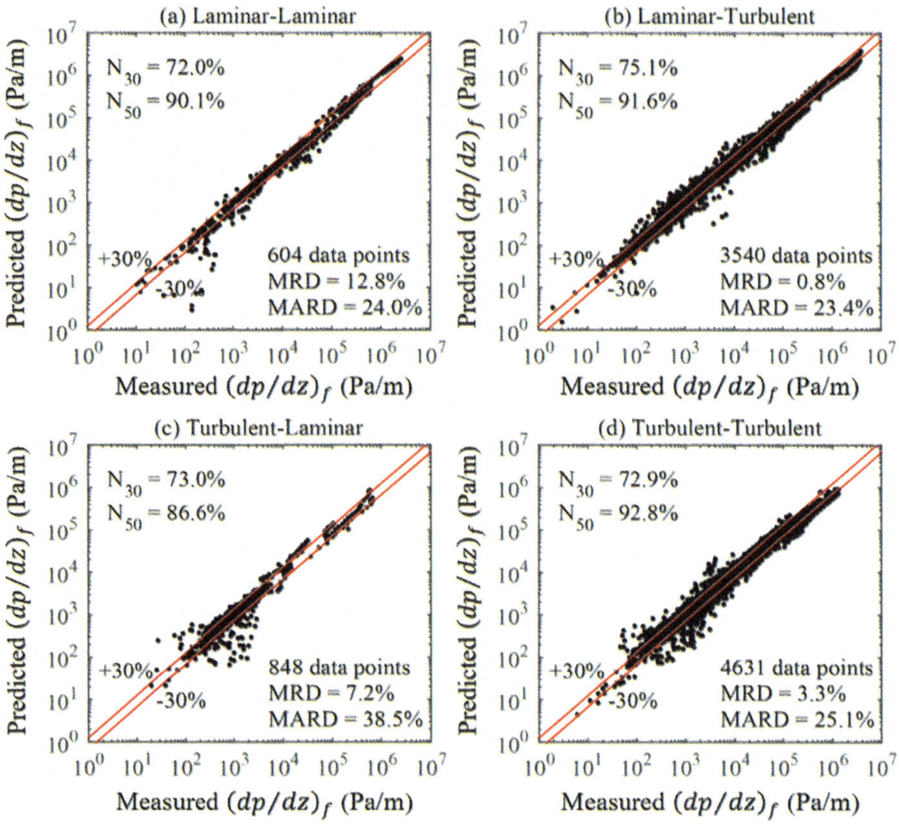

Fig. 16.12 Prediction of Bhagwat (2015) correlation against the entire data for different single-phase flow regimes (all data, 9623 data points): (**a**) laminar-laminar, (**b**) laminar-turbulent, (**c**) turbulent-laminar, (**d**) turbulent-turbulent

performs consistently better than Muller-Steinhagen and Heck (1986) correlation for all four cases of single-phase flow regimes. Finally, the performance of the correlation of Bhagwat (2015) over the entire data (9623 data points) consisting of horizontal, vertical, and upward pipe inclinations is presented graphically in Fig. 16.14. For the consolidated data bank, correlation of Bhagwat (2015) predicts 74% and 92% of data within ±30% and ±50% error bands with a MRD of 3.3% and MARD of 25.6%. In comparison to this, as shown by Fig. 16.15, Muller-Steinhagen and Heck (1986) correlation predicts 63% and 88% of data points within ±30% and ±50% error bands with a MRD of 8.0% and MARD of 28.8%.

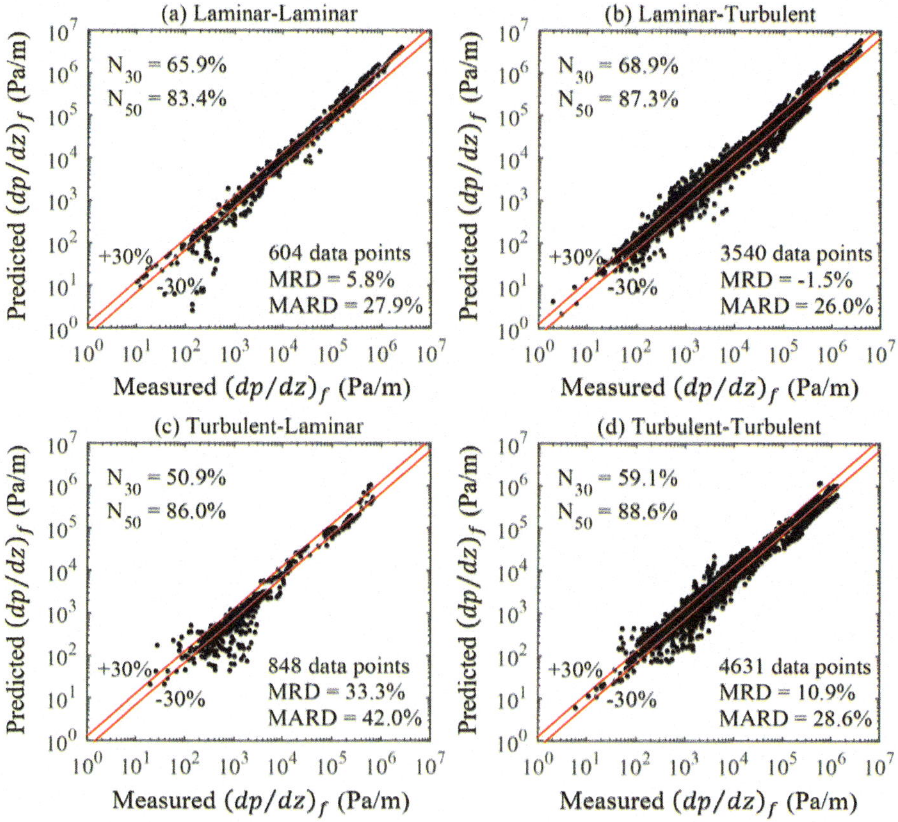

Fig. 16.13 Prediction of Muller-Steinhagen and Heck (1986) correlation against the entire data for different single-phase flow regimes (all data, 9623 data points): (**a**) laminar-laminar, (**b**) laminar-turbulent, (**c**) turbulent-laminar, (**d**) turbulent-turbulent

An Illustrative Example

Use of Xu and Fang (2012) Two-Phase Frictional Pressure Drop Correlation Consider a boiling two-phase flow of R134a through a 6 mm I.D. horizontal copper tube at a mass flux of 800 kg/m^{2}s and a system pressure of 1500 kPa. The two-phase refrigerant enters the 1-m-long tube with a quality of $x = 0.3$ and exits with a quality of $x = 0.7$. Calculate the two-phase frictional pressure drop for this flow using the two-phase frictional pressure drop correlation of Xu and Fang (2012). For simplicity, assume that the quality changes linearly with the pipe length, and use a mean quality of 0.5.

For refrigerant R134a at 1500 kPa, use the following thermophysical properties:

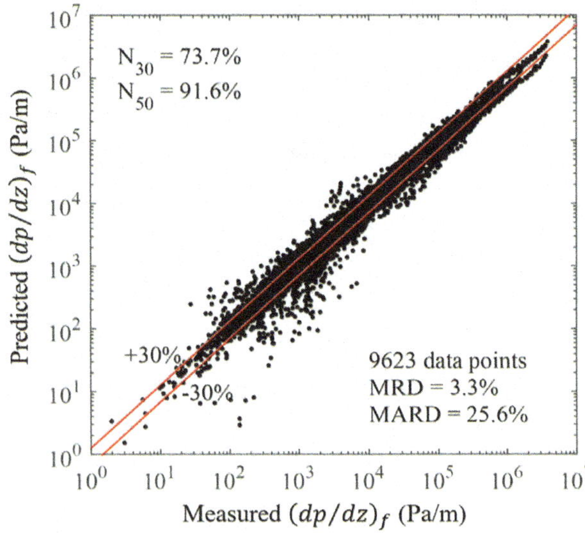

Fig. 16.14 Performance of Bhagwat (2015) correlation against entire data (9623 data points)

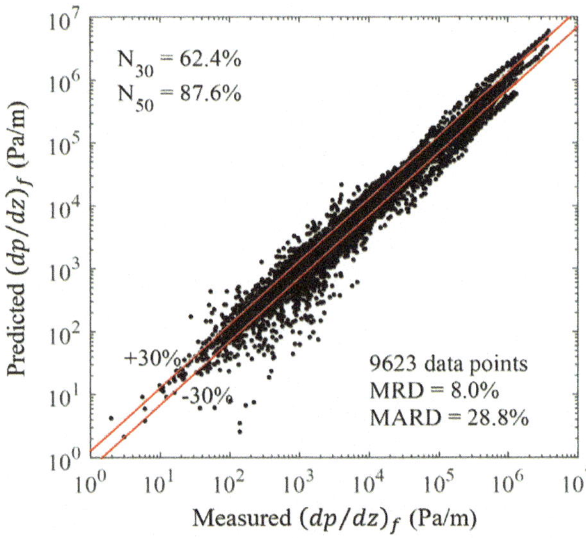

Fig. 16.15 Performance of Muller-Steinhagen and Heck (1986) correlation against entire data (9623 data points)

Liquid density (ρ_L) = 1078 kg/m^3, gas density (ρ_G) = 76.95 kg/m^3, liquid viscosity (μ_L) = 0.0001746 Pa.s, gas viscosity (μ_G) = 0.0000138 Pa.s, and surface tension (σ) = 0.00427 N/m.

Solution

The two-phase frictional pressure drop for Xu and Fang (2012) is determined from Eq. (16.24):

$$\Phi_{LO}^2 = \left[Y^2x^3 + (1-x)^{0.33}\left(1 + 2x\left(Y^2 - 1\right)\right)\right]\left(1 + 1.54(1-x)^{0.5}La\right)$$

where from Eq. (16.23): $Y = \sqrt{\frac{(dp/dz)_{GO}}{(dp/dz)_{LO}}}$ and from Eq. (15.19): $La = \sqrt{\frac{\sigma}{g(\rho_L - \rho_G)}\frac{1}{D_h}}$.

It should be noted that Xu and Fang (2012) recommend use of Fang et al. (2011) correlations to calculate the single-phase fanning friction factors f_{LO} and f_{GO}. The Fang et al. (2011) correlations given by Eqs. (16.25a) and (16.25b) are

$$f_{LO} = 0.0625\left[\log\left(\frac{150.39}{Re_{LO}^{0.98865}} - \frac{152.66}{Re_{LO}}\right)\right]^{-2}$$

$$f_{GO} = 0.0625\left[\log\left(\frac{150.39}{Re_{GO}^{0.98865}} - \frac{152.66}{Re_{GO}}\right)\right]^{-2}$$

The liquid and gas Reynolds numbers (Re_{LO} and Re_{GO}) required in Eqs. (16.25a) and (16.25b) are calculated by the following equations:

$$Re_{LO} = \frac{GD}{\mu_L} = \frac{800 \times 0.006}{0.0001746} = 27491$$

$$Re_{GO} = \frac{GD}{\mu_G} = \frac{800 \times 0.006}{0.0000138} = 347826$$

Thus, from the friction factor correlations of Fang et al. (2011), Eqs. (16.25a) and (16.25b), $f_{LO} = 0.005995$ and $f_{GO} = 0.0035175$.

The single-phase pressure drop for liquid and gas phases are calculated from Eqs. (16.16) and (16.18):

$$\left(\frac{dp}{dz}\right)_{LO} = \frac{2f_{Lo}G^2}{D\rho_L} = \frac{2 \times 0.005995 \times 800^2}{0.006 \times 1078} = 1,186\ \text{Pa/m}$$

$$\left(\frac{dp}{dz}\right)_{GO} = \frac{2f_{Go}G^2}{D\rho_G} = \frac{2 \times 0.0035175 \times 800^2}{0.006 \times 76.95} = 9,752\ \text{Pa/m}$$

Thus, from Eq. (16.23), $Y = \sqrt{\frac{(dp/dz)_{GO}}{(dp/dz)_{LO}}} = \sqrt{\frac{9,752}{1,186}} = 2.867$

And from Eq. (15.19), $La = \sqrt{\frac{\sigma}{g(\rho_L - \rho_G)}\frac{1}{D_h}} = \sqrt{\frac{0.00427}{9.81(1078 - 76.95)}\frac{1}{0.006}} = 0.1099$

The two-phase frictional multiplier (Φ_{LO}^2) is calculated from Eq. (16.24) to be

$$\Phi_{LO}^2 = \left[Y^2x^3 + (1-x)^{0.33}\left(1 + 2x\left(Y^2 - 1\right)\right)\right]\left(1 + 1.54(1-x)^{0.5}La\right)$$
$$= \left[2.867^2 \times 0.5^3 + (1-0.5)^{0.33}\left(1 + 2 \times 0.5\left(2.867^2 - 1\right)\right)\right]$$
$$\left(1 + 1.54(1-0.5)^{0.5} \times 0.1099\right) = 8.455$$

Finally, the two-phase frictional pressure drop is determined from Eq. (16.16) as $\left(\frac{dp}{dz}\right)_f = \Phi_{LO}^2\left(\frac{dp}{dz}\right)_{LO} = 8.455\times\ 1,186 = 10,028$ Pa/m

Discussion The total pressure drop in gas-liquid two-phase flow is the sum of hydrostatic, frictional, and accelerational components as shown in Eq. (13.1). For the horizontal pipe orientation of this problem, the hydrostatic pressure drop is zero, and the frictional pressure drop was calculated to be 10,028 Pa/m. The remaining accelerational component of two-phase pressure drop is due to the expansion of gas phase as the two-phase mixture travels downstream and is trivial for non-boiling two-phase flow and hence can be neglected. However, in the case of boiling two-phase flow (problem under consideration), the pressure drop due to acceleration of the gas phase can offer a significant contribution to the total pressure drop depending upon the pipe diameter and pipe orientation. It should be noted that similar to hydrostatic component of pressure drop (see Eq. 13.2), the calculation of accelerational component of two-phase pressure drop also requires correct estimation of void fraction. The accelerational two-phase pressure drop component is expressed by Eq. (16.37) as given by Ghajar and Bhagwat (2014b):

$$\left(\frac{dp}{dz}\right)_a = \frac{1}{L}\left\{\left[\frac{G_L^2}{\rho_L\ (1-\alpha_G)} - \frac{G_G^2}{\rho_G\ \alpha_G)}\right]_{out} - \left[\frac{G_L^2}{\rho_L\ (1-\alpha_G)} - \frac{G_G^2}{\rho_G\ \alpha_G)}\right]_{in}\right\} \tag{16.37}$$

The void fraction at the inlet and outlet of the pipe needed in Eq. (16.37) for the calculation of the accelerational component of two-phase pressure drop was determined from the void fraction correlation of Woldesemayat and Ghajar (2007) to be $\alpha_{G,\ in} = 0.783$ and $\alpha_{G,\ in} = 0.926$. The final value of accelerational pressure drop calculated from Eq. (16.37) is 4064 Pa/m. Using Eq. (13.1), the total pressure drop for this problem is

$$\left(\frac{dp}{dz}\right)_t = \left(\frac{dp}{dz}\right)_h + \left(\frac{dp}{dz}\right)_f + \left(\frac{dp}{dz}\right)_a = 0 + 10{,}028 + 4{,}064 = 14,092\ \text{Pa/m}$$

Therefore, the accelerational pressure drop contributes to about 29% of the total two-phase pressure drop, while the remaining 71% contribution is due to the frictional two-phase pressure drop. It should be noted that in the case of adiabatic two-phase flow, the quality and void fraction at pipe inlet and exit remain practically unchanged and hence the accelerational two-phase pressure drop may be neglected.

Chapter 17
Modeling of Stratified Flow

Background

All the correlations presented in Chap. 16 are found to work well for all but stratified flow pattern. As mentioned earlier, stratified flow pattern in horizontal and downward pipe inclinations has a peculiar flow structure, and it is practically difficult to accurately model it with existing flow pattern-independent two-phase flow models. Apparently, the stratified flow pattern needs to be modeled using mechanistic flow model approach introduced by Taitel and Dukler (1976) and its derivatives. In addition to Taitel and Dukler (1976) model, also known as "flat surface model," the two-phase literature also reports semi-mechanistic models such as apparent rough surface (ARS) model introduced by Hart et al. (1989), and "double circle model" proposed by Chen et al. (1997), to calculate void fraction and two-phase pressure drop in stratified two-phase flow. It must be mentioned that the ARS and double circle models are developed only for stratified flow in horizontal pipe orientation and need to be scrutinized against stratified flow data in downward pipe inclinations. This chapter describes step-by-step procedure involved in solving the stratified flow model using these three different methods. The schematic of the type (shape) of gas-liquid interface associated with each of these methods is illustrated in Fig. 17.1.

Taitel and Dukler (1976) Model

The Taitel and Dukler (1976) model also known as "flat surface model" is developed for stratified flow at equilibrium, and as its name suggests, it assumes a flat interface between the gas and liquid phase. It also assumes that the gas-liquid interface is smooth and that the interfacial friction factor is equal to the gas phase friction factor. Their model considers the stratified flow of the pattern shown in Fig. 17.2. The

A. J. Ghajar, *Single- and Two-Phase Flow Pressure Drop and Heat Transfer in Tubes*, Mechanical Engineering Series,
https://doi.org/10.1007/978-3-030-87281-6_17

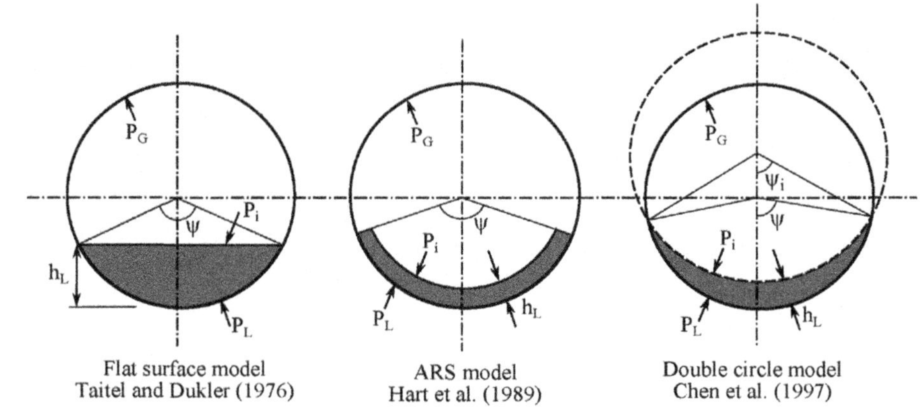

Fig. 17.1 Geometry considerations in three different types of stratified flow models

Fig. 17.2 Schematic of horizontal stratified flow at equilibrium

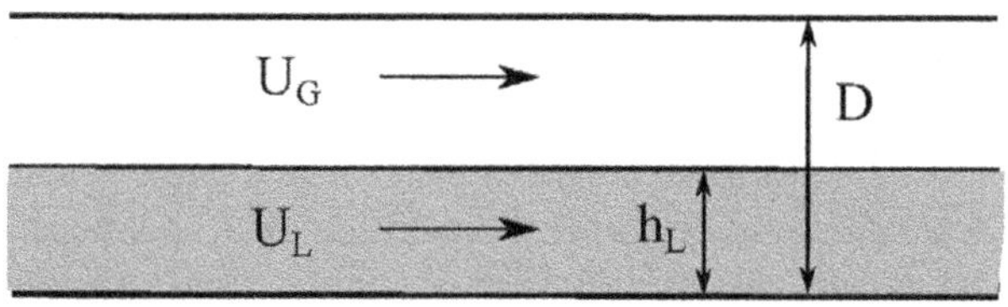

momentum balance equations for each phase (gas and liquid) could be written as shown in Eqs. (17.1) and (17.2). These equations also show the simplified forms of these momentum balance equations under the assumption of negligible acceleration of each phase and constant liquid level (height) along the pipe length. Considering that the pressure drop in the gas phase is equal to the pressure drop in the liquid phase, the momentum balance equations written for individual phases can be combined together to form Eq. (17.3). The individual phase and interfacial shear stress required in Eq. (17.3) are calculated using Eq. (17.4) where the single-phase friction factor for individual phase ($j = G, L$) is calculated from Eq. (17.5):

$$\begin{aligned} A_G\left(\frac{dp}{dz}\right) &= \tau_{wG}P_G + \tau_i P_i + A_G\rho_G g\left(\sin\theta - \cos\theta\frac{d}{dz}(1-h_L)\right) - A\frac{d}{dz}(GxU_G) \\ &= \tau_{wG}P_G + \tau_i P_i + A_G\rho_G g\sin\theta \end{aligned} \tag{17.1}$$

$$\begin{aligned} A_L\left(\frac{dp}{dz}\right) &= \tau_{wL}P_L - \tau_i P_i + A_L\rho_L g\left(\sin\theta - \cos\theta\frac{dh_L}{dz}\right) - A\frac{d}{dz}(G(1-x)U_L) \\ &= \tau_{wL}P_L - \tau_i P_i + A_L\rho_L g\sin\theta \end{aligned} \tag{17.2}$$

$$\tau_{wL}\frac{P_L}{A_L} - \tau_{wG}\frac{P_G}{A_G} - \tau_i P_i\left(\frac{1}{A_L} + \frac{1}{A_G}\right) + (\rho_L - \rho_G)g\sin\theta = 0 \tag{17.3}$$

$$\tau_{wG} = \frac{f_G\rho_G U_G^2}{2}, \tau_{wL} = \frac{f_L\rho_L U_L^2}{2}, \tau_i = \frac{f_i\rho_G(U_G - U_L)^2}{2} \tag{17.4}$$

$$f_j = \begin{pmatrix} 16/Re_j & : (Re_j \le 2300) \\ 0.046Re_j^{-0.2} & : (Re_j > 2300) \end{pmatrix} \tag{17.5}$$

The interfacial friction factor (f_i) is assumed to be equal to the gas phase friction factor (f_G). The gas and liquid phase Reynolds numbers are based on actual phase velocity and the hydraulic pipe diameters (based on area occupied by each phase) expressed as $Re_G = (\rho_G D_G U_G)/\mu_G$ and $Re_L = (\rho_L D_L U_L)/\mu_L$, respectively. The hydraulic diameters are calculated from Eq. (17.6) where the area occupied by gas (A_G) and liquid (A_L) phase is defined by Eq. (17.7). Considering the flat gas-liquid interface geometry of the Taitel and Dukler (1976) model, the wetted perimeter of pipe occupied by the gas (P_G) and liquid (P_L) phase and the interface (P_i) is expressed by Eq. (17.8).

$$D_G = \frac{4A_G}{P_G + P_i}, D_L = \frac{4A_L}{P_L} \tag{17.6}$$

$$A_G = \alpha_G\frac{\pi D^2}{4}, A_L = (1 - \alpha_G)\frac{\pi D^2}{4} \tag{17.7}$$

$$P_L = \frac{\Psi D}{2}, P_G = \pi D - P_L, P_i = \sin\left(\frac{\Psi}{2}\right)D \tag{17.8}$$

The void fraction, the liquid height, and the angle subtended by the flat interface with the pipe centerline are correlated with each other in the form of Eqs. (17.9) and (17.10). The liquid height (h_L) required in calculation of Ψ can be obtained using graphical solution presented in Fig. 14.14. Alternatively, Eqs. (17.1, 17.2, 17.3, 17.4, 17.5, 17.6, 17.7, 17.8, and 17.9) are to be solved iteratively until Eq. (17.3) is satisfied. The void fraction value that satisfies Eq. (17.3) is then used to calculate actual phase velocities, hydraulic diameters, Reynolds number, friction factors, and subsequently the shear stress values. Finally, Eq. (17.1) or (17.2) is solved to obtain the frictional pressure gradient.

$$\alpha_G = 1 - \frac{\Psi - \sin\Psi}{2\pi} \tag{17.9}$$

$$\Psi = 2\arccos\left(1 - \frac{2h_L}{D}\right) \tag{17.10}$$

Apparent Rough Surface Model

The concept of ARS model was first introduced by Hamersma and Hart (1987) and Hart et al. (1989) for wavy stratified two-phase flow in horizontal pipes and is in the form of gas pressure drop equation using two-phase friction factor. This two-phase friction factor (f_{TP}) is based on the gas (f_G) and interfacial (f_i) friction factors weighted by the fraction of pipe circumference (ϕ) wetted by the liquid phase. The following step-by-step approach needs to be followed to determine the values of α_G, f_G, f_i, ϕ, and f_{TP} and finally the two-phase frictional pressure drop using ARS model:

1. First calculate the void fraction using

$$\frac{1-\alpha_G}{\alpha_G} = \frac{U_{SL}}{U_{SG}}\left[1 + \left(108Re_{SL}^{-0.726}\frac{\rho_L}{\rho_G}\right)^{0.5}\right] \tag{17.11}$$

2. Calculate fraction of the pipe circumference (ϕ) wetted by the liquid phase using Eq. (17.12). It must be noted that ϕ is the fraction of circumference and not the angle formed by the pipe centerline with the gas-liquid interface (Ψ). The relation between ϕ and Ψ can be expressed as $\phi = \Psi/2\pi$:

$$\phi = 0.52(1-\alpha_G)^{0.374} + 0.26\left(\frac{\rho_L U_L^2}{(\rho_L - \rho_G)gD}\right)^{0.58} \tag{17.12}$$

3. Calculate interfacial friction factor (f_i) using Eq. (17.13) where ε is interfacial roughness.

$$\left.f_i = \frac{0.0625}{\left[\log\left(\frac{15}{Re_G} + \frac{\varepsilon}{3.715D}\right)\right]^2}\right\} \text{where } \varepsilon = \frac{2.3D(1-\alpha_G)}{4\phi} \tag{17.13}$$

4. From Eq. (17.14), calculate the gas phase friction factor (f_G) based on gas phase Reynolds number as a function of actual gas velocity ($Re_G = (\rho_G U_G D)/\mu_G$):

$$\left.f_G = \frac{0.07725}{(\log(Re_G/7))^2}\right\} 2100 < Re_G < 10^8 \tag{17.14}$$

5. The two-phase friction factor is then calculated using

$$f_{TP} = f_G(1-\phi) + f_i\phi \tag{17.15}$$

6. Finally, the two-phase pressure drop based on two-phase friction factor is calculated from Eq. (17.16). Note that Eq. (17.16) uses actual gas phase velocity (U_G) and not the superficial gas velocity (U_{SG}):

$$\left(\frac{dp}{dz}\right)_f = \frac{2f_{TP}\rho_G U_G^2}{D} \tag{17.16}$$

Double Circle Model

Double circle model is a mechanistic model proposed by Chen et al. (1997), which considers the concave shape of the gas-liquid interface by using the geometrical intersection of a hypothetical circle with the pipe centerline (see Fig. 17.1). Compared to the ARS model, the shape of gas-liquid interface used by double circle model is more realistic, and the overall model is based on mechanistic model of Taitel and Dukler (1976). To begin with, use of Chen et al. (1997) model requires determination of Ψ and Ψ_i in radians. Note that the subtended angle (Ψ) used by Chen et al. (1997) is half of that used by Taitel and Dukler (1976) and ARS models. The following step-by-step procedure is required to solve for void fraction and frictional pressure drop using double circle model:

1. Calculate the angle subtended by liquid phase with the pipe centerline using the relationship $\Psi = \pi\phi$ where the fraction of pipe circumference wetted by liquid phase is obtained by Eq. (17.12) (same as ARS model).
2. Once the angle Ψ is known, the angle Ψ_i is calculated from Eq. (17.17) using iterative technique. The value of void fraction required in this equation is a guess value obtained by solving for the Taitel and Dukler (1976) flat surface model:

$$\Psi_i = \left(\frac{\sin\Psi_i}{\sin\Psi}\right)^2\left(\Psi + \frac{\sin^2\Psi}{\tan\Psi_i} - \frac{\sin 2\Psi}{2} - \pi(1-\alpha_G)\right) \tag{17.17}$$

3. Next, the diameter of offset circle (D_i) is determined from

$$D_i = D\frac{\sin\Psi}{\sin\Psi_i} \tag{17.18}$$

4. The perimeter of pipe occupied by the gas and liquid phase and the gas-liquid interface is determined from Eq. (17.19). The calculation of the hydraulic diameters (D_G and D_L) and the area occupied by each phase (A_G and A_L) is the same as that given by Eqs. (17.6) and (17.7):

$$P_G = (\pi - \Psi)D, P_L = \Psi D, P_i = \Psi_i D_i \tag{17.19}$$

5. Interfacial friction factor is calculated using Eq. (17.20) where the superficial gas velocity during smooth to wavy stratified flow transition is defined by Eq. (17.21). Friction factors for gas (f_G) and liquid (f_L) phase are obtained using procedure similar to Taitel and Dukler (1976) model, i.e., Eq. (17.5).

$$\frac{f_i}{f_G} = 1 + 3.75\left(\frac{1 - \alpha_G}{\phi}\right)^{0.2}\left(\frac{U_{SG}}{U_{SG,t}} - 1\right)^{0.08} \tag{17.20}$$

$$U_{SG,t} = \sqrt{\frac{4\nu_L(\rho_L - \rho_G)g}{0.06\rho_G U_L}} \tag{17.21}$$

6. Solve for Eq. (17.3) by updating the values of void fraction until convergence is obtained.
7. Use converged value of void fraction obtained from step 6 to calculate for two-phase pressure drop using Eqs. (17.1) or (17.2).

Note that for a quick convergence, the initial guess of void fraction required to calculate actual phase velocities, hydraulic diameters, and Ψ_i is calculated from the Taitel and Dukler (1976) model. It must be mentioned that the double circle method is more complex than the Taitel and Dukler (1976) and ARS models since it involves two iterative solutions for α_G and Ψ_i. The following example problem provides an overview of the procedure required to solve for void fraction and pressure drop in two-phase stratified flow.

An Illustrative Example

Consider the horizontal two-phase flow of air and water in a 78 mm I.D. smooth polycarbonate pipe. The gas and liquid phase superficial velocities are 20 and 0.04 m/s, respectively. Assuming the stratified flow pattern to exist, determine the void fraction and two-phase frictional pressure drop using (a) Taitel and Dukler (1976) (flat surface), (b) ARS, and (c) double circle models.

For air (gas) and water (liquid), use the following thermophysical properties:

Liquid density (ρ_L) = 998 kg/m^3, gas density (ρ_G) = 1.2 kg/m^3, liquid viscosity (μ_L) = 0.001 Pa.s, and gas viscosity (μ_G) = 18.5 × 10^{-6} Pa.s.

Solution

(a) Calculation of void fraction and pressure drop using Taitel and Dukler (1976) method: First, calculate the Reynolds number based on superficial velocity of gas and liquid phase. The single-phase friction factors for smooth pipe are found using appropriate friction factor correlation such as Blasius (1913):

$$Re_{SG} = \frac{\rho_G U_{SG} D}{\mu_G} = \frac{1.2 \times 20 \times 0.078}{18.5 \times 10^{-6}} = 1.012 \times 10^5$$

$$f_G = \frac{0.079}{Re_{SG}^{\ 0.25}} = \frac{0.079}{\left(1.012 \times 10^5\right)^{0.25}} = 0.00443$$

$$Re_{SL} = \frac{\rho_L U_{SL} D}{\mu_L} = \frac{998 \times 0.04 \times 0.078}{0.001} = 3113.7$$

$$f_L = \frac{0.079}{Re_{SL}^{\ 0.25}} = \frac{0.079}{(3113.7)^{0.25}} = 0.0106$$

Now calculate single-phase pressure drop (based on superficial phase velocity) for each phase and hence X parameter from Eq. (16.21). The single-phase pressure drop calculation for each phase using Eqs. (16.19) and (16.17) is shown as follows:

$$\left(\frac{dp}{dz}\right)_G = \frac{2 f_G \rho_G U_{SG}^2}{D} = \frac{2 \times 0.00443 \times 1.2 \times 20^2}{0.078} = 54.52\ \text{Pa/m}$$

$$\left(\frac{dp}{dz}\right)_L = \frac{2 f_L \rho_L U_{SL}^2}{D} = \frac{2 \times 0.0106 \times 998 \times 0.04^2}{0.078} = 0.434\ \text{Pa/m}$$

$$\therefore\ X = \sqrt{\frac{(dp/dz)_L}{(dp/dz)_G}} = \sqrt{\frac{0.434}{54.52}} = 0.09$$

For horizontal two-phase flow, Eq. (14.13) gives $Y = 0$, and using the graphical solution in Fig. 14.14, the non-dimensional liquid height is $h_L/D \approx 0.1$. Using this ratio of equilibrium liquid height and pipe diameter, the angle subtended by the liquid phase with pipe centerline and hence the void fraction in stratified flow is calculated using Eqs. (17.10) and (17.9):

$$\Psi = 2\ \text{arc}\ \cos(1 - 0.2) = 73.74^o\ (1.287^r) \Rightarrow \alpha_G = 1 - \frac{\Psi - \sin\Psi}{2\pi} = 0.948$$

This value of void fraction of 0.948 is an initial estimate of the void fraction obtained using the graphical solution. This void fraction may be further updated to satisfy Eq. (17.3). Once the void fraction is determined, the actual phase velocities (see Table 12.1 for U_G and U_L), hydraulic diameters (Eq. 17.6), cross-sectional area occupied by each phase (Eq. 17.7), and wetted perimeter of each phase (Eq. 17.8) are calculated as follows:

$$U_G = \frac{U_{SG}}{\alpha_G} = \frac{20}{0.948} = 21.1\ \mathrm{m/s}, U_L = \frac{U_{SL}}{1-\alpha_G} = \frac{0.04}{1-0.948} = 0.77\ \mathrm{m/s}$$

$$A_G = \alpha_G \frac{\pi D^2}{4} = 0.00453\ \mathrm{m}^2, A_L = (1-\alpha_G)\frac{\pi D^2}{4} = 0.000248\ \mathrm{m}^2$$

$$P_L = 0.0502\ \mathrm{m}, P_G = 0.195\ \mathrm{m}, P_i = 0.0468\ \mathrm{m}, D_G = 0.0749\ \mathrm{m}, D_L = 0.0197\ \mathrm{m}$$

The Reynolds numbers based on actual gas and liquid phase velocities are found to be

$$Re_G = \frac{1.2 \times (20/0.948) \times 0.0749}{18.5 \times 10^{-6}} = 1.025 \times 10^5$$

$$Re_L = \frac{998 \times (0.04/(1-0.948)) \times 0.0197}{0.001} = 15139$$

The single-phase friction factors for each phase are found using Eq. (17.5):

$$f_G = 0.046\left(1.025 \times 10^5\right)^{-0.2} = 0.00457$$

$$f_L = 0.046 \times (15139)^{-0.2} = 0.00671$$

Now, the wall and interfacial shear stress using $f_i = f_G$ is calculated using Eq. (17.4).

$$\tau_{wG} = \frac{0.00457 \times 1.2 \times 21.1^2}{2} = 1.232\ \mathrm{N/m}^2$$

$$\tau_{wL} = \frac{0.00671 \times 998 \times 0.77^2}{2} = 1.985\ \mathrm{N/m}^2$$

$$\tau_i = \frac{0.00457 \times 1.2 \times (21.1-0.77)^2}{2} = 1.133\ \mathrm{N/m}^2$$

Now solve Eq. (17.3) using the known values of shear stress, wetted perimeter, and cross-sectional area occupied by each phase. For these values, Eq. (17.3) is not satisfied (residual $= -121.1$), and hence, the value of h_L/D is updated to find a new value of void fraction, and the remaining steps are repeated. Iterate through these calculations until the residual of Eq. (17.3) is approximately zero. After a few iterations, it is found that Eq. (17.3) is satisfied (residual $= 0.008$) for $h_L/D \approx 0.1117$ and $\alpha_G = 0.938$ ($\Psi = 78.1^o = 1.363^r$). Finally, after the convergence of Eq. (17.3), two-phase frictional pressure drop is obtained by solving either Eq. (17.1) or Eq. (17.2). It should be noted that use of Eq. (17.1) or (17.2) requires calculation of updated values of shear stresses (τ_{wL}, τ_{wG}, τ_i) based on updated friction factors (f_G, f_L, f_i), the cross-sectional area occupied by each phase (A_G, A_L), and wetted perimeters of each phase and the gas-liquid interface (P_G, P_L, P_i). Two-phase pressure drop calculation using Eq. (17.2) is shown as follows:

$$\left(\frac{dp}{dz}\right)_f = \tau_{wL}\frac{P_L}{A_L} - \tau_i\frac{P_i}{A_L} = 1.445\frac{0.0531}{0.000296} - 1.17\frac{0.0491}{0.000296} = 66.1\ \text{Pa/m}$$

(b) *ARS model*: Calculate void fraction using Eq. (17.11) where $Re_{SL} = \rho_L U_{SL} D/\mu_L = 3113.7$:

$$\frac{1-\alpha_G}{\alpha_G} = \frac{0.04}{20}\left[1 + \left(108 \times 3113.7^{-0.726} \times \frac{998}{1.2}\right)^{0.5}\right] \Rightarrow \alpha_G = 0.966$$

Now, calculate the actual gas and liquid phase velocities (see Table 12.1), $U_G = 20.7$ m/s and $U_L = 1.176$ m/s. The fractional wetted perimeter is calculated from Eq. (17.12):

$$\phi = 0.52(1-0.966)^{0.374} + 0.26\left(\frac{998 \times 1.176^2}{(998-1.2) \times 9.81 \times 0.078}\right)^{0.58} = 0.523$$

Calculate gas Reynolds number based on actual gas velocity:

$$Re_G = \frac{\rho_G U_G D}{\mu_G} = \frac{1.2 \times 20.7 \times 0.078}{18.5 \times 10^{-6}} = 1.046 \times 10^5$$

Using Eq. (17.14), the friction factor for single-phase gas flow (f_G) is calculated as follows:

$$f_G = \frac{0.07725}{\left(\log\left(Re_G/7\right)\right)^2} = \frac{0.07725}{\left(\log\left(1.046 \times 10^5\right)/7\right)^2} = 0.0044$$

The interfacial friction factor is calculated using Eq. (17.13) where the interfacial roughness is $\varepsilon = 2.3D(1-\alpha_G)/4\phi = 0.00291$:

$$f_i = \frac{0.0625}{\left[\log\left(\frac{15}{Re_G} + \frac{\varepsilon}{3.715D}\right)\right]^2} = \frac{0.0625}{\left[\log\left(\frac{15}{1.046\times10^5} + \frac{0.00291}{3.715\times0.078}\right)\right]^2} = 0.0156$$

Now, from Eq. (17.15), two-phase friction factor is calculated as

$$f_{TP} = (1-\phi)f_G + \phi f_i = (1-0.523)\times 0.0044 + 0.523\times 0.0156 = 0.01025$$

The two-phase frictional pressure drop is calculated using Eq. (17.16):

$$\left(\frac{dp}{dz}\right)_f = \frac{2f_{TP}\rho_G U_G^2}{D} = \frac{2\times 0.01025\times 1.2\times 20.7^2}{0.078} = 135.1\ \text{Pa/m}$$

(c) *Double circle model*: Initially, the angle subtended by the liquid layer with the pipe centerline is calculated using the relationship $\Psi = \pi\phi = 1.153^r$ where $\phi = 0.367$ is calculated at a void fraction of $\alpha_G = 0.938$ (converged solution of Taitel and Dukler (1976) model) using Eq. (17.12). Next, the angle Ψ_i is calculated using Eq. (17.17). It should be noted that the angles (Ψ and Ψ_i) used in Eq. (17.17) are expressed in radians:

$$\Psi_i = \left(\frac{\sin\Psi_i}{\sin(1.153)}\right)^2\left(1.153 + \frac{\sin^2(1.153)}{\tan\Psi_i} - \frac{\sin(2\times 1.153)}{2} - \pi(1-0.938)\right)$$

The solution to this equation, $\Psi_i = 0.931^r$, is obtained iteratively. Now, the diameter D_i of eccentric circle is found using Eq. (17.18).

$$D_i = D\frac{\sin\Psi}{\sin\Psi_i} = 0.078\times\frac{\sin(1.153)}{\sin(0.931)} = 0.088\ \text{m}$$

The wetted perimeter of gas and liquid phase based on Ψ and that of gas-liquid interface based on Ψ_i are calculated from Eq. (17.19). The calculation of hydraulic diameters is similar to that of Taitel and Dukler (1976) given by Eq. (17.6):

$$P_L = 0.0899\ \text{m}, P_G = 0.155\ \text{m}, P_i = 0.0827\ \text{m}, D_G = 0.0753\ \text{m}, D_L = 0.0132\ \text{m}$$

The Reynolds number of each phase is based on actual phase velocity and hydraulic pipe diameter, and the single-phase friction factors for each phase using Eq. (17.5) are calculated as follows:

$$Re_G = \frac{\rho_G U_G D_G}{\mu_G} = \frac{1.2\times 21.3\times 0.0753}{18.5\times 10^{-6}} = 104036$$

$$Re_L = \frac{\rho_L U_L D_L}{\mu_L} = \frac{998\times 0.645\times 0.0132}{0.001} = 8497$$

$$f_G = 0.046(Re_G)^{-0.2} = 0.046(104036)^{-0.2} = 0.00456$$

$$f_L = 0.046(Re_L)^{-0.2} = 0.046\times(8497)^{-0.2} = 0.00753$$

To find f_i from Eq. (17.20), first, determine the superficial gas velocity corresponding to the smooth to wavy stratified flow transition ($U_{SG,t}$) from Eq. (17.21):

$$U_{SG,t} = \sqrt{\frac{4 \times (0.001/998)(998 - 1.2) \times 9.81}{0.06 \times 1.2 \times 0.465}} = 0.918$$

$$f_i = 0.00456\left[1 + 3.75\left(\frac{1 - 0.938}{0.367}\right)^{0.2}\left(\frac{20}{0.918} - 1\right)^{0.08}\right] = 0.0198$$

Now, the wall and interfacial shear stress is calculated using Eq. (17.4):

$$\tau_{wG} = \frac{0.00456 \times 1.2 \times 21.3^2}{2} = 1.244\ \mathrm{N/m^2}$$

$$\tau_{wL} = \frac{0.00753 \times 998 \times 0.645^2}{2} = 1.563\ \mathrm{N/m^2}$$

$$\tau_i = \frac{0.0198 \times 1.2 \times (21.3 - 0.645)^2}{2} = 5.087\ \mathrm{N/m^2}$$

Equation (17.3) is not satisfied (residual = 1083), and hence, the void fraction value is updated, and all other subsequent steps are repeated. At convergence (residual = 0.002), $\alpha_G = 0.961$, $\Psi = 1.48^r$, and $\Psi_i = 1.407^r$. For the converged value of void fraction, recalculate the gas, liquid, and interfacial shear stress from Eq. (17.4), and then calculate two-phase pressure drop using Eq. (17.1) or (17.2). Note that these equations require updated values of shear stresses based on the converged values of void fraction. Pressure drop from Eq. (17.2) is calculated as follows:

$$\left(\frac{dp}{dz}\right)_f = \tau_{wL}\frac{P_L}{A_L} - \tau_i\frac{P_i}{A_L} = 4.277\frac{0.1156}{0.0001834} - 4.234\frac{0.1109}{0.0001834} = 135.6\ \mathrm{Pa/m}$$

Discussion For similar flow conditions, Badie et al. (2000) measured $\alpha_G = 0.956$ and $(dp/dz)_f = 115$ Pa/m. Note that the void fraction and frictional pressure drop obtained by solving for ARS model and double circle model are comparable to measurements of Badie et al. (2000). The prediction of void fraction using Taitel and Dukler (1976) model is within ±2% of the measured void fraction by Badie et al. (2000). However, the frictional pressure drop calculated using Taitel and Dukler (1976) is significantly less than the experimental value. This is possibly because of the use of $f_i/f_G = 1$ in their model. The use of suitable and more realistic model for $f_i/$

f_G in Taitel and Dukler (1976) may yield better results. For example, the use of $f_i/f_G = 10$ recommended by Crowley et al. (1992) would give a better prediction of frictional pressure drop as 126.5 Pa/m. Conclusively, the success of these three methods in the prediction of void fraction and pressure drop in stratified two-phase flow is subject to the validity and accuracy of the closure relationships used in these models. Different forms of models for the calculation of f_i/f_G could be obtained from Ottens et al. (2001).

Chapter 18
Entrainment

Background

A peculiar phenomenon known as entrainment is observed in annular flow due to the relative motion between the gas and liquid phase. The entrainment process is characterized by the flow of tiny liquid droplets into the central fast-moving gas core. The liquid entrainment is a consequence of the significant shear at the gas-liquid interface that causes tearing of the liquid wave crests in the form of ligaments and depends on the phase flow rates, pipe diameter and orientation, and the surface tension at the gas-liquid interface. The liquid entrainment fraction (E) is defined as the ratio of mass flow rate/flux of liquid drops entering into the gas core to the total mass flow rate/flux of the liquid phase. Correct knowledge of the rate of liquid entrainment or alternatively the fraction of liquid entrainment is crucial in analyzing heat and mass transfer processes in annular flows. In the case of boiling two-phase flows, accurate estimation of the liquid entrainment fraction is necessary in determination of critical heat flux and dryout conditions.

Entrainment Mechanisms

Ishii and Grolmes (1975) studied the different possible mechanisms responsible for liquid entrainment process. They found that liquid entrainment can happen due to any or all of the wave undercut, wave rolling, wave coalescence, and ripple shearing mechanisms. Additionally, the phenomenon of "bubble burst" due to rupture of liquid phase crests and "droplet impingement" due to rolling waves may also contribute to the entrainment process.

The different entrainment mechanisms suggested by Ishii and Grolmes (1975) are illustrated in Fig. 18.1. As shown in Fig. 18.2, critical values of gas and liquid flow

A. J. Ghajar, *Single- and Two-Phase Flow Pressure Drop and Heat Transfer in Tubes*, Mechanical Engineering Series,
https://doi.org/10.1007/978-3-030-87281-6_18

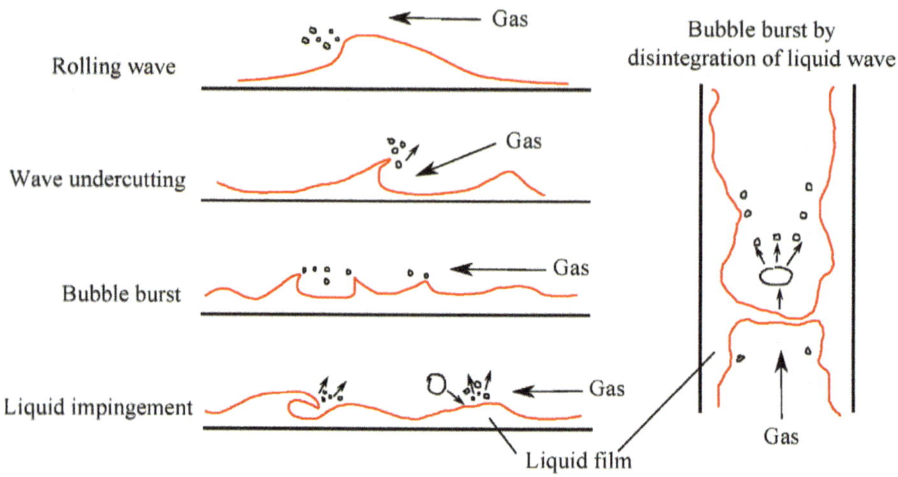

Fig. 18.1 Entrainment mechanisms suggested by Ishii and Grolmes (1975)

Fig. 18.2 Inception criteria for entrainment. (Adapted from Ishii and Grolmes, 1975)

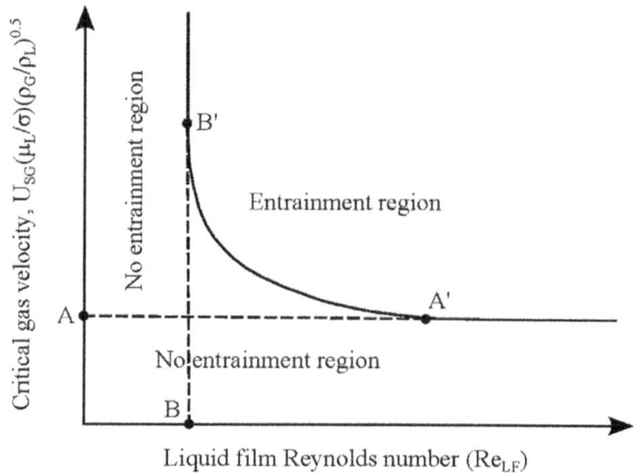

rates are associated with the entrainment process and below which no liquid entrainment is expected to occur. Based on the experimental data, two-phase flow literature offers several correlations to determine the non-dimensional critical gas velocity and the liquid film Reynolds number (Re_{LF}) below which the entrainment phenomenon is negligible or alternatively above which the amount of liquid entrainment is significant (see Fig. 18.2). Note that the Reynolds number based on liquid flowing in the form of film is defined by Ishii and Grolmes (1975) as $Re_{LF} = (4\delta U_{LP}\rho_L)/\mu_L$.

One such criterion in terms of liquid film Reynolds number proposed by Ishii and Grolmes (1975) is represented by Eq. (18.1). This equation sets the absolute limit for inception of entrainment process and represents point B in Fig. 18.2 below which entrainment cannot take place. Ishii and Grolmes (1975) found that at point A', $Re_{LF} = 1635$ and beyond which the entrainment affected region is insensitive to the increase in liquid flow rates. The critical gas velocity for Re_{LF} greater than that by Eq. (18.1) is obtained from Eq. (18.2). The critical gas velocity to the left-hand side of this equation is essentially a non-dimensional superficial gas velocity normalized using liquid phase dynamic viscosity and the surface tension. Note that for Re_{LF} greater than $Re_{LF,C}$ but less than 160, critical gas velocity is independent of liquid viscosity number given by Eq. (18.3):

$$Re_{LF,C} = 154.7\left(\frac{\rho_L}{\rho_G}\right)^{0.75}\left(\frac{\mu_G}{\mu_L}\right)^{1.5} \tag{18.1}$$

$$\frac{U_{SG}\mu_L}{\sigma}\sqrt{\frac{\rho_G}{\rho_L}} \geq \begin{cases} 1.5Re_{LF}^{-0.5} & : Re_{LF,C} < Re_{LF} < 160 \\ 11.78N_\mu^{0.8}Re_{LF}^{-0.33} & : N_\mu \leq 0.066, 160 \leq Re_{LF} \leq 1635 \\ 1.35Re_{LF}^{-0.33} & : N_\mu > 0.066, 160 \leq Re_{LF} \leq 1635 \\ N_\mu^{0.8} & : N_\mu \leq 0.066, Re_{LF} > 1635 \\ 0.1146 & : N_\mu > 0.066, Re_{LF} > 1635 \end{cases} \tag{18.2}$$

$$N_\mu = \frac{\mu_L}{\left(\rho_L\sigma\sqrt{\sigma/(g\Delta\rho)}\right)^{0.5}} \tag{18.3}$$

Liquid Entrainment Fraction Correlations

Note that Eqs. (18.1) and (18.2) predict the onset of entrainment and not the magnitude of liquid entrainment fraction. To predict the liquid entrainment fraction, we need specific correlations.

Ishii and Mishima (1989) Correlation Their proposed correlation is applicable to air-water equilibrium annular flow. As shown in Eq. (18.4), their correlation is a function of modified Weber number given by Eq. (18.5). Although the correlation of Ishii and Mishima (1989) consists of non-dimensional numbers as a function of fluid physical properties, it is primarily based on air-water data, and its validity for other fluid combinations must be scrutinized before use:

$$E = \tanh\left(7.25\times10^{-7}We^{1.25}Re_{SL}^{0.25}\right) \left\} \begin{array}{l} 0.1 \leq p_{sys} \leq 0.4\text{ MPa} \\ 9.5 \leq D \leq 32\text{ mm} \\ 370 \leq Re_{SL} \leq 6400 \end{array}\right. \tag{18.4}$$

$$We = \frac{\rho_G U_{SG}^2 D}{\sigma}\left(\frac{\rho_L - \rho_G}{\rho_G}\right)^{0.33} \tag{18.5}$$

Cioncolini and Thome (2012) Correlation They proposed a method to predict liquid entrainment fraction in gas-liquid annular flows. In comparison to Ishii and Mishima (1989), correlation of Cioncolini and Thome (2012) is based on a more comprehensive data set, 2293 data points for circular pipes and 71 data points for noncircular pipes, and is claimed to be valid for $5 \le D \le 95$ mm and $0.1 \le p_{sys} \le 20$ MPa and for both adiabatic and evaporating flows in vertical upward flow.

Bhagwat and Ghajar (2015b) Correlation They introduced a correction factor (ξ + $120\cos^2\theta$) to the Cioncolini and Thome (2012) correlation to improve its accuracy at high system pressures and horizontal and upward inclined pipe orientations. This correction factor embedded into original correlation of Cioncolini and Thome (2012) is expressed by Eq. (18.6) where ξ is calculated from Eq. (18.7). It should be pointed out that the original correlation of Cioncolini and Thome (2012) used a constant value 279.6 in place of the correction factor (ξ + $120\cos^2\theta$) in Eq. (18.7). The need of a correction factor for the effect of high system pressure on liquid entrainment fraction is explained in the work by Bhagwat and Ghajar (2015b). Note that the use of Eq. (18.7) requires system pressure (p_{sys}) with units of MPa. The gas core Weber number and the gas core density are calculated using Eqs. (18.8) and (18.9), respectively:

$$E = \left(1 + (\xi + 120\cos^2\theta) We_c^{-0.8395}\right)^{-2.209} \quad \left\} \begin{array}{l} 0.1 \le p_{sys} \le 20\ \text{MPa} \\ 5 \le D \le 95\ \text{mm} \\ 10 \le We_c \le 10^5 \end{array} \right. \tag{18.6}$$

$$\xi = \begin{cases} 280 & : 0.1 \le p_{sys} < 10\ \text{MPa} \\ 4637.8 \times p_{sys}^{-1.6} & : p_{sys} \ge 10\ \text{MPa} \end{cases} \tag{18.7}$$

$$We_c = \frac{\rho_c U_{SG}^2 D}{\sigma} \tag{18.8}$$

$$\rho_c = \frac{x + E(1-x)}{(x/\rho_G) + (E(1-x)/\rho_L)} \tag{18.9}$$

Equation (18.6) needs to be solved using iterative technique since the core density required in calculation of the core Weber number depends upon the liquid

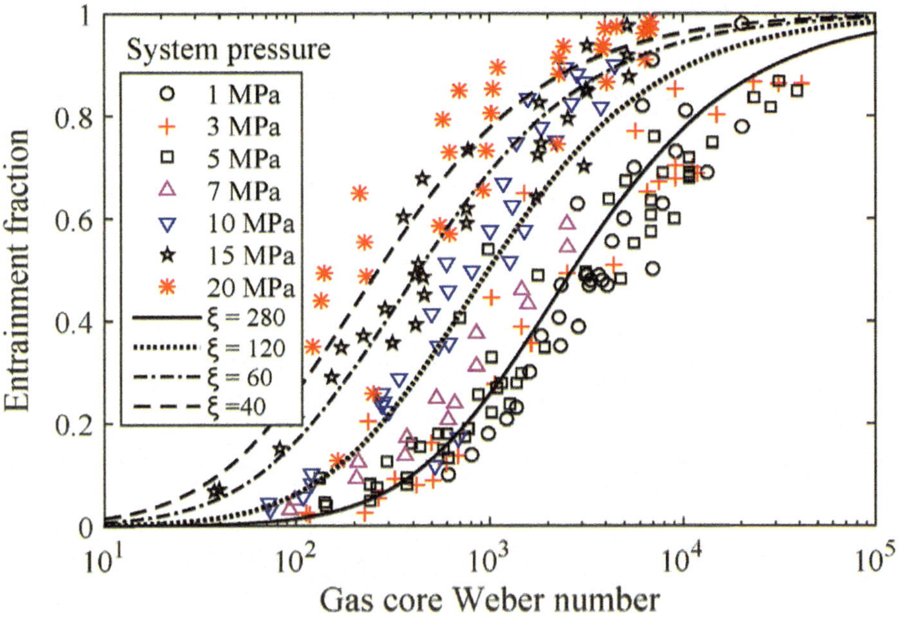

Fig. 18.3 Variation of liquid entrainment fraction as a function of system pressure (trends predicted by Eq. (18.6))

entrainment fraction. Cioncolini and Thome (2012) suggested a two-step predictor and corrector method such that the predictor method predicts the liquid entrainment fraction for an initial guess of core density approximately equal to the gas phase density (i.e., $\rho_c \approx \rho_G$). In the corrector step, the E value from predictor step is used to calculate a new value of gas core density and hence the liquid entrainment fraction. The variation of liquid entrainment fraction as a function of system pressure using the proposed modification by Bhagwat and Ghajar (2015b) to the original Cioncolini and Thome (2012) correlation is illustrated in Fig. 18.3. In this figure, the value of $\xi \approx 280$ is the original fixed value used by Cioncolini and Thome (2012), and the ξ values of 120, 60, and 40 correspond to 10, 15, and 20 MPa of the system pressure, respectively (see Eq. 18.7). Clearly the proposed modification predicts more than 90% of the data points within ±30% deviation for system pressures of 10, 15, and 20 MPa which shifts the trend of E vs We_c with change in ξ values which is a function of system pressure. From Fig. 18.3, it is evident that at low ($We_c \leq 10$) and high ($We_c \geq 10^5$) gas core Weber numbers, the effect of system pressure on liquid entrainment fraction is small. Figure 18.3 is plotted for vertical two-phase flow and can be used as a graphical solution to determine liquid entrainment fraction.

An Illustrative Example

Use of Bhagwat and Ghajar (2015b) Correlation Consider the vertical upward annular two-phase flow of air-water mixture in a 45 mm I.D. smooth pipe. At near atmospheric system pressure, the two-phase mixture flows with a quality (x) and mass flux (G) of 0.25 and 210 kg/m^2, respectively. The physical properties of air and water are as follows: $\rho_G = 1.5$ kg/m^3, $\rho_L = 998$ kg/m^3, and $\sigma = 0.072$ N/m. Determine the liquid entrainment fraction.

Solution

Liquid entrainment fraction (E) can be estimated from Eq. (18.6) using an iterative technique since the core density (ρ_c) required in calculation of the core Weber number (We_c) depends upon the liquid entrainment fraction. Using a two-step predictor and corrector method, first estimate the gas core Weber number (We_c) from Eq. (18.8) using gas phase density (ρ_G) and superficial gas velocity (U_{SG}) to predict the initial estimate of the liquid entrainment fraction. The superficial gas velocity required in determination of We_c is calculated as $U_{SG} = Gx/\rho_G = 210 \times 0.25/1.5 = 35$ m/s. Then find the gas core density (ρ_c) from Eq. (18.9) using the initial estimate of E, and recalculate core Weber number and liquid entrainment fraction in corrector step. For near atmospheric system pressure, $\xi = 280$ from Eq. (18.7).

Predictor Step Initial estimate of gas core Weber number (We_c) using Eq. (18.8).

$$We_c = \frac{\rho_G U_{SG}^2 D}{\sigma} = \frac{1.5 \times 35^2 \times 0.045}{0.072} = 1148.4$$

Initial estimate of the liquid entrainment fraction (E) using Eq. (18.6).

$$E = \left(1 + 280 \times We_c^{-0.8395}\right)^{-2.209} = 0.288$$

Corrector Step Calculate the gas core density (ρ_c) from Eq. (18.9) using the initial estimate of E, and recalculate the gas core Weber number (We_c):

$$\rho_c = \frac{x + E(1-x)}{(x/\rho_G) + (E(1-x)/\rho_L)} = \frac{0.25 + 0.288 \times 0.75}{(0.25/1.5) + (0.288 \times 0.75/998)} = 2.8 \text{ kg/m}^3$$

$$We_c = \frac{\rho_c U_{SG}^2 D}{\sigma} = \frac{2.8 \times 35^2 \times 0.045}{0.072} = 2143.7$$

The final value of liquid entrainment fraction (E) is determined from the recalculated value of the gas core Weber number:

$$E = \left(1 + 280 \times We_c^{-0.8395}\right)^{-2.209} = 0.44$$

Discussion The calculated value of the liquid entrainment fraction (E) can now be used to determine, for example, the liquid film mass flow rate ($\dot{m}_{LF}$) as follows:

$$\dot{m}_{LF} = (1 - E)\dot{m}_L = (1 - 0.44) \times 0.25 = 0.14 \text{ kg/s}$$

where the liquid mass flow rate ($\dot{m}_L$) required in the $\dot{m}_{LF}$ equation is calculated as

$$\dot{m}_L = G(1 - x) \times A = 210(1 - 0.25) \times \frac{\pi}{4} \times 0.045^2 = 0.25 \text{ kg/s}$$

Chapter 19
Modeling of Annular Flow

Background

Annular flow structure exhibits separated flow characteristics and hence could be appropriately modeled using separated flow model approach. The two-phase flow literature reports several annular flow pattern-specific models to determine two-phase flow parameters such as void fraction, pressure drop, and heat transfer. Chapter 15 gives some of the existing correlations in the two-phase flow literature that can satisfactorily predict void fraction in annular flow. This chapter specifically focuses on some salient features of annular flow such as liquid film thickness, liquid film flow rate, interfacial friction, and their interrelationships through the concept of triangular relationship.

Triangular Relationship in Annular Flow

Triangular relationship in the annular flow in essence is a relationship between three principal dependent system parameters, i.e., pressure gradient, liquid film thickness (δ), and the liquid film flow rate ($\dot{m}_{LF}$), such that if any of these two variables are known, the third unknown variable may be calculated. Although the triangular relationship forms a basis for different correlations developed for annular flow, these dependent variables that form a triangular relationship cannot not be practically calculated from the independent system parameters such as phase flow rates, fluid properties, and pipe geometry and thus further require closed-form solutions. The closed-form solutions required to form triangular relationship for annular flow are usually in the form of equations for liquid entrainment fraction (E) and two-phase frictional multiplier (Φ^2) involving two-phase (f_{TP}) or interfacial (f_i) friction factors. This chapter briefly introduces the concept of triangular relationship and illustrates a

A. J. Ghajar, *Single- and Two-Phase Flow Pressure Drop and Heat Transfer in Tubes*, Mechanical Engineering Series,
https://doi.org/10.1007/978-3-030-87281-6_19

step-by-step approach to calculate the system parameters in annular flow. For more details, refer to Hewitt and Hall-Taylor (1970) and Collier and Thome (1996).

In order to solve for the triangular relationship, Hewitt and Hall-Taylor (1970) recommend use of following simplifying assumptions:

1. Film thickness is very small with respect to pipe diameter, i.e., $\delta << D$.
2. Shear stress in liquid film is constant and equal to the wall shear stress.
3. Hydrostatic and accelerational effects on the gas core and liquid film are negligible (compared to frictional pressure drop).

Assuming axis symmetric flow, uniform film thickness, and negligible contribution of hydrostatic and accelerational pressure drops (in comparison to frictional pressure drop), the momentum balance on the liquid film in annular flow leads to Eq. (19.1) for two-phase frictional pressure drop. This two-phase frictional pressure drop is expressed in terms of actual velocity of the liquid flowing in the form of film (in contact with pipe wall) and that the liquid entrainment is negligible. Using the definition of Φ_L^2, Eq. (16.17) could be also expressed as Eq. (19.2):

$$\left(\frac{dp}{dz}\right)_f = \frac{4\tau_{wL}}{D} = \frac{2f_{TP}\rho_L U_L^2}{D} \tag{19.1}$$

$$\Phi_L^2 = \frac{(dp/dz)_f}{(dp/dz)_L} = \frac{1}{(1-\alpha_G)^2}\left(\frac{f_{TP}}{f_L}\right) \tag{19.2}$$

Hewitt and Hall-Taylor (1970) have reported that using the concept of equivalent diameter and definition of void fraction under no entrainment, the Reynolds number of liquid flowing in the form of film is identical to the Reynolds number of liquid phase flowing alone through the pipe. This implies that based on identical Reynolds number defined by Eq. (19.3), the friction factors of liquid flowing in the form of film and that alone through the pipe must be identical, i.e., $f_{TP} = f_L$. Using this relationship, Eq. (19.2) reduces to Eq. (19.4). This relationship however does not acknowledge the existence of liquid entrainment fraction which in certain practical cases may not be ignored. To account for the liquid entrainment fraction, Turner and Wallis (1965) have proposed use of a similar relationship in which the single-phase pressure drop for total liquid flow is replaced by the single-phase pressure drop for that part of the liquid flow which is in the form of film. The two-phase frictional multiplier suggested by Turner and Wallis (1965) is expressed by Eq. (19.5) where the single-phase pressure drop of liquid film flow is obtained from Eq. (19.6):

$$\frac{\rho_L U_L 4\delta}{\mu_L} = \frac{\rho_L U_{SL} D}{\mu_L} \tag{19.3}$$

$$\Phi_L^2 = \frac{1}{(1-\alpha_G)^2} \tag{19.4}$$

$$\Phi_{LF}^2 = \frac{(dp/dz)_f}{(dp/dz)_{LF}} = \frac{1}{(1-\alpha_G)_{LF}^2} \tag{19.5}$$

$$\left(\frac{dp}{dz}\right)_{LF} = \frac{2f_{LF}\rho_L(U_{SL}(1-E))^2}{D} = \frac{2f_{LF}(G(1-x)(1-E))^2}{D\rho_L} \tag{19.6}$$

Equation (19.5) can be considered as an empirical correlation for determination of void fraction in the presence of liquid entrainment process; however, its implementation also requires the knowledge of two-phase frictional pressure drop. For a known or measured value of void fraction and liquid entrainment fraction, Eq. (19.5) may not predict the correct magnitude of two-phase frictional pressure drop due to lack of information on the interfacial roughness and hence the interfacial shear stress. To address this issue, one more closure relationship is required that links the interfacial friction/roughness to the liquid flow rate or any other previously calculated variables such as void fraction (see Chap. 15) or liquid entrainment fraction (see Chap. 18). This relationship could be obtained by considering yet another form of two-phase frictional multiplier in the form of Φ_G^2. Using similar assumptions that are used to obtain Eq. (19.1), a momentum balance for gas core is used to obtain Eq. (19.7). Using Eq. (19.7), the two-phase frictional multiplier (Φ_G^2) assuming only gas phase flowing through pipe can be written as Eq. (19.8) where the interfacial and gas shear stress are expressed by Eqs. (19.9) and (19.10), respectively. Thus, Φ_G^2 could be expressed as a function of void fraction, interfacial friction factor, core density, actual gas velocity, and liquid film velocity as expressed by Eq. (19.11):

$$\left(\frac{dp}{dz}\right)_f = \frac{4\tau_i}{(D-2\delta)} \tag{19.7}$$

$$\Phi_G^2 = \frac{(dp/dz)_f}{(dp/dz)_G} = \frac{4\tau_i/(D-2\delta)}{4\tau_G/D} \tag{19.8}$$

$$\tau_i = f_i\left(\frac{\rho_c(U_G - U_{LF})^2}{2}\right) \tag{19.9}$$

$$\tau_G = f_G\left(\frac{\rho_G U_{SG}^2}{2}\right) \tag{19.10}$$

$$\Phi_G^2 = \frac{1}{\sqrt{\alpha_G}} \frac{f_i}{f_G} \frac{\rho_c}{\rho_G} \frac{(U_G - U_{LF})^2}{U_{SG}^2} = \frac{1}{\sqrt{\alpha_G}} \frac{f_i}{f_G} \frac{\rho_c}{\rho_G} \left[\frac{1}{\alpha_G} - \frac{U_L(1-E)}{U_{SG}} \right]^2 \quad (19.11)$$

Note that for low liquid flow rates, the velocity of liquid flowing in the form of film may be small compared to the fast-moving gas core such that $U_{LF} << U_G$. Also for low liquid flow rates, the liquid entrainment fraction may be negligible leading to $\rho_c/\rho_G \approx 1$. Thus, for such a simplified two-phase flow situation, Eq. (19.11) reduces to Eq. (19.12). Furthermore, for a more simplified assumption of smooth gas-liquid interface, ($f_i \approx f_G$), and Φ_G^2 can be expressed as Eq. (19.13):

$$\Phi_G^2 = \frac{1}{\alpha_G{}^{5/2}} \frac{f_i}{f_G} \quad (19.12)$$

$$\Phi_G^2 = \frac{1}{\alpha_G{}^{5/2}} \quad (19.13)$$

It must be mentioned that the assumption of smooth gas-liquid interface (f_i/f_G) is an unrealistic assumption for annular flow and the calculation of two-phase frictional multiplier and the two-phase frictional pressure drop may deviate significantly from actual conditions. It is evident from Fig. 19.1 that by accounting for the interfacial roughness, the predicted values of Φ_G^2 could be significantly greater than that for the ideal conditions. Going back to Eq. (19.11), the ratio f_i/f_G could be replaced with the empirical correlation in the form of Eq. (19.14) proposed by Wallis (1962). Note that two-phase flow literature reports several empirical relationships to model the ratio of f_i/f_G. Some of these correlations are listed in Table 19.1:

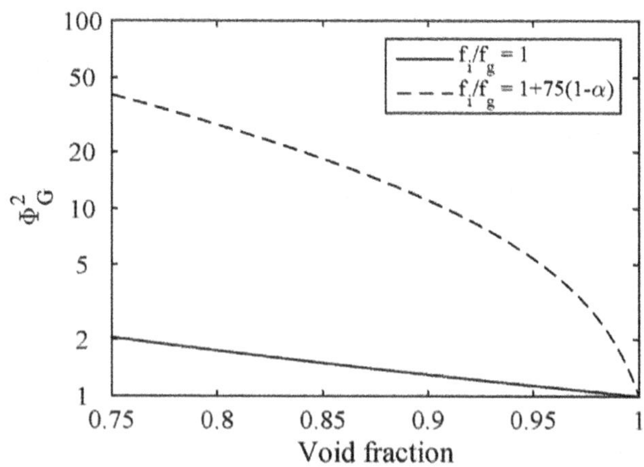

Fig. 19.1 Effect of f_i/f_G on two-phase frictional multiplier (Φ_G^2)

Table 19.1 List of correlations for interfacial friction factors

Source	Correlation
Moeck (1970)	$f_i = 0.005\left(1 + 1458\left(\frac{\delta}{D}\right)^{1.42}\right)$
Laurinat et al. (1984)	$f_i = f_G\left(2 + 2.5 \times 10^{-5}\frac{Re_{LF}}{D}\right)\alpha_G^{5/2}$
Wallis (1962)	$f_i = 0.005(1 + 75(1 - \alpha_G))$
Wongwises and Kongkiatwanitch (2001)	$f_i = 17.172Re_{SG}^{-0.768}\left(\frac{\delta}{D}\right)^{-0.253}$
Fore et al. (2000)	$f_i = 0.005\left(1 + 300\left[\left(1 + \frac{17500}{Re_G}\right)\frac{\delta}{D} - 0.0015\right]\right)$

$$\frac{f_i}{f_G} = 1 + 300\frac{\delta}{D} = 1 + 75(1 - \alpha_G) \tag{19.14}$$

The mass flow rate of liquid phase in the form of film, the actual velocity of the liquid film, and the Reynolds number of the liquid film are defined by Eqs. (19.15), (19.16), and (19.17), respectively:

$$\dot{m}_{LF} = \dot{m}_L(1 - E) \tag{19.15}$$

$$U_{LF} = U_L(1 - E) \tag{19.16}$$

$$Re_{LF} = Re_{SL}(1 - E) \tag{19.17}$$

The density of the gas core (ρ_c) is obtained using Eq. (18.9) based on liquid entrainment fraction. Note that only for negligible liquid entrainment fraction, $\rho_c \approx \rho_G$. Figure 19.2 shows the algorithm to use in triangular relationship for the determination of frictional pressure drop, liquid film thickness, and flow rate of liquid in the form of film. In the calculation of the liquid film thickness (δ) using the triangular relationship, since it was assumed that the film thickness is very small with respect to the pipe diameter ($\delta << D$), Eq. (19.18) is recommended (see Fig. 19.2). For the details on the development of Eq. (19.18), refer to Ghajar and Bhagwat (2017).

$$\left.\alpha_G = \frac{A_G}{A} = \frac{(D - 2\delta)^2}{D^2} = 1 - \frac{4\delta}{D}\right\} \ (\delta << D) \tag{19.18}$$

An Illustrative Example

Consider the vertical upward two-phase flow of air-water mixture in a 45 mm I.D. smooth pipe. At near atmospheric system pressure, the two-phase mixture flows with a quality (x) and mass flux (G) of 0.25 and 210 kg/m^2, respectively. The physical properties of air and water are as follows: $\rho_G = 1.5$ kg/m^3, $\rho_L = 998$ kg/m^3, $\mu_G = 18.5 \times 10^{-6}$ Pa-s, $\mu_L = 0.001$ Pa·s, and $\sigma = 0.072$ N/m. Using triangular relationship for annular flow, determine the film thickness (δ), film flow

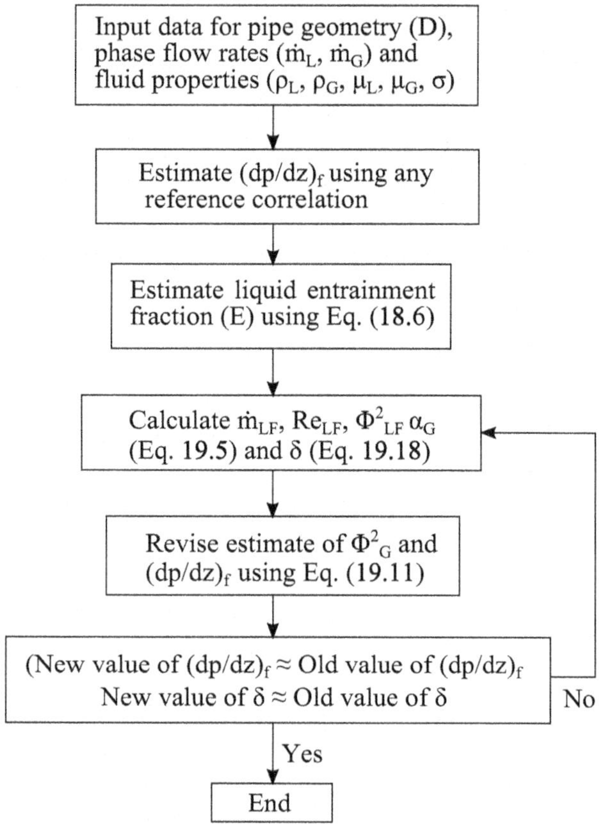

Fig. 19.2 Algorithm to solve triangular relationship in annular flow

rate ($\dot{m}_{LF}$), and pressure drop $(dp/dz)_f$. Assume that adiabatic annular flow at equilibrium exists in the pipe.

Solution

The solution to this problem follows the calculation algorithm illustrated in Fig. 19.2.

1. *Make initial estimate of two phase pressure drop*: Calculate gas and liquid phase Reynolds numbers (see Table 12.1) and single-phase friction factors using appropriate friction factor correlation such as Blasius (1913):

$$Re_{SG} = \frac{GxD}{\mu_G} = \frac{210 \times 0.25 \times 0.045}{18.5 \times 10^{-6}} = 127703$$

$$f_G = \frac{0.079}{Re_{SG}{}^{0.25}} = \frac{0.079}{(127703)^{0.25}} = 0.00418$$

$$Re_{SL} = \frac{G(1-x)D}{\mu_L} = \frac{210 \times (1-0.25) \times 0.045}{0.001} = 7088$$

$$f_L = \frac{0.079}{Re_{SL}{}^{0.25}} = \frac{0.079}{(7088)^{0.25}} = 0.0086$$

Calculate the frictional pressure drop for gas and liquid phase assuming each phase flowing alone through the pipe:

$$\left(\frac{dp}{dz}\right)_G = \frac{2f_G(Gx)^2}{D\rho_G} = \frac{2 \times 0.00418(210 \times 0.25)^2}{0.045 \times 1.5} = 341.3 \text{ Pa/m}$$

$$\left(\frac{dp}{dz}\right)_L = \frac{2f_L(G(1-x))^2}{D\rho_G} = \frac{2 \times 0.0086(210 \times 0.75)^2}{0.045 \times 998} = 9.5 \text{ Pa/m}$$

The parameter (X) of Lockhart and Martinelli (1949) and the two-phase frictional multiplier (Φ_G^2) are now obtained from Eqs. (16.21) and (16.20), respectively. For turbulent-turbulent region of gas and liquid flow $C = 20$ from Table 16.1:

$$X = \sqrt{\frac{(dp/dz)_L}{(dp/dz)_G}} = \sqrt{\frac{9.5}{341.3}} = 0.167$$

$$\Phi_G^2 = 1 + CX + X^2 = 1 + 20 \times 0.167 + 0.167^2 = 4.37$$

Now, the two-phase frictional pressure drop is calculated as

$$\left(\frac{dp}{dz}\right)_f = \Phi_G^2 \times \left(\frac{dp}{dz}\right)_G = 4.37 \times 341.3 = 1491.5 \text{ Pa/m}$$

2. *Estimate liquid entrainment fraction (E)*: Liquid entrainment fraction can be estimated using Eq. (18.6). First estimate the gas core Weber number using gas phase density and predict the initial estimate of liquid entrainment fraction. The superficial gas velocity required in determination of We_c is calculated as $U_{SG} = Gx/\rho_G = 210 \times 0.25/1.5 = 35$ m/s. Then find the gas core density using initial estimate of E, and recalculate Weber number and liquid entrainment fraction in corrector step. For near atmospheric system pressure, $\xi = 280$ from Eq. (18.7).

$$We_c = \frac{\rho_G U_{SG}^2 D}{\sigma} = \frac{1.5 \times 35^2 \times 0.045}{0.072} = 1148.4$$

$$E = \left(1 + 280 \times We_c^{-0.8395}\right)^{-2.209} = 0.288$$

$$\rho_c = \frac{x + E(1-x)}{(x/\rho_G) + (E(1-x)/\rho_L)} = \frac{0.25 + 0.288 \times 0.75}{(0.25/1.5) + (0.288 \times 0.75/998)} = 2.8\ \text{kg/m}^3$$

$$We_c = \frac{\rho_c U_{SG}^2 D}{\sigma} = \frac{2.8 \times 35^2 \times 0.045}{0.072} = 2143.7$$

$$E = \left(1 + 280 \times We_c^{-0.8395}\right)^{-2.209} = 0.44$$

The updated value of core density (ρ_c) based on entrainment in corrector step is $\rho_c = 3.47$ kg/m^3. Based on the predicted entrainment value ($E = 0.44$) and using Eq. (19.15), the liquid film flow rate can be calculated as follows. The liquid mass flow rate required in this equation is $\dot{m}_L = G(1-x) \times A = 210(1-0.25) \times \pi/4 \times 0.045^2 = 0.25$ kg/s.

$$\dot{m}_{LF} = (1-E)\dot{m}_L = (1-0.44) \times 0.25\ (\text{kg/s}) = 0.14\ \text{kg/s}$$

3. *Estimate liquid film thickness (δ)*: To determine the liquid film thickness, first calculate the two-phase frictional multiplier (Φ_{LF}^2) based on liquid film flow rate given by Eq. (19.5). To determine Φ_{LF}^2, first calculate the friction factor (f_{LF}) for liquid film flow. The Reynolds number of liquid film is calculated from Eq. (19.17) such that $Re_{LF} = Re_{SL} \times (1 - E) = 3969$, and the corresponding friction factor is calculated from the Blasius (1913) friction factor correlation:

$$f_{LF} = \frac{0.079}{Re_{SLF}{}^{0.25}} = \frac{0.079}{(3969)^{0.25}} = 0.00995$$

$$\left(\frac{dp}{dz}\right)_{LF} = \frac{2f_{LF}(G(1-x)(1-E))^2}{D\rho_L} = \frac{2 \times 0.00995 \times (210 \times 0.75 \times 0.56)^2}{0.045 \times 998}$$
$$= 3.45 \text{ Pa/m}$$

$$\Phi_{LF}^2 = \frac{(dp/dz)_f}{(dp/dz)_{LF}} = \frac{1491.5}{3.45} = 432.3$$

$$\Phi_{LF}^2 = \frac{1}{(1-\alpha_G)^2} = 432.3 \Rightarrow (1-\alpha_G) = \frac{4\delta}{D} = 0.0481$$

$$\therefore \alpha_G = 0.952, \text{and from Eq.(19.18)}, \delta = 0.541 \text{ mm}$$

4. *Update the value of two-phase pressure drop*: Use Eq. (19.11) along with Eq. (19.14) to calculate the new value of two-phase frictional multiplier (Φ_G^2) and compare it with the initial calculated value of $\Phi_G^2 = 4.37$. Here, we first find $U_{SL} = G(1-x)/\rho_L = 210 \times 0.75/998 = 0.158$ m/s and $U_L = U_{SL}/(1-\alpha_G) = 0.158/0.0481 = 3.28$ m/s:

$$\Phi_G^2 = \frac{1}{\sqrt{\alpha_G}} \frac{f_i}{f_G} \frac{\rho_c}{\rho_G} \left[\frac{1}{\alpha_G} - \frac{U_L(1-E)}{U_{SG}}\right]^2$$

$$\Phi_G^2 = \frac{1}{\sqrt{0.952}}(1 + 75 \times 0.0481) \times \frac{3.47}{1.5}\left(\frac{1}{0.952} - \frac{3.28 \times (1-0.44)}{35}\right)^2 = 10.88$$

Now, the updated value of two-phase frictional pressure drop is calculated as

$$\left(\frac{dp}{dz}\right)_f = \Phi_G^2 \times \left(\frac{dp}{dz}\right)_G = 10.88 \times 341.3 = 3713.3 \text{ Pa/m}$$

5. *Repeat steps (3) and (4) to get consistent values of δ and $(dp/dz)_f$*: Since new value of Φ_G^2 and frictional pressure drop is greater than the initial estimate, recalculate Φ_{LF}^2 value to obtain new estimates of δ and Φ_G^2. After 12 iterations, consistent values of liquid film thickness, two-phase frictional multiplier (Φ_G^2), and hence the two-phase frictional pressure drop are obtained. The final values of parameters in triangular relationship are:

   Liquid film thickness (δ) = 0.398 mm.
   Liquid film flow rate ($\dot{m}_{LF}$) = 0.14 kg/s.
   Two-phase pressure drop $(dp/dz)_f$ = 2745 Pa/m.
   In addition, based on the iterated value of Φ_{LF}^2, $\alpha_G = 0.965$.

Discussion For similar experimental conditions, Nguyen (1975) measured $\alpha_G = 0.951$ and $(dp/dz)_f = 2440$ Pa/m. Note that the converged values of void fraction and pressure drop obtained by solving triangular relationship are comparable with measured data of Nguyen (1975). Moreover, the film thickness ($\delta = 0.398$ mm) obtained using triangular relationship is comparable to the predicted ($\delta = 0.431$ mm) film thickness by empirical correlation of Henstock and Hanratty (1976). It must be mentioned that the final converged values obtained using triangular relationship depend upon the closure relationships for liquid entrainment fraction (E), interfacial friction factor (f_i), and hence the two-phase frictional multiplier (Φ_G^2). There is no unanimity over the use of a particular set of closure equations, and hence use of different set of equations may yield slightly different solutions of triangular relationship.

Chapter 20
Non-boiling Two-Phase Heat Transfer

Background

Knowledge of non-boiling two-phase heat transfer is important in several applications in oil and gas and chemical engineering industry. In particular, during the production of two-phase hydrocarbon fluids from oil reservoirs and its transportation to the surface processing facilities, the temperature of hydrocarbons drops drastically and is favorable for hydrate formation and wax deposition. Wax deposition can result in problems including reduction of inner tube diameter causing blockage, increase in surface roughness of tube leading to restricted flow line pressure, and decrease in production, which can lead to various mechanical problems. In such situations, the correct knowledge of heat transfer coefficients in two-phase flow is of utmost importance for the purpose of flow assurance in oil and gas industry.

The non-boiling heat transfer coefficient in gas-liquid two-phase flow similar to void fraction and two-phase pressure drop is found to be influenced by several parameters such as pipe orientation and flow rates of individual phases. These influences on the two-phase heat transfer coefficient for different pipe orientations and liquid and gas flow rates will be briefly discussed next. Detailed information on these influences can be obtained from Ghajar and Tang (2010) and Bhagwat and Ghajar (2016, 2017).

Effect of Pipe Inclination and Phase Flow Rates on Heat Transfer Coefficient

As shown in Fig. 20.1, in upward pipe inclinations, the two-phase heat transfer coefficient (h_{TP}) increases linearly with increase in the pipe inclination from horizontal and attains a maximum value at $\theta = +75^\circ$. At low gas and liquid flow rates,

A. J. Ghajar, *Single- and Two-Phase Flow Pressure Drop and Heat Transfer in Tubes*, Mechanical Engineering Series,
https://doi.org/10.1007/978-3-030-87281-6_20

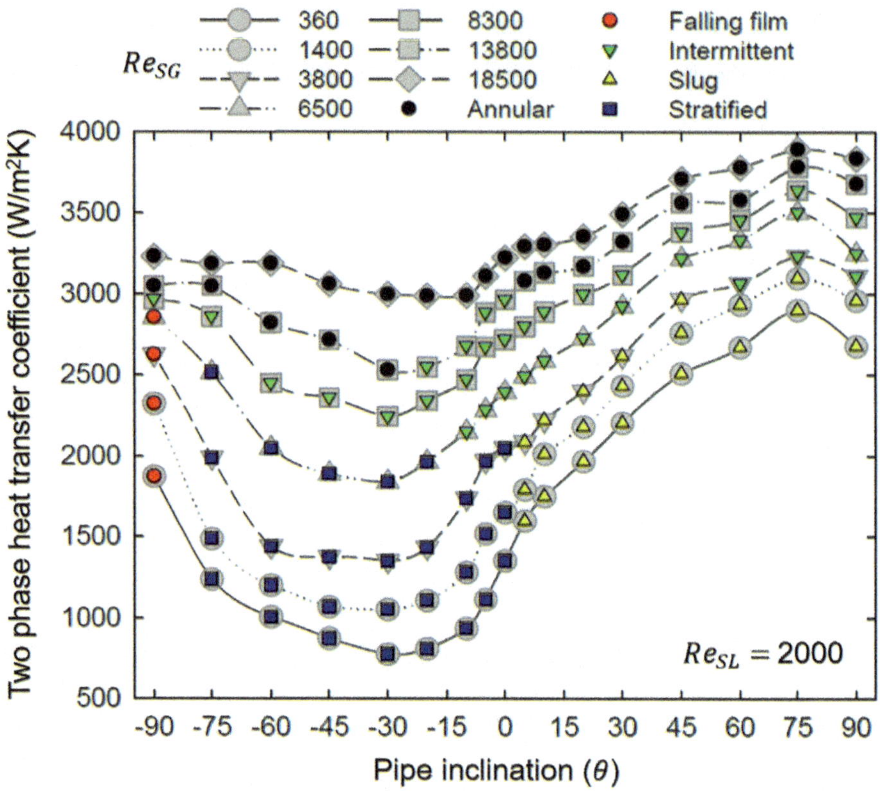

Fig. 20.1 Effect of change in pipe inclination on two-phase heat transfer coefficient at Re_{SL}. (Data measured at the Two-Phase Flow Lab, OSU, Stillwater, OK)

h_{TP} increases by ≈55% as pipe is inclined from horizontal toward vertical upward position. For the case of low liquid flow rates and high gas flow rates, this difference between h_{TP} at horizontal and vertical upward flow is reduced to ≈6%. Note that at moderate liquid flow rates and high gas flow rates, the difference between h_{TP} at horizontal and vertical upward pipe inclination is more substantial. This is because of the combined effect of increase in sweeping action of the disturbance waves at the gas-liquid interface and gradual approach toward uniform cross-sectional distribution of the flow with increase in the liquid flow rate. Both of these conditions favor and aid heat transfer process in two-phase flow.

Similar to the total two-phase pressure drop trends (see Fig. 16.1a), the two-phase heat transfer coefficient at +75° is higher than that at +60° and +90° inclinations. This anomaly is possibly due to the phenomenon of liquid accumulation at pipe bottom wall that obstructs the motion of gas phase and eventually gives rise to the blow through action of the gas phase. This entire process results into better vigorous interaction between the two phases and hence increases the two-phase heat transfer coefficient. Similar observation is reported in literature by Spedding et al. (2000).

In comparison to two-phase total pressure drop (see Fig. 16.1a), in downward pipe inclinations, the two-phase heat transfer coefficient responds sharply to the change in downward pipe inclination. As shown in Fig. 20.1, at $Re_{SL} = 2000$ and the lowest value of $Re_{SG} = 360$, the two-phase heat transfer coefficient initially decreases with decrease in pipe inclination from horizontal and attains a minimum value in the vicinity of $\theta = -30^{\circ}$ and then increases as the pipe is inclined toward vertical downward position. For low gas and liquid flow rates, stratified flow pattern (without coupling between gas-liquid phases except at the interface) prevails in downward pipe inclinations. At $Re_{SL} = 2000$ and varying Re_{SG}, the average decrease in h_{TP} at $\theta \approx -30^{\circ}$ with reference to the horizontal flow is about 30%. This inflection in the trend of h_{TP} vs θ appears to be the consequence of slippage at the gas-liquid interface which in turn is a function of pipe inclination. For low gas and liquid flow rates, stratified flow pattern prevails in downward pipe inclinations. The physical structure of stratified flow patterns in downward pipe inclinations is such that there is no coupling between the two phases except the interfacial interaction. At low gas flow rates, the gravity-driven nature of the flow increases the residence time of gas phase in the test section and as a result decreases the two-phase heat transfer coefficient. It is interesting to note that the general trends of void fraction show inflection at $\theta \approx -45^{\circ}$ (see Fig. 15.1); however, the inflection in two-phase heat transfer coefficient variation is at $\theta \approx -30^{\circ}$. Based on the flow visualization, there appears to be no physical reason behind this discrepancy, and thus more detailed experiments such as measurements of liquid film thickness are deemed necessary for better understanding of this deviation. Similar to upward pipe inclinations, effect of downward pipe inclination on h_{TP} diminishes with increase in gas and liquid flow rates (inertia-driven region of two-phase flows). For instance, at $Re_{SL} = 6000$, the highest and average decrease in two-phase heat transfer coefficients are 11% and 3% compared to 47% and 30% at $Re_{SL} = 2000$. As shown in Fig. 20.1, at steeper downward pipe inclinations and low liquid and gas Reynolds numbers, higher heat transfer coefficient is due to better wetting of the pipe periphery due to unstable gas-liquid interface (and hence frequent splashing). This will be explained further in the next section.

Circumferential Variation of Heat Transfer Coefficient with Pipe Inclination

Since the experimental setup described in Chap. 13 measures local values of two-phase heat transfer coefficient, it is also of interest to check the circumferential variation of h_{TP} for fixed values of gas and liquid flow rates/velocities and varying pipe orientations. Figure 20.2 shows circumferential values of two-phase heat transfer coefficient for horizontal and various upward pipe inclinations. Each local h_{TP} is the average of all axial two-phase heat transfer coefficients (at seven axial stations) measured at a fixed circumferential location. It is observed that due to lack

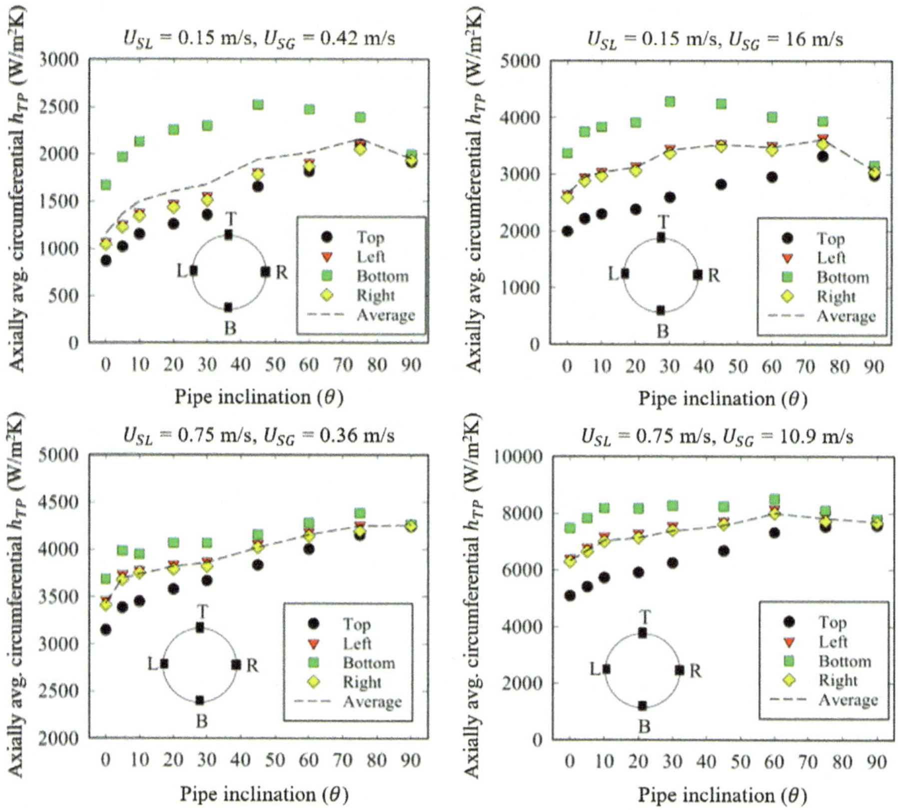

Fig. 20.2 Circumferential variation of two-phase heat transfer coefficient as a function of pipe orientation for fixed values of gas and liquid flow rates. (Data measured at the Two-Phase Flow Lab, OSU, Stillwater, OK)

of symmetry in upward pipe inclinations, the heat transfer coefficient at pipe bottom is always higher than the top and side walls of the pipe. As the pipe inclination approaches vertical upward position, the difference between the circumferential h_{TP} decreases. In the case of vertical upward flow, all four values of h_{TP} (at top, bottom, and two sides of the pipe) are comparable because of the flow symmetry. For low values of U_{SL} and U_{SG}, the average value (axial as well as circumferential) of two-phase heat transfer coefficient is influenced by the heat transfer coefficients at the pipe top and side walls. This situation corresponds to slug/wavy slug flow where bottom wall of the pipe is always in contact with the thick liquid film, while the top and side walls are wetted intermittently due to alternate flow of gas and liquid slugs. With increase in the gas flow rate, for example, at $U_{SG} = 16$ m/s, entire pipe cross section is wetted continuously resulting into the shift of average heat transfer coefficient toward the mean of local h_{TP} at bottom and top wall of the pipe. However, there is still a difference between top and bottom wall heat transfer coefficient (for non-vertical flow) since the liquid film thickness at bottom wall is thicker compared to that at the top wall of the pipe, and hence $h_{TP,\ Bottom} > h_{TP,\ Top}$.

As it was mentioned earlier, at steeper downward pipe inclinations and low liquid and gas Reynolds numbers, higher heat transfer coefficients are observed (see Fig. 20.1). The increase in h_{TP} for steeper downward pipe inclinations could be explained based on the variation in the physical structure of the stratified flow pattern. For these steeper (near vertical and vertical downward) pipe orientations, the flow pattern is stratified and falling film flow, respectively. At $\theta = -60^{\circ}$, the visual observations show that the gas-liquid interface of stratified flow pattern is unstable such that the liquid phase splashes frequently on the pipe top wall and momentarily bridges the pipe cross section. Barnea et al. (1982) also reported that at steeper pipe inclinations, liquid lumps are torn away from the unstable gas-liquid interface all the way to the top wall of the pipe. They also mentioned that at low to moderate liquid flow rates, the gas-liquid interface in steeper pipe orientations tend to become concave, and the liquid film climbs the tube periphery with increase in the liquid flow rate and pipe orientation. Additionally, at higher gas flow rates, the work of Fukano and Ousaka (1989) reveals that the disturbance waves create a pumping action on the liquid layer causing it to rise in circumferential direction and also carry it in the downstream direction. Thus, the liquid phase moves in both circumferential and axial direction and provides mixing and secondary flows. This implies that, compared to near horizontal downward pipe inclinations, a greater fraction of the pipe circumference (wetted perimeter) is in contact with the liquid phase for near vertical downward pipe inclinations and hence allows higher rates of two-phase heat transfer. Maximum value of the two-phase heat transfer coefficient is observed for the case of vertical downward flow. In the case of vertical downward flow, entire pipe circumference is wetted by a thin axisymmetric liquid film (falling film flow), whereas the stratified flow in near vertical downward pipe inclinations experiences asymmetric and thick liquid film that partially wets the pipe circumference. Thus, the presence of thin liquid film offers lesser thermal resistance and hence results into highest heat transfer coefficient. The three different forms of stratified flow with variation in the liquid film thickness and the corresponding circumferential variation of two-phase heat transfer coefficient for different downward pipe inclinations are depicted in Fig. 20.3. Each local h_{TP} is the average of all axial two-phase heat transfer coefficients (at seven axial stations) measured at a fixed circumferential location. The variation in circumferential heat transfer coefficients agrees with the aforementioned circumferential variation of liquid film/layer thickness as a function of pipe inclination. For $\theta = -5^{\circ}$, the local two-phase heat transfer coefficient is highest at pipe bottom wetted by a liquid film, while that at the pipe top wall is smallest due to the absence of liquid film. As the pipe inclination approaches vertical downward position, all the local two-phase heat transfer coefficients are comparable indicating the presence of axisymmetric two-phase flow.

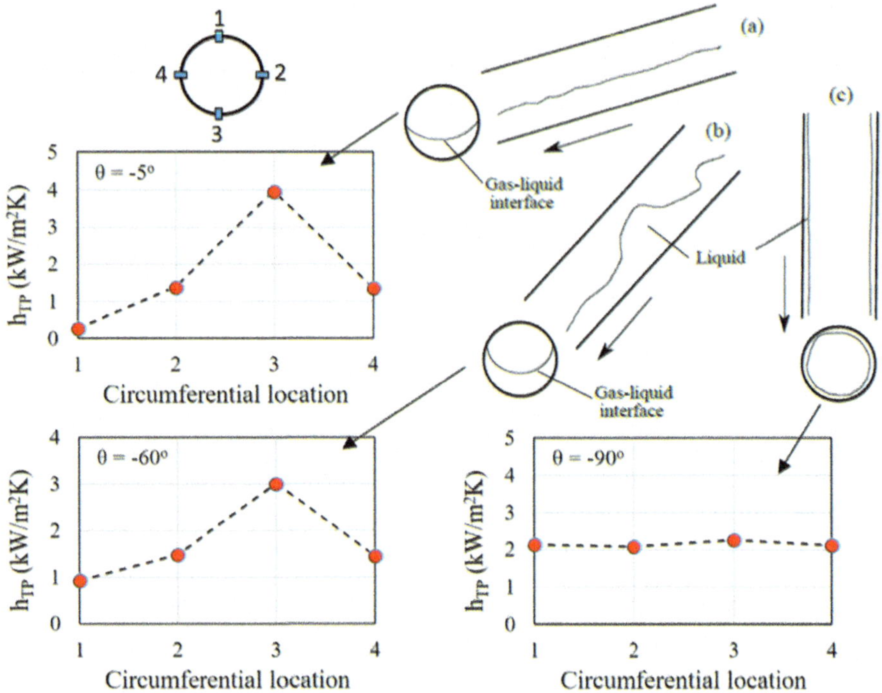

Fig. 20.3 Variation in liquid film thickness and corresponding local two-phase heat transfer coefficients for (**a**) near horizontal, (**b**) steep, and (**c**) vertical downward pipe inclinations ($Re_{SL} = 2000$, $Re_{SG} = 280$). (Data measured at the Two-Phase Flow Lab, OSU, Stillwater, OK)

Modeling of Non-boiling Two-Phase Heat Transfer

In two-phase flow literature, majority of the documented non-boiling two-phase heat transfer modeling methods are limited to specific pipe orientations, fluid combinations, and most importantly flow patterns. The subsea and surface pipelines carrying mixture of oil and gas together are usually laid in undulated form having near horizontal inclinations, and the existence of a specific flow pattern is uncertain. For such a scenario, the use of flow pattern and pipe orientation-specific two-phase heat transfer correlations may predict significant deviations from the actual conditions. To address this issue, Ghajar and coworkers in a series of papers have extensively studied the phenomenon of non-boiling two-phase heat transfer in pipes at different orientations; see, for example, Kim et al. (1999, 2000) and Kim and Ghajar (2006). Ghajar and Tang (2009) proposed a robust correlation for two-phase heat transfer coefficient (h_{TP}) as presented by Eq. (20.1). Their correlation is based on 986 experimental data points for different pipe orientations, fluid combinations, and flow patterns:

$$h_{TP} = h_L F_p \left[1 + 0.55 \left(\frac{x}{1-x} \right)^{0.1} \left(\frac{1-F_p}{F_p} \right)^{0.4} \left(\frac{Pr_G}{Pr_L} \right)^{0.25} \left(\frac{\mu_L}{\mu_G} \right)^{0.25} I^{0.25} \right] \quad (20.1)$$

$$F_p = (1 - \alpha_G) + \alpha_G F_s^2 \quad (20.2)$$

$$F_s = \frac{2}{\pi} \arctan \left(\sqrt{\frac{\rho_G (U_G - U_L)^2}{gD(\rho_L - \rho_G)}} \right) \quad (20.3)$$

$$I = 1 + Eo \mid \sin\theta \mid \quad (20.4)$$

The flow pattern factor (F_p) given by Eq. (20.2) makes the Ghajar and Tang (2009) correlation independent of the flow patterns and depends on shape factor F_s and void fraction (α_G). The shape factor (F_s) given by Eq. (20.3) considers the effect of the shape of gas-liquid interface on two-phase heat transfer. The shape factor (F_s) is applicable for slip ratios equal or greater than one ($S \geq 1$), which is common in gas-liquid flow, and represents the shape changes of the gas-liquid interface by the force acting on the interface due to the relative momentum and gravitational forces. The void fraction (α_G) in Eq. (20.2) should be calculated using the simple correlations of Woldesemayat and Ghajar (2007), using Eqs. (15.11) and (15.12). The expression for the inclination factor (I), given by Eq. (20.4), includes representation of the relative forces acting on the liquid phase in the flow direction due to the momentum and buoyancy forces. The inclination factor accounts for the effect of pipe orientation on h_{TP} and is defined in terms of Eötvös number:

$$Eo = \frac{(\rho_L - \rho_G) g D^2}{\sigma} \quad (20.5)$$

The single-phase heat transfer coefficient (h_L) required in Eq. (20.1) is based on in situ Reynolds number. All thermophysical properties required are calculated at bulk mean temperature except for the μ_w that is calculated at the pipe wall temperature. Note that the in situ liquid phase Reynolds number used in Eq. (20.6) for h_L is defined in terms of the square root of liquid holdup as shown in Eq. (20.7):

$$h_L = 0.027 Re_L^{0.8} Pr_L^{0.33} \left(\frac{k_L}{D} \right) \left(\frac{\mu_L}{\mu_w} \right)^{0.14} \quad (20.6)$$

$$Re_L = \frac{4\dot{m}_L}{\pi D \mu_L \sqrt{1-\alpha_G}} = \frac{G(1-x)D}{\mu_L \sqrt{1-\alpha_G}} \tag{20.7}$$

Comparison of Ghajar and Tang (2009) Correlation with Experimental Data The general two-phase heat transfer correlation of Ghajar and Tang (2009), Eq. (20.1), is verified against 986 experimental data points (176 for horizontal flow, 555 for inclined flow, and 255 for upward vertical flow) from 5 different sources with different experimental facilities. The data covered different pipe inclination angles, gas-liquid combinations (air-water, air-silicone, helium-water, Freon-12-water), and all flow patterns with the exception of stratified flow. Their correlation is applicable over the following range of parameters: $750 \leq Re_{SL} \leq 1.27 \times 10^5$, $14 \leq Re_{SG} \leq 21 \times 10^5$, $0.01 \leq Pr_G/Pr_L \leq 0.15$, $0.0036 \leq \mu_G/\mu_L \leq 0.026$, $910\ \text{kg/m}^3 \leq \rho_L \leq 1210\ \text{kg/m}^3$, and $0^\circ \leq \theta \leq +90^\circ$. Figure 20.4 shows the comparison of their correlation with the entire set of experimental data. Their correlation predicted 90% of the experimental data within ±25%, confirming the robustness of Eq. (20.1) for prediction of two-phase heat transfer coefficients.

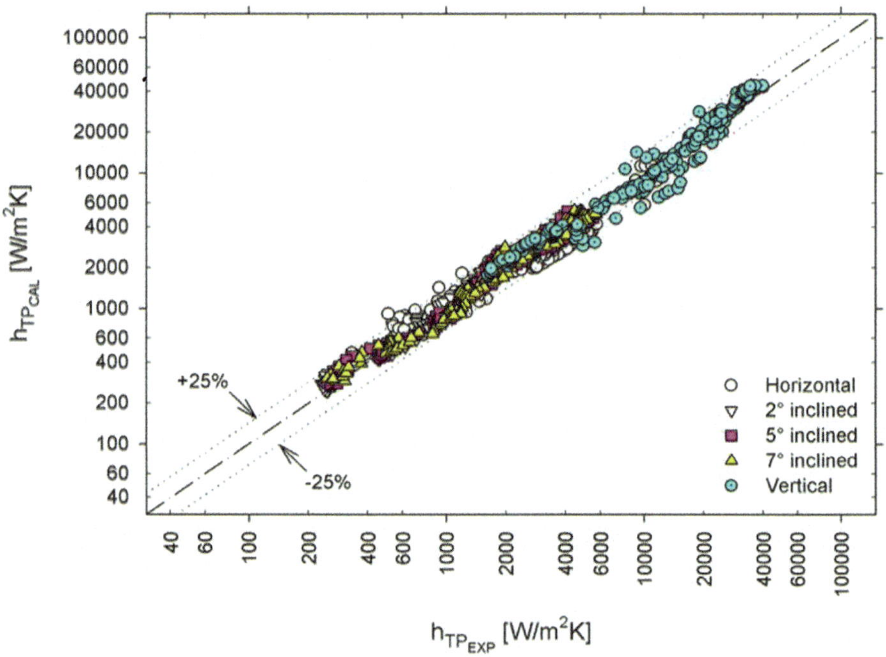

Fig. 20.4 Comparison of the predictions by Eq. (20.1) with all 986 experimental data points for different inclination angles, gas-liquid combinations, and flow patterns

Application of Reynolds Analogy to Non-boiling Two-Phase Heat Transfer

The concept of Reynolds analogy that relates the friction coefficient or the frictional pressure drop with heat transfer in single-phase flow can also be extended to the case of two-phase flow. Using the concept of Reynolds analogy, Tang and Ghajar (2011) have developed a correlation to predict non-boiling two-phase heat transfer coefficient from the knowledge of two-phase frictional pressure drop in horizontal and upward inclined flows. Their correlation correlates the two-phase to single-phase heat transfer coefficient ratio to the two-phase frictional multiplier (ratio of two-phase to single-phase pressure drop). The physical form of their correlation is given by Eq. (20.8). Tang (2011) suggested that the exponents of F_p and Φ_L could vary between 0.1 and 0.5 depending upon the pipe orientation and fluid properties; however, the general structure of Reynolds analogy-based correlation remains unaltered. In Eq. (20.8), the exponent (0.3) of F_p and Φ_L is based on the overall performance of Tang and Ghajar (2011) model for various pipe inclinations and fluid combinations. The same experimental data set used for the development of Eq. (20.1) was used for the development of Eq. (20.8). The Reynolds analogy-based two-phase heat transfer correlation of Tang and Ghajar (2011), Eq. (20.8), predicted 85% of the experimental data within ±30%:

$$h_{TP} = h_L {F_p}^{0.3} \left(\frac{\dot{m}_L}{\dot{m}}\right) \left(\frac{\rho_L}{\rho_M}\right)^{0.5} \Phi_L^{0.3} \tag{20.8}$$

$$\rho_M = \rho_L(1-\alpha_G) + \rho_G \alpha_G \tag{20.9}$$

$$\left(\frac{dp}{dz}\right)_f = \Phi_L^2 \left(\frac{dp}{dz}\right)_L \text{ where } \left(\frac{dp}{dz}\right)_L = \frac{2 f_L G^2 (1-x)^2}{D \rho_L} \tag{20.10}$$

In Eq. (20.8), the two-phase mixture density (ρ_M) is calculated from Eq. (20.9) using void fraction. The void fraction required in calculation of ρ_M is calculated using Woldesemayat and Ghajar (2007) correlation (Eqs. 15.11 and 15.12). F_p is calculated from Eq. (20.2), and h_L is calculated from Eq. (20.6) using the Reynolds number (Re_{SL}) based on superficial liquid velocity (see Table 12.1). Note that Φ_L used in Eq. (20.8) is the square root of the two-phase frictional multiplier calculated using Eq. (20.10). In Eq. (20.10), $(dp/dz)_L$ and $(dp/dz)_f$ are the single-phase liquid and two-phase frictional pressure drop, respectively. The friction factor (f_L) required for the single-phase liquid pressure drop calculation can be calculated using any appropriate friction factor correlation such as correlation of Blasius (1913) or Churchill (1977b). In case the two-phase frictional pressure drop $(dp/dz)_f$ is not known/measured, then suitable correlation valid for given two-phase flow conditions such as the correlations recommended by Ghajar and Bhagwat (2017) may be used to determine Φ_L.

For vertical downward flow, Bhagwat et al. (2012) pointed out that the two-phase heat transfer coefficient is less than that measured in vertical upward flow at similar mass flow rates, while the two-phase pressure drop is considerably higher in the former orientation than the latter. Thus, Tang and Ghajar (2011) Reynolds analogy-based heat transfer correlation, Eq. (20.8), with fixed exponents does not give satisfactory performance in the prediction of two-phase heat transfer coefficient for vertical downward flow. Bhagwat et al. (2012) proposed the following form of the Reynolds analogy-based correlation that can be used to predict non-boiling heat transfer coefficient for vertical downward flow:

$$h_{TP} = h_L \Phi_L^{0.55} \tag{20.11}$$

In Eq. (20.11), the single-phase heat transfer coefficient (h_L) and the two-phase frictional multiplier (Φ_L) are calculated the same way as that of Eq. (20.8). Equation (20.11) is based on the 165 data points measured at the OSU two-phase flow lab in a vertically downward-oriented 0.0127 m stainless steel pipe using air-water fluid combination. This correlation predicts more than 90% of the measured data points within ±20% error bands and is applicable for $1400 \leq Re_{SL} \leq 29{,}000$ and $90 \leq Re_{SG} \leq 18{,}000$.

It should be noted the physical structure of both Ghajar and Tang (2009) general heat transfer correlation, Eq. (20.1), and Tang and Ghajar (2011) Reynolds analogy-based heat transfer correlation, Eq. (20.8), is modular and robust, and hence the application of these correlations can be potentially extended to a wider range of two-phase flow conditions by modifying the empirical multiplying factors and exponents. Both of these correlations, Eqs. (20.1) and (20.8), require accurate information of void fraction. Ghajar and Bhagwat (2013, 2014b) have evaluated several void fraction correlations available in the literature against air-water data (5162 data points for horizontal, upward, and downward inclinations) and refrigerant-vapor (645 data points for horizontal flow). They found the performance of Woldesemayat and Ghajar (2007) correlation (see Eqs. (15.11) and (15.12)) or a more recent correlation by Bhagwat and Ghajar (2014) for a wide range of gas-liquid two-phase flow (see Eqs. (15.13, 15.14, 15.15, 15.16, 15.17, 15.18, and 15.19)) are acceptable for general application with requisite accuracy. For more specific flow conditions, other top-performing void fraction correlations shortlisted by Ghajar and Bhagwat (2013, 2014b) may be used.

An Illustrative Example

Use of the General and Reynolds Analogy-Based Heat Transfer Correlations
Consider the vertical upward two-phase flow of air and silicone oil in a 12 mm

I.D. stainless steel pipe having a surface roughness of 0.002 mm. The mass flow rate of gas and liquid phase is 0.0015 kg/s and 0.9 kg/s, respectively. The fluid thermophysical properties may be taken as follows: $\rho_G = 1.2$ kg/m ^{3}, $\rho_L = 920$ kg/m ^{3}, $\mu_G = 18.4 \times 10^{-6}$ Pa-s, $\mu_L = 0.005$ Pa-s, $\mu_w = 0.004$ Pa-s, $\sigma = 0.02$ N/m, $k_L = 0.12$ W/mK, $Pr_G = 0.71$, and $Pr_L = 64$. For a measured void fraction of $\alpha_G = 0.5$, determine the two-phase heat transfer coefficient for the flow of air and silicone oil using (1) the general heat transfer correlation of Ghajar and Tang (2009), Eq. (20.1), and (20.2) Reynolds analogy-based heat transfer correlation of Tang and Ghajar (2011), given by Eq. (20.8).

Solution

1. Calculation of h_{TP} using Ghajar and Tang (2009) general heat transfer correlation:

Referring to Table 12.1, the superficial gas (U_{SG}) and liquid (U_{SL}) velocities and the two-phase flow quality (x) are first calculated as follows:

$$U_{SG} = \frac{4\dot{m}_G}{\pi D^2 \rho_G} = \frac{4 \times 0.0015}{\pi \times 0.012^2 \times 1.2} = 11.05 \text{ m/s}$$

$$U_{SL} = \frac{4\dot{m}_L}{\pi D^2 \rho_L} = \frac{4 \times 0.9}{\pi \times 0.012^2 \times 920} = 8.65 \text{ m/s}$$

$$x = \frac{\dot{m}_G}{\dot{m}_G + \dot{m}_L} = \frac{0.0015}{0.0015 + 0.9} = 0.00166$$

From the given information of void fraction, the actual velocity of each phase is calculated as follows. In case the void fraction is not known, Woldesemayat and Ghajar (2007) explicit void fraction correlation, Eqs. (15.11) and (15.12), or Bhagwat and Ghajar (2014) implicit void fraction correlation, Eqs. (15.13, 15.14, 15.15, 15.16, 15.17, 15.18, and 15.19), can be used for the calculation of the void fraction. Refer to the example problem in Chap. 15 for the details of the void fraction calculations using either one of the correlations:

$$U_G = \frac{U_{SG}}{\alpha_G} = \frac{11.05}{0.5} = 22.1 \text{ m/s}$$

$$U_L = \frac{U_{SL}}{1 - \alpha_G} = \frac{8.65}{1 - 0.5} = 17.3 \text{ m/s}$$

Now, the shape factor is calculated from Eq. (20.3):

$$F_s = \left[\frac{2}{\pi}\arctan\left(\sqrt{\frac{\rho_G(U_G - U_L)^2}{gD(\rho_L - \rho_G)}}\right)\right]$$

$$= \left[\frac{2}{\pi}\arctan\left(\sqrt{\frac{1.2 \times (22.1 - 17.3)^2}{9.81 \times 0.012 \times (920 - 1.2)}}\right)\right] = 0.298$$

The flow pattern factor is obtained from Eq. (20.2):

$$F_p = (1 - \alpha_G) + \alpha_G F_S^2 = (1 - 0.5) + 0.5 \times 0.298^2 = 0.544$$

The inclination factor for vertical flow is calculated by Eq. (20.4):

$$I = 1 + \frac{(\rho_L - \rho_G)gD^2}{\sigma} \mid \sin\theta \mid = 1 + \frac{(920 - 1.2) \times 9.81 \times 0.012^2}{0.02} = 65.9$$

Next, the in situ Reynolds number based on void fraction is calculated from Eq. (20.7):

$$Re_L = \frac{4\dot{m}_L}{\pi D \mu_L \sqrt{1 - \alpha_G}} = \frac{4 \times 0.9}{\pi \times 0.012 \times 0.005 \times \sqrt{1 - 0.5}} = 27009.5$$

The single-phase heat transfer coefficient is then calculated using Eq. (20.6):

$$h_L = 0.027 Re_L^{0.8} Pr_L^{0.33}\left(\frac{k_L}{D}\right)\left(\frac{\mu_L}{\mu_w}\right)^{0.14}$$

$$= 0.027 \times 27009.5^{0.8} \times 64^{0.33} \times \left(\frac{0.12}{0.012}\right)\left(\frac{0.005}{0.004}\right)^{0.14} = 3856\ \mathrm{W/m^2 - K}$$

Using the general two-phase heat transfer correlation of Ghajar and Tang (2009) given by Eq. (20.1), h_{TP} is calculated as

$$h_{TP} = h_L F_p\left[1 + 0.55\left(\frac{x}{1 - x}\right)^{0.1}\left(\frac{1 - F_p}{F_p}\right)^{0.4}\left(\frac{Pr_G}{Pr_L}\right)^{0.25}\left(\frac{\mu_L}{\mu_G}\right)^{0.25} I^{0.25}\right]$$

$$= 3856 \times 0.544\left[1 + 0.55\left(\frac{0.00166}{1 - 0.00166}\right)^{0.1}\left(\frac{1 - 0.544}{0.544}\right)^{0.4}\left(\frac{0.71}{64}\right)^{0.25} \times \left(\frac{0.005}{18.4 \times 10^{-6}}\right)^{0.25} 65.9^{0.25}\right]$$

$$= 4224\ \mathrm{W/m^2 - K}$$

2. Calculation of h_{TP} using Tang and Ghajar (2011) Reynolds analogy-based heat transfer correlation: To use the Reynolds analogy-based heat transfer correlation, we first need to find the two-phase frictional pressure drop and hence the two-phase frictional multiplier (Φ_L^2). The following steps calculate the two-phase pressure drop using method of Lockhart and Martinelli (1949).

The Reynolds number of each phase and associated friction factors are calculated as shown below. Since the two-phase flow is through steel pipe having a roughness of 0.002 mm, we need to use appropriate correlation such as Colebrook (1939) or Churchill (1977b) to account for the effect of pipe wall surface roughness on friction factor. Here the Colebrook (1939) fanning friction factor correlation was used:

$$Re_{SG} = \frac{\rho_G U_{SG} D}{\mu_G} = \frac{1.2 \times 11.05 \times 0.012}{18.4 \times 10^{-6}} = 8,648 \Rightarrow f_G = 0.0081$$

$$Re_{SL} = \frac{\rho_L U_{SL} D}{\mu_L} = \frac{920 \times 8.65 \times 0.012}{0.005} = 19,099 \Rightarrow f_L = 0.0063$$

The Colebrook (1939) fanning friction factor correlation with $\varepsilon/D = 1.67 \times 10^{-4}$ that was used for the calculation of gas phase friction factor (f_G) and liquid phase friction factor (f_L) is as shown here:

$$\frac{1}{\sqrt{f}} = -4.0\ log\left(\frac{\varepsilon/D}{3.7} + \frac{1.256}{Re\ \sqrt{f}}\right)$$

Now the single-phase pressure drop due to flow of gas and liquid phase is calculated by

$$\left(\frac{dp}{dz}\right)_G = \frac{2 f_G \rho_G U_{SG}^2}{D} = \frac{2 \times 0.0081 \times 1.2 \times 11.05^2}{0.012} = 197.8\ \text{Pa/m}$$

$$\left(\frac{dp}{dz}\right)_L = \frac{2 f_L \rho_L U_{SL}^2}{D} = \frac{2 \times 0.0063 \times 920 \times 8.65^2}{0.012} = 72,278.5\ \text{Pa/m}$$

The Lockhart and Martinelli (1949) parameter X given by Eq. (16.21) is calculated using single-phase pressure drop as shown here:

$$X = \sqrt{\frac{(dp/dz)_L}{(dp/dz)_G}} = \sqrt{\frac{72,278.58}{197.8}} = 19.11$$

Since single-phase flow of both phases is in turbulent region, use $C = 20$ from Table 16.1, and the two-phase frictional multiplier using method of Lockhart and Martinelli (1949) is calculated from Eq. (16.20) as shown here:

$$\Phi_L^2 = 1 + \frac{C}{X} + \frac{1}{X^2} = 1 + \frac{20}{19.11} + \frac{1}{19.11^2} = 2.05 \Rightarrow \Phi_L = 1.43$$

Two-phase mixture density based on void fraction is calculated using Eq. (20.9) as follows:

$$\rho_M = \alpha_G \rho_G + (1 - \alpha_G)\rho_L = 0.5 \times 1.2 + (1 - 0.5) \times 920 = 460.6\ \text{kg/m}^3$$

Now using Eq. (20.8), two-phase heat transfer coefficient can be found. The value of h_L based on Re_{SL} is calculated from Eq. (20.6). The value of $F_p = 0.544$ was calculated in Part (1) of this problem:

$$\begin{aligned} h_L &= 0.027 Re_{SL}^{0.8} Pr_L^{0.33} \left(\frac{k_L}{D}\right) \left(\frac{\mu_L}{\mu_w}\right)^{0.14} \\ &= 0.027 \times 19099^{0.8} \times 64^{0.33} \times \left(\frac{0.12}{0.012}\right) \left(\frac{0.005}{0.004}\right)^{0.14} = 2922.6\ \text{W/m}^2 - \text{K} \end{aligned}$$

Finally, the two-phase heat transfer coefficient using Reynolds analogy-based correlation is found using Eq. (20.8):

$$\begin{aligned} h_{TP} &= h_L {F_p}^{0.3} \left(\frac{\dot{m}_L}{\dot{m}}\right) \left(\frac{\rho_L}{\rho_M}\right)^{0.5} \Phi_L^{0.3} \\ &= 2922.6 \times 0.544^{0.3} \left(\frac{0.9}{0.9015}\right) \left(\frac{920}{460.6}\right)^{0.5} 1.43^{0.3} \\ &= 3824\ \text{W/m}^2 - \text{K} \end{aligned}$$

Discussion Rezkallah and Sims (1987) made two-phase heat transfer measurements for similar experimental conditions and found that $h_{TP} \approx 3900$ W/m ^{2}-K. The general heat transfer correlation of Ghajar and Tang (2009), Eq. (20.1), overpredicted the measured value of h_{TP} by about 8%, and the Reynolds analogy-based heat transfer correlation of Tang and Ghajar (2011), Eq. (20.8), underpredicted the measured value of h_{TP} by about 2%. The general heat transfer correlation and Reynolds analogy-based heat transfer correlation are sensitive to the void fraction (through F_P and ρ_M). Note that in case the two-phase frictional pressure drop is not known, the accuracy of the Reynolds analogy concept depends on the choice of two-phase frictional multiplier correlation appropriate to the two-phase flow under consideration.

References

Abdulkadir, M., Hernandez-Perez, V., Sharaf, S., Lowndes, I. S., & Azzopardi, B. J. (2010). Experimental investigation of phase distributions of two-phase air–silicone oil flow in a vertical pipe. *World Academy of Science, Engineering and Technology, 37*, 52–59.

Abduvayat, P., Manabe, R., & Arihara, N. (2003). Effects of pressure and pipe diameter on gas-liquid two-phase flow behavior in pipelines. *SPE Annual Technical Conference, SPE, 84229.*

Badie, S., Hale, C. P., Lawrence, C. J., & Hewitt, G. F. (2000). Pressure gradient and holdup in horizontal two-phase gas liquid flows with low liquid loading. *International Journal of Multiphase Flow, 26*, 1525–1543.

Barnea, D. (1986). Transition from annular and from dispersed bubble flow - unified models for the whole range of pipe inclinations. *International Journal of Multiphase Flow, 12*, 733–744.

Barnea, D. (1987). A unified model for predicting flow pattern transitions for the whole range of pipe inclinations. *International Journal of Multiphase Flow, 13*, 1–12.

Barnea, D., Shoham, O., & Taitel, Y. (1982). Flow pattern transition for downward inclined two-phase flow: Horizontal to vertical. *Chemical Engineering Science, 37*, 735–740.

Beggs, H. D. (1972). *An experimental study of two-phase flow in inclined pipes*, Ph.D. Thesis, University of Tulsa, Tulsa, Oklahoma.

Beggs, H. D., & Brill, J. P. (1973). A study of two-phase flow in inclined pipes. *Journal of Petroleum Technology, 25*(5), 607–617.

Bendiksen, K. H. (1984). An experimental investigation of the motion of long bubbles in inclined tubes. *International Journal of Multiphase Flow, 10*, 467–483.

Bhagwat, S. M. (2015). *Experimental measurements and modeling of void fraction and pressure drop in upward and downward inclined non-boiling gas-liquid two-phase flow*, Ph.D. Thesis, Oklahoma State University, Stillwater, Oklahoma.

Bhagwat, S. M., & Ghajar, A. J. (2014). A flow pattern independent drift flux model based void fraction correlation for a wide range of gas-liquid two-phase flow. *International Journal of Multiphase Flow, 59*, 186–205.

Bhagwat, S. M., & Ghajar, A. J. (2015a). An empirical model to predict the transition between stratified and non-stratified gas-liquid two-phase flow in horizontal and downward inclined pipes. *Heat Transfer Engineering, 36*(18), 1489–1498.

Bhagwat, S. M., & Ghajar, A. J. (2015b). Modified liquid entrainment fraction correlation for varying pipe orientation and system pressure. *International Journal of Multiphase Flow, 74*, 1–4.

A. J. Ghajar, *Single- and Two-Phase Flow Pressure Drop and Heat Transfer in Tubes*, Mechanical Engineering Series,
https://doi.org/10.1007/978-3-030-87281-6

Bhagwat, S. M., & Ghajar, A. J. (2016). Experimental investigation of non-boiling gas-liquid two-phase flow in upward inclined pipes. *Experimental Thermal and Fluid Science, 79*, 301–318.

Bhagwat, S. M., & Ghajar, A. J. (2017). Experimental investigation of non-boiling gas-liquid two-phase flow in downward inclined pipes. *Experimental Thermal and Fluid Science, 89*, 219–237.

Bhagwat, S. M., Mollamahmutoglu, M., & Ghajar, A. J. (2012). Experimental investigation and empirical analysis of non-boiling gas-liquid two-phase heat transfer in vertical downward pipe orientation. *Proceedings of ASME 2012 Summer Heat Transfer Conference, 2*: 349–359.

Blasius, H. (1913). Das Anhlichkeitsgesetz bei Reibungsvorgangen in Flussikeiten. *Gebiete Ingenieurw, 134*.

Bowers, C. D., & Hrnjak, P. S. (2010). *Determination of void fraction in separated two-phase flows using optical techniques, international refrigeration and air-conditioning conference* (pp. 2293–2302). Purdue University.

Brognaux, L. B., Webb, R. L., & Chamra, L. M. (1997). Single phase heat transfer in micro-fin tubes. *International Journal of Heat and Mass Transfer, 40*(18), 4345–4357.

Celeta, G. P. (2008). *Single- and two-phase flow heat transfer in micropipes*, 5th European thermal-sciences Conference (The Netherlands).

Çengel, Y. A., & Ghajar, A. J. (2020). *Heat and mass transfer – Fundamentals & applications* (6th ed., pp. 498–500). McGraw-Hill Education.

Chen, X. T., Cai, X. D., & Brill, J. P. (1997). Gas liquid stratified wavy flow in horizontal pipelines. *Journal of Energy Resources Technology, 119*, 209–216.

Cheng, S. C., Wong, Y. L., & Groeneveld, D. C. (1988). CHF prediction for horizontal flow. In *International symposium on phase change heat transfer* (pp. 211–215).

Chisholm, D. (1967). A theoretical basis for the Lockhart-Martinelli correlation for two-phase flow. *International Journal of Heat and Mass Transfer, 10*, 1767–1778.

Chisholm, D. (1973). Pressure gradients due to the friction during the flow of evaporating two-phase mixtures in smooth tubes and channels. *International Journal of Heat and Mass Transfer, 16*, 347–358.

Churchill, S. W. (1977a). Comprehensive correlating equations for heat, mass and momentum transfer in fully developed flow in smooth tubes. *Industrial and Engineering Chemistry, 16*, 109–116.

Churchill, S. W. (1977b). Friction factor equation spans all fluid-flow regimes. *Chemical Engineering Journal, 7*, 91–92.

Churchill, S. W., & Usagi, R. (1972). A general expression for the correlation of rates of transfer and other phenomena. *AIChE, 18*, 1121–1128.

Cioncolini, A., & Thome, J. R. (2012). Entrained liquid fraction prediction in adiabatic and evaporating annular two-phase flow. *Nuclear Engineering and Design, 243*, 200–213.

Colebrook, C. F. (1939). Turbulent flow in pipes, with particular reference to the transition between the smooth and rough pipe laws. *Journal of the Institution of Civil Engineers, 11*, 1938–1939.

Collier, J. G., & Thome, J. R. (1996). *Convective boiling and condensation* (3rd ed.). Oxford Science Publications.

Cook, W. (2008). *An experimental apparatus for measurement of pressure drop and void fraction and non-boiling two-phase heat transfer and flow visualization in pipes for all inclinations*, M.S. Thesis, Oklahoma State University, Stillwater, Oklahoma.

Crawford, T. J., Weinberger, C. B., & Weisman, J. (1985). Two-phase flow patterns and void fractions in downward flow part I: Steady state flow patterns. *International Journal of Multiphase Flow, 11*, 761–782.

Crowley, C. J., Wallis, G. B., & Barry, J. J. (1992). Validation of one-dimensional wave model for the stratified to slug flow regime transition with consequences for wave growth and slug frequency. *International Journal of Multiphase Flow, 18*, 249–271.

Das, G., Das, P. K., Purohit, N. K., & Mitra, A. K. (2002). Liquid holdup in concentric annuli during cocurrent gas–liquid upflow. *Canadian Journal of Chemical Engineering, 80*, 153–157.

Dittus, F. W., & Boelter, L. M. K. (1930). Heat transfer in automobile radiators of the tubular type. *University of California Publications in Engineering, 2*, 443–461.

Dix, G. E. (1971). *Vapor void fractions for forced convection with subcooled boiling at low flow rates*, Report NEDO-10491, General Electric Co.

Ducoulombier, M., Colasson, S., Bonjour, J., & Haberschill, P. (2011). Carbon dioxide flow boiling in a single microchannel - part I: Pressure drops. *Experimental Thermal and Fluid Science, 35*, 581–596.

Fang, X. D., Xu, Y., & Zhou, Z. R. (2011). New correlations of single-phase friction factor for turbulent pipe flow and evaluation of existing single-phase friction factor correlations. *Nuclear Engineering and Design, 241*, 897–902.

Fore, L. B., Beus, S. G., & Bauer, R. C. (2000). Interfacial friction in gas-liquid annular flow: Anologies to full and transitional roughness. *International Journal of Multiphase Flow, 26*, 1755–1769.

Fukano, T., & Furukawa, T. (1998). Prediction of the effects of liquid viscosity on interfacial shear stress and frictional pressure drop in vertical upward gas-liquid annular flow. *International Journal of Multiphase Flow, 24*, 587–603.

Fukano, T., & Ousaka, A. (1989). Prediction of the circumferential distribution of film thickness in horizontal and near-horizontal gas–liquid annular flows. *International Journal of Multiphase Flow, 15*, 403–419.

Ghajar, A. J. (2019). Chapter 1: Heat transfer and pressure drop in the transition region of smoth horizontal circular tubes with different inlet configurations. In E. M. Sparrow, J. P. Abraham, J. M. Gorman, & W. J. Minkowycz (Eds.), *Advances in heat transfer* (Vol. 51, pp. 1–53). Academic Press/Elsevier.

Ghajar, A. J., & Bhagwat, S. M. (2013). Effect of void fraction and two-phase dynamic viscosity models on prediction of hydrostatic and frictional pressure drop in vertical upward gas-liquid two-phase flow. *Heat Transfer Engineering, 34*(13), 1044–1059.

Ghajar, A. J., & Bhagwat, S. M. (2014a). Non-boiling gas-liquid two-phase flow phenomenon in near horizontal, upward and downward inclined pipe orientations. *International Journal of Mechanical, Aerospace, Industrial, Mechatronic Engineering, 8*, 1039–1053.

Ghajar, A. J., & Bhagwat, S. M. (2014b). Chapter 4: Flow patterns, void fraction and pressure drop in gas-liquid two-phase flow at different pipe orientations. In *Frontiers and progress in multiphase flow* (pp. 157–212). Springer.

Ghajar, A. J., & Bhagwat, S. M. (2017). Chapter 3: Gas-liquid flow in ducts. In E. E. Michaelides, C. T. Crowe, & J. D. Schwarzkopf (Eds.), *Handbook of multiphase flow* (2nd ed., pp. 287–356). CRC Press/Taylor & Francis.

Ghajar, A. J., & Kim, J. (2006). Calculation of local inside-wall convective heat transfer parameters from measurements of the local outside-wall temperatures along an electrically heated circular tube. In M. Kutz (Ed.), *Heat transfer calculations* (pp. 23.3–23.27). McGraw-Hill.

Ghajar, A. J., & Madon, K. F. (1992). Pressure drop measurements in the transition region for a circular tube with different inlet configurations. *Experimental Thermal and Fluid Science, 5*, 129–135.

Ghajar, A. J., & Tam, L. M. (1994). Heat transfer measurements and correlations in the transition region for a circular tube with three different inlet configurations. *Experimental Thermal and Fluid Science, 8*, 79–90.

Ghajar, A. J., & Tam, L. M. (1995). Flow regime map for a horizontal pipe with uniform wall heat flux and three inlet configurations. *Experimental Thermal and Fluid Science, 10*, 287–297.

Ghajar, A. J., & Tang, C. C. (2009). Advances in void fraction, flow pattern maps and non-boiling heat transfer two-phase flow in pipes with various inclinations. *Advanced Multiphase Flow Heat Transfer, 1*, 1–52.

Ghajar, A. J., & Tang, C. C. (2010). Importance of non-boiling two phase flow heat transfer in pipes for industrial applications. *Heat Transfer Engineering, 31*(9), 711–732.

Ghajar, A. J., & Zurigat, Y. H. (1991). Microcomputer-assisted heat transfer measurements/analysis in a circular tube. *International Journal of Applied Engineering Education, 7*(2), 125–134.

Ghajar, A. J., Tang, C. C., & Cook, L. L. (2010). Experimental investigation of friction factor in the transition region for water flow in mini tubes and micro tubes. *Heat Transfer Engineering, 31*(8), 646–657.

Gnielinski, V. (1976). New equations for heat and mass transfer in turbulent pipe and channel flow. *International Chemical Engineering, 16*, 359–368.

Godbole, P. V., Tang, C. C., & Ghajar, A. J. (2011). Comparison of void fraction correlations for different flow patterns in upward vertical two-phase flow. *Heat Transfer Engineering, 32*(10), 843–860.

Gokcal, B. (2008). *An experimental and theoretical investigation of slug flow for high oil viscosity in horizontal pipes*, Ph.D. Thesis. University of Tulsa, Tulsa, Oklahoma.

Gokcal, B., Al-Sarkhi, A., & Sarica, C. (2009). Effects of high oil viscosity on drift velocity for horizontal and upward inclined pipes. *SPE Projects, Facilities & Construction, 4*, 32–40.

Gomez, L., Shoham, O., Schmidt, Z., Choshki, R., & Northug, T. (2000). Unified mechanistic model for steady state two-phase flow: Horizontal to upward vertical flow. *SPE, 5*, 339–350.

Hamersma, P. J., & Hart, J. (1987). A pressure drop correlation for gas-liquid pipe flow with a small liquid holdup. *Chemical Engineering Science, 42*, 1187–1196.

Haramathy, T. Z. (1960). Velocity of large drops and bubbles in media of infinite or restricted extent. *AICHE Journal, 6*, 281–288.

Hart, J., Hamersma, P. J., & Fortuin, J. M. H. (1989). Correlations predicting frictional pressure drop and liquid holdup during horizontal gas-liquid pipe flow with a small liquid holdup. *International Journal of Multiphase Flow, 15*, 947–964.

Hashemi, A., Kim, J. H., & Sursock, J. P. (1986). *Effect of diameter and geometry on two-phase flow regimes and carryover in model PWR hot leg*, Eighth international heat transfer conference, pp. 2443–2451.

Henstock, W. H., & Hanratty, J. (1976). The interfacial drag and the height of the wall layer in annular flows. *AICHE Journal, 22*, 990–999.

Hewitt, G. F., & Hall-Taylor, N. S. (1970). *Annular two-phase flow* (1st ed.). Pergamon Press.

Hewitt, G. F., & Lacey, P. M. C. (1965). The breakdown of the liquid film in annular two-phase flow. *International Journal of Heat and Mass Transfer, 8*, 781–791.

Hewitt, G. F., & Roberts, D. N. (1969). *Studies of two-phase flow patterns by simultaneous x-ray and flash photography*, Technical Report AERE-M 2159, Atomic Energy Research Establishment.

Hewitt, G., Lacey, P. M. C., & Nicholls, B. (1965). *Transitions in film flow in vertical tube*, Proceedings of two-phase flow symposium, Exeter, U. K., Paper B4 2.

Hewitt, G. F., Martin, C. J., & Wilkes, N. S. (1985). Experimental and modeling studies of churn-annular flow in the region between flow reversal and the pressure drop minimum. *Physico-chemical Hydrodynamics, 6*, 69–86.

Hibiki, T., & Ishii, M. (2003). One-dimensional drift flux model and constitutive equations for relative motion between phases in various two-phase flow regimes. *International Journal of Heat and Mass Transfer, 46*, 4935–4948.

Hills, J. H. (1976). The operation of a bubble column at high throughputs part 1: Gas holdup measurements. *Chemical Engineering Journal, 12*, 89–99.

Hlaing, N. D., Sirivat, A., Siemanond, K., & Wilkes, J. O. (2007). Vertical two-phase flow regimes and pressure gradients: Effect of viscosity. *Experimental Thermal and Fluid Science, 31*, 567–577.

Inoue, Y. (2001). *Measurement of interfacial area concentration of gas–liquid two-phase flow in a large diameter pipe*, M.S. Thesis, Graduate School of Energy Science, Kyoto University, Japan.

Inoue, A., Kurosu, T., Aoki, T., Yagi, M., Misutake, T., & Morooka, S. (1995). Void fraction distribution in BWR fuel assembly and evaluation of subchannel code. *Journal of Nuclear Science and Technology, 32*, 629–640.

Ishii, M. (1977). *One-dimensional drift flux model and constitutive equations for relative motion between phases in various two-phase flow regimes* (pp. 77–47). Argonne National Laboratory.

Ishii, M., & Grolmes, M. A. (1975). Inception criteria for droplet entrainment in two-phase cocurrent film flow. *AICHE Journal, 21*, 308–318.

Ishii, M., & Mishima, K. (1989). Droplet entrainment correlation in annular two-phase flow. *International Journal of Heat and Mass Transfer, 32*, 1835–1846.

Jensen, M. K., & Vlakancic, A. (1999). Technical note –experimental investigation of turbulent heat transfer and fluid flow in internally finned tubes. *International Journal of Heat and Mass Transfer, 42*, 1343–1351.

Jeyachandra, B. C. (2011). *Effect of pipe inclination on flow characteristics of high viscosity oil gas two-phase flow*, Ph.D. Thesis, University of Tulsa, Tulsa, Oklahoma.

Kaji, M., & Azzopardi, B. J. (2010). The effect of pipe diameter on the structure of gas-liquid flow in vertical pipes. *International Journal of Multiphase Flow, 36*, 303–313.

Kandlikar, S. G. (2002). Fundamental issues related to flow boiling in minichannels and microchannels. *Experimental Thermal and Fluid Science, 26*, 389–407.

Kandlikar, S. G., & Grande, W. J. (2002). *Evolution of microchannel flow passages - thermohydraulic performance and fabrication technology*, Proceedings of IMECE2002 ASME International Mechanical Engineering Congress & Exposition (New Orleans, Louisiana).

Kataoka, I., & Ishii, M. (1987). Drift flux model for large diameter pipe and new correlation for pool void fraction. *International Journal of Heat and Mass Transfer, 30*, 1927–1939.

Keinath, B. (2012). *Void fraction, pressure drop and heat transfer in high pressure condensing flows through microchannels*, Ph.D. Thesis, Georgia Institute of Technology.

Kim, J., & Ghajar, A. J. (2006). A general heat transfer correlation for non-boiling gas-liquid flow with different flow patterns in horizontal pipes. *International Journal of Multiphase Flow, 32*, 447–465.

Kim, D., Ghajar, A. J., Dougherty, R. L., & Ryali, V. K. (1999). Comparison of 20 two-phase heat transfer correlations with seven sets of experimental data, including flow pattern and tube inclination effects. *Heat Transfer Engineering, 20*(1), 15–40.

Kim, D., Ghajar, A. J., & Dougherty, R. L. (2000). Robust heat transfer correlation for turbulent gas-liquid flow in vertical pipes. *Journal of Thermophysics Heat Transfer, 14*, 574–578.

Kline, S. J., & McClintock, F. A. (1953). Describing uncertainties in single sample experiments. *Mechanical Engineering, 75*, 3–8.

Krishnamoorthy, C., Rao, R. P., & Ghajar, A. J. (2007). *Single-phase heat transfer in micro-tubes: A critical review*, Proceedings of the ASME-JSME Thermal Engineering Summer Heat Transfer Conference, Paper No. HT2007-32408.

Krishnamoorty, C., & Ghajar, A. J. (2007). *Single-phase friction factor in micro-tubes: A critical review of measurement, instrumentation and data reduction techniques from 1991-2006*, Proceedings of ASME ICNMM2007 5th International Conference on Nanochannels, Microchannels, and Minichannels, Paper No. ICNMM2007-30022.

Laurinat, J. E., Hanratty, T. J., & Dallman, J. C. (1984). Pressure drop and film height measurements for annular gas-liquid flow. *International Journal of Multiphase Flow, 10*, 341–356.

Lips, S., & Meyer, J. P. (2012). Experimental study of convective condensation in an inclined smooth tube. Part II: Inclination effect on pressure drop and void fraction. *International Journal of Heat and Mass Transfer, 55*, 405–412.

Liu, W., Tamai, H., & Takase, K. (2013). Pressure drop and void fraction in steam–water two-phase flow at high pressure. *Journal of Heat Transfer, 135*, 1–13.

Lockhart, R. W., & Martinelli, R. C. (1949). Proposed correlation of data for isothermal two-phase, two component flow in pipes. *Chemical Engineering Progress, 45*, 39–48.

Mandhane, J. M., Gregory, G. A., & Aziz, K. (1974). A flow pattern map for gas-liquid flow in horizontal pipes. *International Journal of Multiphase Flow, 1*, 537–553.

Marchaterre, J. F. (1956). *The effect of pressure on boiling density in multiple rectangular channels*, Report ANL-5522, Argonne National Labs.

Marchaterre, J. F., Petrick, M., Lottes, P. A., Weatherland, R. J., & Flinn, W. S. (1960). *Natural and forced circulation boiling studies*, Report ANL-5735, Argonne National Labs.

Martinelli, R. C., & Boelter, L. M. K. (1942). The analytical prediction of superposed free and forced viscous convection in a vertical pipe. *University of California Publications in Engineering, 5*(2), 23–58.

Metais, B., & Eckert, E. R. G. (1964). Forced, mixed and free convection regimes. *ASME Journal of Heat Transfer, 10*, 295–296.

Meyer, J. P., & Everts, M. (2019). Chapter 3: A review of the recent developments in laminar, transitional, quasi-turbulent, turbulent forced and mixed convective flow through horizontal tubes. In E. M. Sparrow, J. P. Abraham, J. M. Gorman, & W. J. Minkowycz (Eds.), *Advances in heat transfer* (Vol. 51, pp. 131–205). Academic Press/Elsevier.

Mishima, K., & Ishii, M. (1980). Theoretical prediction of onset of horizontal slug flow. *Journal of Fluids Engineering, 102*, 441–445.

Mishima, K., & Ishii, M. (1984). Flow regime transition criteria for upward two-phase flow in vertical tubes. *International Journal of Heat and Mass Transfer, 27*, 723–737.

Moeck, E. O. (1970). *Annular dispersed two-phase flow and critical heat flux*, Ph.D. Thesis, McGill University, Montreal, Quebec, Canada.

Mukherjee, H. (1979). *An experimental study of inclined two-phase flow*, Ph.D. Thesis, University of Tulsa, Tulsa, Oklahoma.

Muller-Steinhagen, H., & Heck, K. (1986). A simple friction pressure drop correlation for two-phase flow in pipes. *Chemical Engineering and Processing, 20*, 297–308.

Nguyen, V. T. (1975). *Two-phase gas-liquid cocurrent flow: an investigation of holdup, pressure drop and flow patterns in a pipe at various inclinations*, Ph.D. Thesis, University of Auckland, New Zealand.

Nicklin, D. J., Wilkes, J. O., & Davidson, J. F. (1962). Two-phase flow in vertical tubes. *Transactions. Institute of Chemical Engineers, 40*, 61–68.

Oshinowo, O. (1971). *Two-phase flow in a vertical tube coil*, Ph.D. Thesis, University of Toronto, Canada.

Ottens, M., Hoefsloot, H. C. J., & Kamersma, P. J. (2001). Correlations predicting liquid holdup and pressure gradient in steady state nearly horizontal cocurrent gas-liquid pipe flow. *Institution of Chemical Engineers, Trans IChemE, 79*, 581–592.

Owen, D. G. (1986). *An experimental and theoretical analysis of equilibrium annular flow*, Ph.D. Thesis, University of Birmingham, UK.

Quiben, J. M., & Thome, J. R. (2007). Flow pattern based two-phase frictional pressure drop model for horizontal tubes. Part I: Diabtic and adiabatic experimental study. *International Journal of Heat and Fluid Flow, 28*, 1049–1059.

Rezkallah, K. S., & Sims, G. E. (1987). An examination of correlations of mean heat transfer coefficients in two-phase and two-component flow in vertical tubes. *AIChE Symposium Series, 83*, 109–114.

Rouhani, S. Z., & Axelsson, E. (1970). Calculation of void volume fraction in the subcooled and quality boiling regions. *International Journal of Heat and Mass Transfer, 13*, 383–393.

Sacks, P. S. (1975). *Measured characteristics of adiabatic and condensing single component two-phase flow of refrigerant in a 0.377 inch diameter horizontal tube*, ASME Winter Annual Meeting, Houston, TX, 75WA/HT-24, pp. 1–12.

Schlegel, J., Hibiki, T., & Ishii, M. (2010). Development of a comprehensive set of drift flux constitutive models for pipes of various hydraulic diameters. *Progress in Nuclear Energy, 52*, 666–677.

Shah, R. K. (1978). A correlation for laminar hydrodynamic entry length solutions for circular and noncircular ducts. *Journal of Fluids Engineering, 100*, 177–179.

Shannak, B. A. (2008). Frictional pressure drop of gas liquid two-phase flow in pipes. *Nuclear Engineering and Design, 238*, 3277–3284.

Shedd, T. A. (2010). Void fraction and pressure drop measurements for refrigerant R410a flows in small diameter tubes. *Technical Report AHRTI*, 20110–20101.

Shoukri, M., Hassan, I., & Gerges, I. (2003). Two-phase bubbly flow structure in large diameter vertical pipe. *Canadian Journal of Chemical Engineering, 81*, 205–211.

Sieder, E. N., & Tate, G. E. (1936). Heat transfer and pressure drop in liquids in tubes. *Industrial and Engineering Chemistry, 28*, 1429–1435.

Spedding, P. L., Che, J. J. J., & Nguyen, N. T. (1982). Pressure drop in two-phase gas-liquid flow in inclined pipes. *International Journal of Multiphase Flow, 8*, 407–431.

Spedding, P. L., Woods, G. S., Raghunathan, S. R., & Watterson, J. K. (2000). Flow pattern, holdup and pressure drop in vertical and near vertical two and three phase upflow. *Chemical Engineering Research and Design, 78*(Part A), 404–418.

Steinke, M. E., & Kandlikar, S. G. (2004). Review of single-phase heat transfer enhancement techniques for application in microchannels, minichannels and micro devices. *International Journal of Heat and Technology, 22*(2), 3–11.

Taitel, Y., & Dukler, A. E. (1976). A model for predicting flow regime transitions in horizontal and near horizontal gas-liquid flow. *AICHE Journal, 22*, 47–55.

Taitel, Y., Barnea, D., & Dukler, A. E. (1980). Modeling flow pattern transitions for steady upward gas-liquid flow in vertical tubes. *AICHE Journal, 26*, 345–354.

Tam, L. M. (1995). *An experimental investigation of heat transfer and pressure drop in the transition region for a horizontal tube with different inlets and uniform wall heat flux*, Ph.D. Thesis, Oklahoma State University.

Tam, H. K. (2013). *Experimental investigation of simultaneous heat transfer and pressure drop measurements for plian and micro-fin tubes*, Ph.D. Thesis, University of Macau.

Tam, L. M., & Ghajar, A. J. (1997). Effect of inlet geometry and heating on the fully developed friction factor in the transition region of a horizontal tube. *Experimental Thermal and Fluid Science, 15*, 52–64.

Tam, L. M., & Ghajar, A. J. (1998). The unusual behavior of local heat transfer coefficient in a circular tube with a bell-mouth inlet. *Experimental Thermal and Fluid Science, 16*, 187–194.

Tam, L. M., & Ghajar, A. J. (2006). Transitional heat transfer in plain horizontal tubes. *Heat Transfer Engineering, 27*(5), 23–38.

Tam, L. M., Tam, H. K., & Ghajar, A. J. (2011). *Simultaneous heat transfer and pressure drop measurements for a horizontal micro-tube*, Proceedings of ASME/JSME 2011 8th Thermal Engineering Joint Conference, Paper No. AJTEC2011-44218.

Tam, H. K., Tam, L. M., Ghajar, A. J., Sc, T., & Zhang, T. (2012). Experimental investigation of heat transfer, friction factor, and optimal fin geometries for the internally micro-fin tubes in the transition and turbulent regions. *Journal of Enhanced Heat Tranfer, 19*(5), 457–476.

Tam, H. K., Tam, L. M., & Ghajar, A. J. (2013). Effect of inlet geometries and heating on the entrance and fully developed friction factors in the laminar and transition regions of a horizontal tube. *Experimental Thermal and Fluid Science, 44*, 680–696.

Tam, L. M., Tam, H. K., Ghajar, A. J., Ng, W. S., & Wu, C. K. (2014). *The effect of inner surface roughness and heating on friction factor in horizontal mini-tubes*, Proceedings of 15th International Heat transfer Conference, Paper No. IHTC15-8732.

Tam, H. K., Tam, L. M., Ghajar, A. J., & Wang, Q. (2017). *Experimental investigation of the heat transfer in a horizontal mini-tube with three different inlet configurations*, Proceedings of the 2nd Thermal and Fluid Engineering Conference (TFEC2017) and the 4th International Workshop on Heat Transfer (IWHT2017), Paper No. TFEC-IWHT2017-17541.

Tam, H. K., Tam, L. M., Ghajar, A. J., Wang, Q., Chen, X. W., & Lei, W. C. (2019). *Experimental investigation of single-phase heat transfer in horizontal mini-tubes with different bending inlets*, 5th International Workshop on Heat/Mass Transfer Advances for Energy Conservation and Pollution Control, Paper No. IWHT2019-105.

Tam, H. K., Xu, Y. X., Tam, L. M., & Ghajar, A. J. (2021). *Effect of heating on the fully developed friction factors in horizontal mini-tubes*, Proceedings of the 5th-6th Thermal and Fluid Engineering Conference, Paper No. TFEC-2021-32064.

Tang, C. C. (2011). *A study of heat transfer in non-boiling two-phase gas liquid flow in pipes for horizontal, slightly inclined, and vertical orientations*, Ph.D. Thesis, Oklahoma State University, Stillwater, Oklahoma.

Tang, C. C., & Ghajar, A. J. (2011). *A mechanistic heat transfer correlation for non-boiling two-phase flow in horizontal, inclined, and vertical pipes*, Proceedings of AJTEC 2011 ASME/JSME 8th Thermal Engineering Joint Conference, Paper No. AJTEC2011-44114.

Terekhov, V. I., & Pakhomov, M. A. (2008). The effect of bubbles on the structure of flow and the friction in downward turbulent gas-liquid flow. *High Temperature, 46*, 854–860.

Turner, J. M., & Wallis, G. B. (1965). *The separate cylinders model of two-phase flow*. Technical Report Report No. NYO-3114-6, Thayer's School of Engineering, Dartmouth College, Hanover, NH.

Usui, K., & Sato, K. (1989). Vertically downward two-phase flow: (I) void distribution and average void fraction. *Journal of Nuclear Science and Technology, 26*(7), 670–680.

Wallis, G. B. (1962). *The onset of droplet entrainment in annular gas-liquid flow*. General Electric Report No. 62, GL 127, Schenectady, New York.

Wallis, G. B. (1969). *One-dimensional two-phase flow*. McGraw-Hill.

Webb, R. L., Narayanamurthy, R., & Thors, P. (2000). Heat transfer and friction characteristics of internal helical-rib roughness. *Journal of Heat Transfer, 122*(1), 134–142.

Weisman, J., & Kang, S. Y. (1981). Flow pattern transitions in vertical and upward inclined lines. *International Journal of Multiphase Flow, 7*, 271–291.

Woldesemayat, M. A., & Ghajar, A. J. (2007). Comparison of void fraction correlations for different flow patterns in horizontal and upward inclined pipes. *International Journal of Multiphase Flow, 33*, 347–370.

Wongs-ngam, J., Nualboonrueng, T., & Wongwises, S. (2004). Performance of smooth and micro-finned tubes in high mass flux region of R-134a during condensation. *Journal of Heat Mass Transfer, 40*, 425–435.

Wongwises, S., & Kongkiatwanitch, W. (2001). Interfacial friction factor in vertical upward gas-liquid annular two-phase flow. *International Journal of Heat and Mass Transfer, 28*, 323–336.

Wongwises, S., & Pipathttakul, M. (2006). Flow pattern, pressure drop and void fraction of gas–liquid two-phase flow in an inclined narrow annular channel. *Experimental Thermal and Fluid Science, 30*, 345–354.

Xiong, R., & Chung, J. N. (2006). *Size effect on adiabatic gas-liquid two-phase flow map and void fraction in micro-channels*, Proceedings of International Mechanical Engineering Congress and Exposition, Chicago, IL.

Xu, Y., & Fang, X. (2012). A new correlation of two-phase frictional pressure drop for evaporating two-phase flow in pipes. *International Journal of Refrigeration, 35*, 2039–2050.

Yang, Y. F., Tam, H. K., Tam, L. M., & Ghajar, A. J. (2021). *Numerical investigation of the forced and mixed convection heat transfer inside horizontal macro- and mini-tubes*, 6th International Workshop on Heat/Mass Transfer Advances for Energy Conservation and Pollution Control.

Yijun, J., & Rezkallah, K. (1993). A study on void fraction in vertical co-current upward and downward two-phase gas-liquid flow - I: Experimental results. *Chemical Engineering Communications, 126*, 221–243.

Zdaniuk, G. J., Chamra, L. M., & Mago, P. J. (2008). Experimental determination of heat transfer and friction in helically finned tubes. *Experimental Thermal and Fluid Science, 32*(3), 761–775.

Zhang, W., Hibiki, T., & Mishima, K. (2010). Correlations of two-phase frictional pressure drop and void fraction in mini-channel. *International Journal of Heat and Mass Transfer, 53*, 433–465.

Zuber, N., & Findlay, J. (1965). Average volume concentration in two-phase systems. *ASME Journal of Heat Transfer, 87*, 453–468.

Index

A. J. Ghajar, *Single- and Two-Phase Flow Pressure Drop and Heat Transfer in Tubes*, Mechanical Engineering Series,
https://doi.org/10.1007/978-3-030-87281-6

The manufacturer's authorised representative in the EU is Springer Nature Customer Service Centre GmbH, Europaplatz 3, 69115 Heidelberg, Germany. If you have any concerns regarding our products, please contact ProductSafety@springernature.com

Printed and bound by CPI Group (UK) Ltd, Croydon, CR0 4YY

30/09/2026

02225491-0007